TRANSACTIONS

OF THE

AMERICAN PHILOSOPHICAL SOCIETY

HELD AT PHILADELPHIA
FOR PROMOTING USEFUL KNOWLEDGE

NEW SERIES—VOLUME XXXIV, PART II

THE ARC SPECTRUM OF IRON (Fe I)

PART I. ANALYSIS OF THE SPECTRUM

Based on the work of many investigators and including unpublished studies by MIGUEL A. CATALÁN

HENRY NORRIS RUSSELL and CHARLOTTE E. MOORE
Princeton University Observatory

PART II. THE ZEEMAN EFFECT

DOROTHY W. WEEKS
Wilson College

THE AMERICAN PHILOSOPHICAL SOCIETY
INDEPENDENCE SQUARE
PHILADELPHIA 6

DECEMBER, 1944

LANCASTER PRESS, INC., LANCASTER, PA.

PREFACE

The present work deals with a spectrum which has been the object of active study for more than seventy years, and carries one phase of the investigation—the term-analysis of the structure—about as far as existing material appears to permit. The spectrum, though complicated, is found to be highly orderly, and almost all the principal terms predicted by theory are now known. Present methods of observation in the laboratory, however, do not succeed in bringing out the spectrum as completely as it is exhibited in the sun, and future observation, especially for the discovery of better laboratory sources, is still highly promising.

The writers take pleasure in expressing their gratitude to many colleagues who have generously made new and unpublished material available. Special thanks are due to Doctor Miguel A. Catalán for the communication of many spectral terms; to Messrs. Arthur S. King and Harold D. Babcock for laboratory and solar spectroscopic material obtained at Mount Wilson; and to Professor George R. Harrison for observations of the Zeeman effect secured at the Massachusetts Institute of Technology.

Henry Norris Russell
Charlotte E. Moore
Dorothy W. Weeks

THE ARC SPECTRUM OF IRON (Fe I)

PART I

ANALYSIS OF THE SPECTRUM

HENRY NORRIS RUSSELL AND CHARLOTTE E. MOORE

I. INTRODUCTION

(1) GENERAL CONSIDERATIONS

The neutral atom of iron provides the classical example of a complex atomic spectrum. The ubiquity of the metal, and the richness and ease of production of its spectrum, have led to the general adoption of its lines as standards,[1] so that their wave-lengths are now more accurately known than those for any other element for which data have been published. Full data are available on temperature classification [2] and Zeeman effect, and a good beginning has been made [3] in the determination of the line intensities, transition-probabilities, and f-values. The spectrum is therefore of unusual interest both to the practical spectroscopist and to the theoretical physicist, and most of all to the astrophysicist. The high abundance of iron causes even its faint arc lines to appear in the cooler stars, and the stronger ones in those as hot as Sirius. Many lines predicted by the analysis of Fe I, and not yet observed in the laboratory, agree so closely, in wave-length and estimated intensity, with unidentified lines in the solar spectrum, that there can be no doubt of their presence there (see § 22).

The great number of iron lines and the wide range in intensity and excitation among them make them especially valuable for the determination of curves of growth, and for the investigation of stellar atmospheres.

A complete term-analysis of this spectrum is therefore of prime importance to astronomers and physicists alike. Practically complete analyses of Fe II [4] and Fe III [5] have already been published. One of comparable extent for Fe I has been delayed partly by the richness of the spectrum but more by political conditions (§ 2). It is at last possible to present it here.

(2) PREVIOUS INVESTIGATIONS

Shortly after Catalán's discovery of multiplets, twenty of them were identified in Fe I by F. M. Walters, Jr.[6] in 1923. In 1924 he published a list of 52 multiplets involving 26 spectral terms [7] of multiplicities 3, 5, and 7—though the last were not connected with the others. This was accomplished almost at the same time by the independent work of Laporte,[8] who classified about 600 lines and derived an ionization potential of 8.15 volts—which he revised to 8.06 in 1926 [9]—adding a few more terms. These results were shown by Hund [10] to be in complete agreement with his general theory of the structure of atomic spectra. Additional terms, almost all discovered by Walters, and communicated by him to colleagues, were announced by Meggers [11] and by Moore and Russell.[12]

Notable extensions of the analysis were accomplished (again independently in the main) by Burns and Walters [13] and by Catalán.[14] The latter accounted for 2350 lines by combinations among 304 energy levels, identified 51 terms, and determined the ionization potential as 7.83 volts. Many fairly strong lines remained unclassified, and 135 of the energy levels were not grouped into terms.

The only published addition to this appears to be Green's identification [15] of a few high terms, but it was generally known that Dr. Catalán was continuing the analysis. In 1936, in response to a request from the writers for his new terms for use in the "Multiplet Table," he generously sent a list of the low even terms, and stated that the remaining terms would be sent as soon as he could copy them. This list included singlet terms, making this the first spectrum (and up to the present, the only one) in which terms of four different multiplicities are known.

---

[1] *Trans. Internat. Astron. Union* **3**: 86, 1928; **4**: 60, 1932; **5**: 84, 1935; **6**: 79, 1938.

[2] King, A. S., *Mount Wilson Contr.* No. 66, 1913; No. 247, 1922; and No. 496, 1934; *Astrophys. Jour.* **37**: 239, 1913; **56**: 318, 1922; **80**: 124, 1934.

[3] King, R. B., and A. S. King, *Mount Wilson Contr.* No. 528, 1935; and No. 581, 1938: *Astrophys. Jour.* **82**: 377, 1935; **87**: 24, 1938.

[4] Dobbie, *Annals Solar Physics Observ.* (Cambridge) **5**, pt. 1, 1938.

[5] Edlén and Swings, *Astrophys. Jour.* **95**: 532, 1942.

[6] *Jour. Washington Acad. Sci.* **13**: 243, 1923.

[7] *Jour. Optical Soc. America* **8**: 245, 1924.

[8] *Zeitschr. f. Physik* **23**, 135, 1924; **26**: 1, 1924.

[9] *Proc. Nat. Acad. Sci.* **12**: 496, 1926.

[10] *Linienspektren*: 163. Berlin, Springer, 1927.

[11] *Astrophys. Jour.* **60**: 60, 1924.

[12] *Mount Wilson Contr.* No. 365, 1928; *Astrophys. Jour.* **68**: 151, 1928.

[13] *Publ. Allegheny Observ.* **6**: 159, 1929.

[14] *Anales Soc. Española Fisica y Quimica* **28**: 1239, 1930.

[15] *Phys. Rev.* **55**: 1209, 1939.

Shortly afterward, the Spanish War broke out. Communications were interrupted, and it was not known whether Dr. Catalán's laboratory and papers were accessible to him—or still in existence. To aid in continuing his work, a line-list of Fe I, containing all data available to the writers and unpublished Zeeman data by Babcock, was prepared at Princeton and sent to him. More than four years later Dr. Catalán sent a list of spectral terms and Zeeman g-values containing a great amount of new material. Many of the g-values appear to have been derived from Babcock's data. Communication is still very difficult, and he has been unable to send his long list of classified lines. This has had to be reconstructed from the term-values.

Additional data became available in 1941 when Professor G. R. Harrison, of the Massachusetts Institute of Technology, entrusted for discussion to Professor Dorothy W. Weeks, of Wilson College, a complete set of records of the Zeeman effect of iron, made on his automatic measuring machine, from spectrograms taken with the great Bitter magnet. This material is much more extensive and more fully resolved than any which had previously existed. Miss Weeks' results, including patterns for 1038 lines and the determination of g-values for 392 energy levels, are reported by her in Part II of this memoir.

For this work, a comprehensive analysis of the spectrum was prerequisite, and Miss Weeks and the writers were able to collaborate to great mutual advantage. The Zeeman patterns for previously unclassified lines were of great value in suggesting identifications, and the g-values which were derived for many known levels were often conclusive, either in confirming previous term-assignments or in suggesting new ones. At the same time the writers reviewed the existing analysis of the spectrum and were able to find a number of new levels and terms. Many of these were identified as terms predicted by Hund's theory, and nothing was found inconsistent with it.

The progress of the analysis of this spectrum has been curiously uneven. The numbers of energy levels detected and published in successive intervals are given in table 1. An attempt has been made in table A (p. 134) to assign each separate energy level to the investigator who first presented good evidence for its existence (irrespective of later assignment of a term-classification).

A list of the total would be misleading, since Walters and Laporte in 1924 worked and published independently and almost simultaneously, and Catalán in 1930 did not have access to the work of Burns and Walters in 1929. The principal contributors to the published analyses have been Walters, Catalán, and Laporte. The most important unpublished contribution is Catalán's. The analysis which is here presented is thus the co-operative work of many investigators, and the work of the writers has been largely editorial. Dr. Catalán would appear as a joint author of this work were it not for the difficulties of communication which have prevented the continuous exchange of results which the writers would otherwise have desired. As things are, they have been obliged to take the responsibility for the final assignment of term designations, the configuration analysis, and the rejection of doubtful levels. A few of Catalán's designations have been altered, mainly on the basis of new and more precise Zeeman data.

The present work is therefore one more example of the cordial international co-operation among spectroscopists which has been so largely responsible for the rapid advance of the analysis of complex spectra. It was not terminated until it was obvious that the law of diminishing returns was in active operation and that very little more in the way of discovery of additional energy levels would result from extensive effort.

The present lists contain 4860 classified lines, arising from combinations among 464 energy levels. Thanks to the complete Zeeman data, all but 19 of these have been grouped into 146 terms, which combine to give 1342 multiplets.

In so complicated a spectrum as this there can be no hope of classifying all the faint lines. There is good reason, however, to believe that the analysis is substantially complete. Almost all the stronger lines have been classified. Taking as a criterion the lines recorded by King in his temperature classification,

TABLE 1

PROGRESS OF ANALYSIS OF THE SPECTRUM

Dates	Low Even	High Even	Odd	Total
1923–24	16	20	94	130
1924–28	18	12	16	46
1929–30	3	62	57	122
1930–44			17	17
Total published	37	94	184	315
Catalán unpublished	12	9	68	89
Present writers	13	19	28	60
Grand total	62	122	280	464

which number 1753 and lie between 10469 and 2298A, we find that 46 remain unclassified, of which one is of intensity 20 and the next 8 on a scale in which the strongest lines reach 1500. The configuration analysis in §§ 10–14 shows also that almost all the terms which theory predicts as likely to give even moderately prominent lines have been identified.

Though this familiar spectrum has been so much observed, it is probable that a very large number of additional arc lines of iron can be detected when better sources for faint lines are devised (§ 25).

II. OBSERVATIONAL DATA

(3) WAVE-LENGTHS

A complete and accurate list of wave-lengths and wave-numbers is the prime necessity for an analysis. There is a wealth of material for Fe I, but this is very far from homogeneous. The principal lines have been measured with very high precision—especially the International Secondary and Tertiary Standards,[16] which depend on three or more independent sets of interferometer measures. Many other lines have been measured by interference methods, and a great many more with gratings of adequate dispersion. There remain, however, many faint lines which have never been accurately measured in the laboratory—some only to 0.1A. Many of these agree with the predictions of the term-analysis within their limits of error and are undoubtedly due to iron.

A detailed and critical list of wave-lengths, based on homogeneous material and standards, is much to be desired, but there is no hope of this during the war.

The wave-lengths which appear in table B (p. 139) represent, therefore, a compilation from all available sources. Reasonable care has been taken to adopt determinations from those sources which, in general, were judged to be most reliable, but it should be emphasized that the list cannot be considered a definitive one. Wave-lengths of extreme precision are essential in the first attempts to break into complex spectra like those of some of the rare earths, which contain few outstanding lines; but the extension of an analysis in which many accurate values of terms and term-differences have already been found makes less severe demands. The existing material has proved generally adequate for the purpose, although the probable error of the worst measure in the table is at least twenty times that of the best.

The long list of sources at the end of table B is arranged roughly in order of accuracy—beginning with the standards, and ending with the rough measures just mentioned. The detailed order of the reference letters has, however, been adopted partly as a matter of convenience (e. g., good measures in the infra-red come early in the list); and the writers explicitly disclaim the assumption that an earlier letter in the alphabet always, or even usually, indicates that they regard the measures referred to as better than some others which may come a few places lower.

The numerous "predicted" lines of Fe I which have been taken from The Revised Rowland Table[17] are discussed in §§ 18–23 and listed in table C (p. 170) along with many others taken from Babcock's extension in the infra-red.[18]

[16] *Trans. Internat. Astron. Union* **3**: 86, 1928; **4**: 60, 1932; **5**: 84, 1935; **6**: 79, 1938.

[17] *Publ. Carnegie Inst. Washington* No. 396, 1928.

[18] Unpublished material.

(4) INTENSITIES

The recorded intensities of the lines in the great majority of spectra are in a state of primeval chaos. Different observers have used radically different scales. The older estimates were usually made on a scale from 1 for the weakest lines to 10 for the strongest. In some recent work (e. g., on the rare earths) a far more open scale has been adopted—from 1 to 1000 or even to 10000. Even the last probably falls short of representing the actual range.

In a few cases where an experienced observer, who has a fairly stable scale of estimation, has photographed the whole range of the spectrum, and made his estimates approximately homogeneous by comparison of overlapping plates, the recorded intensities give a good idea of the relative strengths of the lines on the photographs. Allowances must be made for the varying sensitiveness of the plates and for self-reversal and masking of weak lines by strong neighbors, but the experienced reader can obtain a fairly reliable idea of the actual intensity relations.

In the present case, however, one has to deal with a hodge-podge of estimates on all sorts of scales. The attempt to reduce these to an even roughly homogeneous system is hopeless. An experienced observer making rapid eye estimates on a good set of plates could produce a much better list of intensities.

The intensities found in table B have been taken from what seemed to be as good a source as any (not always the same as for the wave-length). The estimates by King, Meggers, and Kiess, which are on the open scale,[19] are given without parentheses; those of other observers, which are almost all on the narrow scale, in parentheses.

In the course of the analysis, where all known multiplets were written out in detail, the writers obtained a rough idea of the meaning of the estimates of different workers, and they believe that errors in the multiplet analysis resulting from the raggedness of the intensity estimates have been avoided.

All users of table B should, however, be explicitly warned that the tabulated intensities afford only a rough general indication, and should not be used for any quantitative purpose without special study.

(5) TEMPERATURE CLASSIFICATION

These data, which are also of primary importance in the analysis, are taken entirely from King's work.[20] The differences in openness of the intensity scale noted above have very little effect on the temperature classification. The tendency toward assignment of a higher temperature class in the ultra-violet is well known.

[19] Even here several different scales are involved; e. g., King's estimates in his first two papers differ systematically where they overlap, and his later scale in the deep-red is much more open.

[20] *Mount Wilson Contr.* No. 66, 1913; No. 247, 1922; No. 496, 1934; *Astrophys. Jour.* **37**, 239, 1913; **56**: 318, 1922; **80**: 124, 1934.

(6) ZEEMAN EFFECT

The observations and results are discussed by Miss Weeks in Part II.

III. THE ANALYSIS OF THE SPECTRUM

(7) ENERGY LEVELS. ACCURACY OF DETERMINATIONS

The main outlines of the structure of the spectrum have been known since 1924. Burns and Walters in 1929[21] determined the relative positions of 199 energy levels to 0.001 cm^{-1} with the aid of lines measured with the interferometer. These values have been adopted without change except for those levels whose values have been recommended by the International Astronomical Union,[22] although some of the standard wave-lengths upon which they depend have been slightly altered. The remaining 265 levels—most of which were discovered by Catalán—have been determined by the usual process of approximation, finding values for each new level from all the transitions which connect it with previously determined levels, working backward from these to derive improved values for such of the first group as have not already been taken as final, and so on. These values are given to 0.01 cm^{-1}.

The number of combinations from which a level has been determined ranges from more than thirty to one or two in a few instances. In the latter case, the levels were not accepted as real unless their existence was confirmed by position in multiplets, Zeeman effect, or both. Most of them are leading components of terms which have high J-values and give strong lines.

A complete study of the accuracy of the tabular level values would involve one of the accuracy of the various wave-lengths, and is not here attempted. A general idea may be obtained from the residuals of the individual determinations of each level from the mean. For 1118 determinations of 155 odd levels the mean residual, regardless of sign, is ± 0.036 cm^{-1}, and for 325 of 48 high even levels, ± 0.045 cm^{-1}. If these residuals arose solely from the errors of the levels which are being determined, the correct values of the average error could be found by dividing them by $\sqrt{(n-1)/n}$, where n is the number of observations contained in the mean. The average value of the divisor for the individual odd levels is 0.884, and for the even ones 0.873, giving ± 0.041 and ± 0.052 as the average error of a determination, or a general mean of ± 0.043 cm^{-1}. This should be increased to allow for the error of the other level-values from which these were derived, but a large proportion of the others were accurate interferometer values, so that the estimate of ± 0.05 cm^{-1} as the average error of a determination of a level from a single line may be adopted. Results derived from lines in the red will be more accurate, and from ultra-violet lines less so—assuming equal accuracy of wave-lengths. This refinement may await a definitive list of iron wave-lengths. Meanwhile, it may safely be assumed that the level-values given to two places in table A have average errors ranging from ± 0.02 to ± 0.04 cm^{-1}. Twenty-three cases in which the uncertainty is greater are marked by colons. These include:

(*a*) 15 values based on a single well-measured line. (The reality of the level is usually confirmed by other poorly measured lines.)

(*b*) 7 values in which the observations are discordant, with average residual exceeding 0.08 cm^{-1}.

(*c*) one value based on four lines measured to ± 0.1A ($u^5P_3^\circ$). For these levels the uncertainty may reach ± 0.1 cm^{-1}.

Table A (p. 134) contains those energy levels (464 in number) which have been finally accepted as real. It is arranged in order of term-types, the even terms preceding the odd; singlets, triplets; etc.

The first column gives the adopted electron configurations (when assignable with reasonable probability—see §§ 8–15); the second, the term-designations; the third, the energy levels above the ground state a^5D_4; and the fourth, the differences between successive components of a multiple term—positive when, as usual, the order is "inverted." The fifth column gives references to the author who appears to have been the first to present good evidence of the existence of each energy level, and the last, the definitive g-values determined by Miss Weeks, which are identical with the "corrected" values of table E (p. 203).

The unclassified even levels are listed after the classified even levels and similarly for the odd. The notation is that generally adopted.

IV. SPECTRAL STRUCTURE. IONIZATION POTENTIAL

(8) THE LOW LEVELS

The theoretical interpretation of the spectrum given in Hund's classic monograph has been fully confirmed by all subsequent work. The principal low terms arise from the even electronic configurations $3d^6\ 4s^2$ and $3d^7\ 4s$, and are thoroughly intermingled. These combine with very numerous odd terms coming from $3d^6$ 4s 4p and $3d^7$ 4p; and these again with terms of high even configurations, in which 4p is replaced by 5s or 4d.

The low even terms which may theoretically be anticipated are as follows:

Configuration	Quintets	Triplets	Singlets
d^6s^2	D	P D F G H (P F)	S D F G I (S D G)
d^7s	P F	P F, P D F G H (D)	P D F G H (D)
d^8		(P) F	(S D G)

[21] *Publ. Allegheny Observ.* **6**: 159, 1929. See also **8**: 39, 1931.
[22] *Trans. Internat. Astron. Union* **4**: 65, 1932.

The terms inclosed in parenthesis may be expected to be so high as to be almost unobservable.

The quintet terms and the lowest triplet, a^3F, were assigned to these configurations by Laporte in 1924. The assignment of the others on the basis of combination intensities would not be easy, as many cross-connections are strong; but all ambiguity is removed by comparison with the low terms in Fe II and Fe III, which have been thoroughly investigated by Dobbie[23] and by Edlén and Swings.[24] In the latter, the terms derived from d^6 lie much lower than the others, and have been identified with certainty. Their relative positions and those of certain levels of Fe I are given in table 2, the levels being measured from the lowest, a^5D_4, in both cases. The parallelism of the two sets of values is remarkable. The terms a^1I, b^1G, and b^1D in Fe I were discovered with the aid of this relation, the search being fairly easy, with their relative positions so closely predicted. The ^1S and ^1F terms were not found despite careful searching.

TABLE 2

Fe III, d^6			Fe I, d^6s^2				
Term	J	Level	Term	J	Level	I−III	I−0.96III
a^5D	4	0	a^5D	4	0	0	0
	3	436		3	416	− 20	− 3
	2	739		2	704	− 35	− 6
	1	932		1	888	− 44	− 7
	0	1027		0	978	− 49	− 8
a^3P	2	19405	a^3P	2	18378	−1027	−249
	1	20688		1	19552	−1136	−308
	0	21208		0	20038	−1170	−322
a^3H	6	20051	a^3H	6	19390	− 661	+141
	5	20301		5	19621	− 680	+132
	4	20482		4	19788	− 694	+125
a^3F	4	21462	b^3F	4	20641	− 821	+ 37
	3	21700		3	20875	− 825	+ 43
	2	21857		2	21039	− 818	+ 56
a^3G	5	24559	b^3G	5	23784	− 775	+207
	4	24941		4	24119	− 822	+176
	3	25142		3	24339	− 803	+203
a^3D	3	30858	b^3D	3	29372	−1486	−252
	2	30716		2	29357	−1359	−130
	1	30726		1	29320	−1406	−177
a^1I	6	30356	a^1I	6	29313	−1043	+171
a^1G	4	30886	b^1G	4	29799	−1087	+147
a^1S	0	34812					
a^1D	2	35804	b^1D	2	34637	−1167	+265
a^1F	3	42897					
b^3F	4	50276					
	3	50295					
	2	50185					
b^3P	2	50412					
	1	49577					
	0	49148					

[23] *Proc. Roy. Soc.* A **151**: 703, 1935; *Annals Solar Physics Observ.* (Cambridge) **5**, pt. 1, 1938.

[24] *Astrophys. Jour.* **95**: 532, 1942.

TABLE 3

Fe II d^7			Fe I d^7s							
Term	J	Level	Term	J	Level	I−II	Term	J	Level	I−II
a^4F	4½	1873	a^5F	5	6928	5055	a^3F	4	11976	10103
	1½	3118		1	8155	5037		2	12969	9851
a^4P	2½	13474	a^5P	3	17550	4076	b^3P	2	22838	9364
	½	13905		1	17927	4022		0	23052	9147
a^2G	4½	15845	a^3G	6	21716	5871	a^1G	4	24575	8730
	3½	16369		4	22249	5880				
a^2P	1½	18361	c^3P	2	24336	5975	a^1P	1	27543	9182
	½	18887		0	25091	6204				
a^2H	5½	20340	b^3H	6	26106	5766	a^1H	5	28820	8480
	4½	20806		4	26628	5822				
a^2D	2½	20517	a^3D	3	26225	5708	a^1D	2	28605	8088
	1½	21308		1	26406	5098				
b^2F	3½	31999	d^3F	4	37046	5047	a^1F	3	40534	8535
	2½	31812		2	36940	5128				
d^2D	2½	48039								
	1½	47675								

The comparison between d^7 of Fe II and d^7s of Fe I involves two terms in the latter for each one in the former, with multiplicities greater and less by one. The results are shown in table 3. For brevity, only the components of highest and lowest J are tabulated. Here again the agreement, both in position and in separation, is conclusive. Dobbie assigns the configuration d^7 in Fe II to the term c^2F at 44915 and d^6s to b^2F, but decisive evidence for interchanging the two appears upon comparing d^6 of Fe III with d^6s of Fe II,

TABLE 4

Fe III d^6	Fe II d^6s						II−III	Fe I d^6s^2
Desig	Desig	Level	II−III	Desig	Level	II−III	Wtd. Mean	Desig
a^5D	a^6D	0	0	a^4D	7955	7955	3182	a^5D
a^3P	b^4P	20831	1426	b^2P	25787	6382	3078	a^3P
a^3H	a^4H	21252	1201	b^2H	26170	6119	2840	a^3H
a^3F	b^4F	22637	1175	a^2F	27315	5853	2734	b^3F
a^3G	a^4G	25429	870	b^2G	30389	5830	2523	b^3G
a^3D	b^4D	31483	625	b^2D	36253	5395	2215	b^3D
a^1I	a^2I	32876	2520					a^1I
a^1G	c^2G	33467	2581					b^1G
a^1S	a^2S	37227	2415					
a^1D	c^2D	38164	2360					b^1D
a^1F	c^2F	44915	2018					
	(b^2F)	(31999)	(−10898)					

as shown in table 4, where only the leading components of the terms are given. The designations of the related terms in Fe III and Fe I are added for reference. For the doublets derived from singlets the differences II–III are closely in line with the means, weighted in accordance with the multiplicities, in the other cases. These are found in the last column but one of the table. This comparison shows also that d^2D of Fe II does not come from d^6s, as Dobbie tentatively suggests, but from d^7, where there is a place for it (table 3).

The nature of c^3F is settled by the Zeeman data. It may be assigned with confidence as the lowest term from d^8. For elements of neighboring atomic number, in which the lowest terms from d^n are well-determined, the differences between these and the lowest terms of $d^{n-1}s$ are as follows for the leading components:

Cr I	$d^6\ {}^5D - d^5s\ {}^7S$	35399
Co I	$d^9\ {}^2D - d^8s\ {}^4F$	24016
Ni I	$d^{10}\ {}^1S - d^9s\ {}^3D$	14525

The corresponding term in Fe I should be from 25000 to 30000 above a^5F_5. For c^3F_4 (32873) this difference is 25955. The over-all separations of the terms are 243 and 973 for Cr I and Co I. The value 892 for c^3F also falls into line.

The next lowest term of the d^8 configuration should be ^1D, followed by ^3P. The separations of $^1D_2$ and $^3P_2$ from $^3F_4$ are 6868, 8215 for d^2s^2 in Ti I, and 13521 (raised by perturbations) and 15609 for d^8s^2 in Ni I. It is probable that these terms in Fe I lie about 10000 and 13000 cm^{-1} above c^3F.

The list of low even terms of Fe I is complete up to the level where combinations would be very difficult to find—except for two singlet terms from d^6s^2.

(9) RELATED HIGHER TERMS. IONIZATION POTENTIAL

The strongest lines of Fe I arise from terms based on the three lowest terms of Fe II and form a distinctive group, as follows (table 5):

TABLE 5

Limit Fe II		Added Electron			
		4s	4p			5s
d^6s	a^6D		$z^7P°$	$z^7D°$	$z^7F°$	e^7D
		a^5D	$z^5P°$	$z^5D°$	$z^5F°$	e^5D
			$y^5P°$	$x^5D°$	$x^5F°$	g^5D
d^6s	a^4D		$z^3P°$	$z^3D°$	$z^3F°$	e^3D
d^7	a^4F	a^5F	$y^5D°$	$y^5F°$	$z^5G°$	e^5F
		a^3F	$y^3D°$	$y^3F°$	$z^3G°$	e^3F

These terms and their mutual relations have long been recognized. They include all the lower terms belonging to each group (low even, odd, high even) and are almost completely separated from the higher terms of each group except for g^5D, e^3D, which are intermingled with others. Because of this isolation, they should be little perturbed, and comparison of the 4s and 5s terms should give a good approximation to the ionization potential. In making this, it is possible to improve considerably on a simple Rydberg formula by assuming that the difference in the denominators n* for 4s and 5s is not exactly unity, but has a value interpolated between the results for homologous terms in the spectra of elements of neighboring atomic numbers, in which the limits and values of n* are accurately determined by longer series.

For the configuration $d^{n-1}4s$ and $d^{n-1}5s$, each low term has a high one uniquely in series with it, but for $d^{n-2}4s^2$ and $d^{n-2}4s5s$, both the higher terms such as e^7D, e^5D are, in a sense, in series with a^5D. The differences Δn^* are smaller for the change involving an increase in multiplicity, but both run smoothly with atomic numbers, and may be used as is illustrated in table 6.

TABLE 6

VALUES OF Δn^* (5s−4s)

Low Configuration		$d^{n-1}s$	$d^{n-2}s^2$	
Change in Multiplicity		0	+2	0
At. No.	Element			
24	Cr	1.075		
25	Mn		0.960	1.080
26	Fe			
27	Co		0.999	1.108
28	Ni	1.077	1.007	1.120
29	Cu	1.066	1.013	1.142
30	Zn		1.025	

The value of Δn^* for $d^{n-1}s$ is for the higher multiplicity. Ill-determined or perturbed terms are omitted, and some which do not exist are left blank.

The values of Δn^* in Fe I may now be estimated as 1.072 for a^5F, e^5F, 0.982 for a^5D, e^7D, and 1.094 for a^5D, e^5D. The corresponding limits are easily found with the aid of a detailed Rydberg Table,[25] changing approximate values until the desired Δn^* is obtained. The results for the component of greatest J-value in each term are as follows (table 7):

TABLE 7

Terms	n*	n*	Δn^*	Limit	Ionization
a^5D_4, e^7D_5	1.3170	2.2950	0.9820	63650	63650
a^5D_4, e^5D_4	1.3135	2.4075	1.0940	63630	63630
a^5F_5, e^5F_5	1.3699	2.4419	1.0723	65408	63535

The limit given in the last column but one is $a^6D_{4\frac{1}{2}}$ in Fe II (the ground-level) for the first two lines, but

[25] *Rydberg Interpolation Table.* Princeton Univ. Observ., 1934.

$a^4F_{4\frac{1}{2}}$ (which is higher by 1873) for the last line. Subtracting this gives the value of $a^6D_{4\frac{1}{2}} - a^5D_4$, which measures the ionization potential. The three determinations are remarkably concordant with mean 63605.

An independent and probably a still better determination can be obtained from the higher ^7D terms which were found during the revision of the spectrum. The term g^7D evidently arises from $(a^6D)6s$. Search for the 7s term found the level here called h^7D_5. The remaining components were not located, but this is determined by three good combinations with the 4p triad, and is trustworthy. Fitting a Ritz formula to e^7D_5, g^7D_5, h^7D_5 (by adjustment of the limit) gives:

$$n^7D_5 = 63732 - T; \quad T = \frac{R}{n - 1.6454 - 3.07 \times 10^{-6}T}$$

This corresponds to an ionization potential of 7.862 volts, 0.016 volts higher than the first determination. The value 63700, corresponding to 7.858 volts, may be adopted. This is 0.02 volt higher than the previously accepted value and is probably reliable to 0.01 volt.

V. ELECTRON CONFIGURATIONS

(10) The High Even Terms

The terms arising from configurations involving 5s electrons have been discussed. A 4d electron added to the three principal limit-terms should give pentads, of which the following terms have been identified:

a^6D; e^7G, e^7F, f^7D, e^7P, e^7S; e^5G, f^5F, f^5D, e^5P, e^5S
a^4F; e^5H, f^5G, g^5F, h^5D, f^5P; e^3H, e^3G, f^3F, f^3D, e^3P
a^4D; g^5G, ... i^5D ...

All but one of the 42 levels arising from a^6D and of the 38 from a^4F have been found in each case. They form closely packed groups—the first between 50342 and 52067 and the second from 53061 to 55726. A large majority of the levels can be conclusively identified from the multiplet intensities and g-values, but the assignments are uncertain for the levels of small J. These complicated groups were first unravelled by Catalán. His conclusion that e^5S_2 and e^7F_3 are practically coincident is confirmed by the writers, but the present arrangement of the $(a^4F)4d$ group differs from his. In Co I, Ni I, and Cu I, where components of the limit-terms are widely separated, there is definite evidence that in the configura-

TABLE 8

Term Values and Denominators

Limits	$a^6D_{4\frac{1}{2}}$ 63700			$a^4F_{4\frac{1}{2}}$ 65573, $a^4F_{3\frac{1}{2}}$ 66130			$a^4D_{3\frac{1}{2}}$ 71655		
Electron	Desig	Term	n*	Desig	Term	n*	Desig	Term	n*
4s	a^5D_4	63700	1.312	a^5F_5	58645	1.368	a^5D_1	71655	1.238
				a^3F_4	53597	1.431			
5s	e^7D_5	20884	2.292	e^5F_5	18567	2.431	g^5D_4	20305	2.325
	e^5D_4	19023	2.402	e^3F_4	17612	2.496	e^3D_3	20361	2.321
6s	g^7D_5	9899	3.330						
7s	h^7D_5	5803	4.349						
4d	e^7S_3	12130	3.008	f^5P_3	12413	2.973			
	e^7P_4	13225	2.881	h^5D_4	12418	2.973			
	f^7D_5	13322	2.870	g^5F_5	12512	2.961	i^5D_4	13957	2.804
	e^7F_6	13358	2.866	f^5G_6	12404	2.974			
	e^7G_7	13048	2.905	e^5H_7	12298	2.987	g^5G_6	13653	2.835
	e^5S_2	12551	2.957	e^3P_2	11250	3.123			
	e^5P_3	11863	3.041	f^3D_3	12382	2.977			
	f^5D_4	13277	2.875	f^3F_4	11447	3.096			
	f^5F_5	12597	2.951	e^3G_5	12391	2.976			
	e^5G_6	13177	2.886	e^3H_6	12289	2.988			
4p	$z^7P_4^\circ$	39989	1.657	$y^5D_4^\circ$	32477	1.838	$y^5P_3^\circ$	34888	1.774
	$z^7D_5^\circ$	44349	1.573	$y^5F_5^\circ$	31878	1.855	$x^5D_4^\circ$	32029	1.851
	$z^7F_6^\circ$	41050	1.635	$z^5G_6^\circ$	30729	1.890	$x^5F_5^\circ$	31398	1.870
	$z^5P_3^\circ$	34644	1.780	$y^3D_3^\circ$	27398	2.001	$z^3P_2^\circ$	37708	1.706
	$z^5D_4^\circ$	37800	1.704	$y^3F_4^\circ$	28887	1.949	$z^3D_3^\circ$	40333	1.650
	$z^5F_5^\circ$	36825	1.726	$z^3G_5^\circ$	30194	1.906	$z^3F_4^\circ$	40348	1.649
5p	$u^5P_3^\circ$	12008	3.023						
	$t^5D_4^\circ$	12623	2.948						
	$u^5F_5^\circ$	12683	2.941						

tion d^{n-2}s·d the leading components of all the ten terms belonging to a pentad are derived from the component of highest J-value of the limiting terms and all lie at almost the same level—while in d^{n-1}·d the two leading components of the terms of higher multiplicity both go to the limit with greatest J and lie close together, while the leading components of the terms of lower multiplicity go to the next component of the limit and are near the level of the third components of the terms of higher multiplicity. Catalán has assumed that the former rule applies to both sets of pentads in Fe I. For those derived from d^{6}s(^{6}D) the Zeeman data confirm this, but for those from d^7(^{4}F) the other arrangement gives a better representation of the intensities and also of the g-values. For the smaller values of J the mutual perturbations of the levels are great. The g's differ widely from the theoretical values (§ 29), and large perturbations of the intensities and levels are to be anticipated. There can be no doubt that these levels belong to the pentad as a whole, but the assignment of individual term-designations has little significance.

Of the third pentad, with limit a^{4}D, only the terms g^{5}G (discovered by Miss Weeks from the Zeeman effect) and part of i^{5}D have been identified. There are unclassified lines in the appropriate region of the spectrum which may come from other levels belonging to this pentad, but not enough to locate them.

The term-values for the leading components of the terms based on the three lowest terms of Fe II, referred to these as limits, and the corresponding values of n* are given in table 8. The triplets of (a^{4}F)4d are referred to a^{4}F$_{3\frac{1}{2}}$ as limit, for reasons described above. The limits are referred to the ground-level of Fe I as origin. Table 8 shows that few more high even terms should be observable even though more complete line lists were available. The group (a^{4}D)4d could probably be filled up. The ^{5}D$_4$ term from (a^{6}D)6s should have n* about 3.45, and be at 63700–9200 or 54500, and its combinations might be found. The ^{5}F, ^{3}F of (a^{4}F$_4$)6s may be expected near 56000 but should give faint lines and those from (a^{4}D)6s still fainter ones.

A 5d electron should give n* about 3.9 and levels about 7000 below the limit. The unidentified levels 1, 2, and 3, which combine like septets, may belong to the pentad group with limit a^{6}D.

Terms having limits in Fe II higher than a^{4}D are hardly to be expected. The next lowest limits are a^{4}P$_{2\frac{1}{2}}$ at 13477 and a^{2}P$_{1\frac{1}{2}}$ at 18361 above a^{6}D$_{4\frac{1}{2}}$. The addition of any even electron but 5s to them would produce states lying above the principal ionization level, subject to auto-ionization and giving faint and diffuse lines, if any. The terms arising from (a^{4}P)5s have been searched for unsuccessfully.

The energy of binding of a 4s electron to the various states of Fe II to produce the known low even terms and the corresponding values of n* are given in table 9.

TABLE 9

VALUES OF n* FOR 4s ELECTRON

Fe II		Fe I			Fe II		Fe I		
d^7	Level	d^{7}s	Term	n*	d^{6}s	Level	d^{6}s^2	Term	n*
a^{4}F	65573	a^{5}F	58645	1.368	a^{6}D	63700	a^{5}D	63700	1.313
		a^{3}F	53597	1.431	a^{4}D	71655		71655	1.238
a^{4}P	77174	a^{5}P	59624	1.357	b^{4}P	84531	a^{3}P	66153	1.288
		b^{3}P	54336	1.421	b^{2}P	89487		71109	1.242
a^{2}G	79545	a^{3}G	57829	1.378	a^{4}H	84952	a^{3}H	65562	1.294
		a^{1}G	54970	1.413	b^{2}H	89870		70480	1.248
a^{2}P	82061	c^{3}P	57725	1.379	b^{4}F	86337	b^{3}F	65696	1.292
		a^{1}P	54518	1.419	a^{2}F	91015		70374	1.249
a^{2}H	84040	b^{3}H	57934	1.376	a^{4}G	89129	b^{3}G	65345	1.296
		a^{1}H	55220	1.410	b^{2}G	94089		70305	1.249
a^{2}D	84217	a^{3}D	57992	1.376	b^{4}D	95183	b^{3}D	65811	1.291
		a^{1}D	55612	1.405	b^{2}D	99953		70581	1.247
b^{2}F	95699	d^{3}F	58653	1.368	a^{2}I	96576	a^{1}I	67263	1.277
		a^{1}F	55165	1.410	c^{2}G	97167	b^{1}G	67368	1.276
d^{5}s^2					a^{2}S	100927			
a^{6}S	87017				c^{2}D	101864	b^{1}D	67227	1.278
					c^{2}F	108614			

TABLE 10

PREDICTED ODD TERMS ARISING FROM 4p

Limits in Fe II		Septets			Quintets							Triplets								Singlets							
Config	Desig	P	D	F	S	P	D	F	G	H	I	S	P	D	F	G	H	I	K	S	P	D	F	G	H	I	K
d^6s	a^6D	x	x	x		x	x	x																			
d^7	a^4F						x	x	x					x	x	x											
d^6s	a^4D					x	x	x					x	x	x												
d^7	a^4P				x	x	x					x	x	x													
d^7	a^2G														x	x	x						x	x	x		
d^7	a^2P											x	x	x						x	x	x					
d^7	a^2H															x	x	x						x	x	x	
d^7	a^2D												x	x	x						x	x	x				
d^6s	b^4P				x	x	x					x	x	x													
d^6s	a^4H								x	x	x					x	x	x									
d^6s	b^4F						x	x	x					x	x	x											
d^5s^2	a^6S	x				x																					
d^6s	a^4G							x	x	x					x	x	x										
d^6s	b^2P											x	x	x						x	x	x					
d^6s	b^2H															x	x	x						x	x	x	
d^6s	a^2F													x	x	x						x	x	x			
d^6s	b^2G														x	x	x						x	x	x		
d^6s	b^4D					x	x	x					x	x	x												
d^7s	b^2F													x	x	x						x	x	x			
d^6s	a^2I																x	x	x						x	x	x
d^6s	c^2G														x	x	x						x	x	x		
d^6s	b^2D												x	x	x						x	x	x				
d^6s	a^2S												x								x						
d^6s	c^2D												x	x	x						x	x	x				
d^6s	c^2F													x	x	x						x	x	x			
Totals		2	1	1	2	6	7	6	4	2	1	4	10	14	14	12	8	4	1	2	6	8	9	8	6	3	1
Observed		2	1	1	2	7	7	6	4	1		2	5	8	7	10	7	3			1	4	4	5	3	1	
Exc. 7 Highest Limits		2	1	1	2	6	7	6	4	2	1	4	7	10	9	9	6	3		2	3	4	4	5	4	2	

Each d^7 limit has two associated d^7s terms and the first six d^6s^2 terms have two associated d^6s limits. The means of n^* for the pairs, weighted according to multiplicity, range from 1.378 to 1.392 in the first group and 1.273 to 1.284 in the second. The values of n^* for the d^6s^2 terms with only doublet limits lie in the latter range. The general means are: for $d^7 \rightarrow d^7s$ 1.385 with average deviation ± 0.004 and for $d^6s \rightarrow d^6s^2$ 1.277, A.D. ± 0.003. The greater binding energy for d^6s^2 corresponds to the completion of a pair of 4s electrons.

(11) THE HIGHER ODD TERMS

The addition of a 4p electron to the various limits gives rise to numerous triads of terms. The observed odd terms should include these (except for some of the highest limits) and a few terms involving a 5p electron. For the known 5p triad from a^6D the mean value of n^* is 2.971—greater by 1.25 than for the corresponding 4p triad. For the triads from a^4F, n^* should be about 3.11 for the quintets and 3.20 for the triplets, leading to levels near 54000 and 55000. These have not been found. Apart from these, the observed odd terms are almost certainly from 4p.

The predicted terms are listed in table 10.

The numbers of predicted and observed terms of each type are given at the bottom of the table. The deficiency of the latter indicates that many terms with high-lying limits have been missed. Omitting the seven highest limits gives the predictions in the lowest line, which are not far from the observed numbers.

The mean levels at which the various triads may be expected can probably best be estimated by assuming that the difference Δn^* between the mean $\overline{n^*}$ for the terms of a 4p triad and those for the related 4s term are the same as in the cases already discussed. This gives the results in table 11.

TABLE 11

DIFFERENCES OF n^*

Limit	High Multiplicity		Low Multiplicity	
	$\overline{n^*}$	Δn^*	$\overline{n^*}$	Δn^*
d^7 a^4F	1.861	+0.493	1.952	+0.521
d^6s a^6D	1.622	+0.310	1.737	+0.425
d^6s a^4D	1.832	+0.594	1.668	+0.430

For the remaining d^7 limits the values of n* (4s) are almost identical with their means—1.372 for the terms of higher multiplicity in a related pair and 1.413 for those of lower multiplicity. Adding the appropriate Δn^* gives 1.865 and 1.934, corresponding to term-values of 31516 and 29338. Proceeding similarly with d^6s, the differences in level between the limit and the 4p triad come out as follows (table 12):

TABLE 12

APPROXIMATE TERM VALUES—4p TRIADS

Limit	High Mult.	Low Mult.
d^7	31500	29300
d^6s High Mult.	42500	38000
d^6s Low Mult.	32400	39000

The term-values for d^6s apply to the groups of four triads related to the low triplet terms of d^6s^2. The singlet d^6s^2 terms have but two related triads. For these the values 42500 and 39000 have been adopted, as the other pair of term-values are affected by the mutual repulsion of terms of the same multiplicity.

The assignment of the terms to specific triads is complicated by considerable perturbations shown by the g-values (Part II, § 29) which indicate a good deal of sharing of identity among the levels, especially those with small J-values. The guiding principles of the present arrangement have been as follows:

(*a*) *Level.* The terms of a given triad are likely to be within 2000 cm^{-1} of the levels determined as above (which it is useless to predict within 1000 cm^{-1}).

(*b*) *Separation.* In general, limit-terms and low related terms of a wide internal separation are associated with odd terms of wide separation.

(*c*) *Intensity.* The terms of a triad will in general combine most strongly with the related low even term having the same limit in Fe II. Each d^7s term in Fe I has one related d^7p triad; most d^6s^2 terms have four—two of the same multiplicity and one each of multiplicity higher and lower by 2. The latter give intersystem combinations with the low term, which may be relatively faint. Combinations of d^7p terms with d^7s terms having different limits are usually stronger than those with d^6sp terms, and vice versa. On account of the Boltzmann factor, the population of high states is less than that of lower states, and their combinations with the same low term tend to be fainter. The combinations of a d^7s term with its related d^7p triad are always strong. Those of a d^6s^2 term with the lower of the related d^6sp triads of the same multiplicity are much weaker; with the higher triad they are stronger, but hardly comparable with d^7s–d^7p. This is clearly shown by King's determinations of f-values for Fe I and Ti I [26] and is a general property of the spectra of the iron group, as is shown by the incidence of ultimate lines.[27] Further evidence of this has appeared in the present work. The combinations of the singlet d^7s terms are much stronger than those of the d^6s^2 singlets, so that the latter were among the last terms to be detected in the analysis.

[26] King, R. B., and A. S. King, *Mount Wilson Contr.* No. 581, 1938; *Astrophys. Jour.* **87**: 24, 1938.

(12) ODD SEPTETS AND QUINTETS

The predicted septet terms have been identified with certainty. The term $y^7P°$ is fully confirmed by the Zeeman effect. Its components are arranged, as was to be expected, in the normal order, with the smallest J lowest, while almost all terms of Fe I are inverted. The term-value 46596 ($n^* = 1.535$) is unusually great for 4p. Very few examples of the configuration $d^{n-3}\,s^2p$ are known in other spectra.

The quintet terms may be assigned with considerable confidence. By applying the approximate term-values of table 12 to the limits in table 9, the estimated mean levels of the triads in table 13 are obtained. For the limit a^6S the separation for d^6sp has been taken as a rough guide. The d^7p triad with limit a^4P should combine more strongly with a^5P, and the d^6sp with b^4P. This puts $v^5P°$, $u^5D°$ in the former, and leaves $x^5P°$, $w^5D°$ for the latter—all near the predicted levels. The relative levels then place $y^5S°$, $z^5S°$.

The $^5P°$ term from a^6S should be related to $y^7P°$, much like the $^5P°$, $^7P°$ terms from $3d^54p$ in Mn II. This $^7P°$ term has normal separations (−264, −176) while the $^5P°$ term is inverted (+114, +72). In Fe I $y^7P°$ has separations −215, −155 and $w^5P°$ +176, +97—much smaller than the other $^5P°$ terms. The difference $^5P_3° - {}^7P_4°$ is 4564 in Mn II and 5715 in Fe I. This identification is conclusive.

Their levels suffice to assign $y^5G°$, $z^5H°$ to a^4H. The related $^5I°$ could be detected only by combinations with a^5F, which should be very faint. The triad with limit b^4F is reasonably filled by $v^5D°$, $w^5F°$, $x^5G°$. The higher one with limit a^4G evidently contains $v^5F°$ and $w^5G°$. The $^5H°$ term has not been found. Two lines at 38628.18 and 38713.63 cm^{-1} of intensities 8 III and 6 III by their behavior in the furnace come from a^5F or a level a little higher. If they are transi-

TABLE 13

ODD QUINTET TERMS

Limit	Est. Level	Adopted Terms					
a^4P	46000	$y^5S°$	44512	$v^5P°$	47967	$u^5D°$	46721
b^4P	42000	$z^5S°$	40895	$x^5P°$	42532	$w^5D°$	43499
a^4H	42000	$y^5G°$	42784	$z^5H°$	43321	$^5I°$	
b^4F	44000	$v^5D°$	44415	$w^5F°$	44243	$x^5G°$	45608
a^6S	44000:			$w^5P°$	46137		
a^4G	46000	$v^5F°$	47606	$w^5G°$	47363	$^5H°$	
b^4D	53000	$t^5P°$	53388	$^5D°$	(53891)	$^5F°$	(54013)

[27] Meggers, W. F., *Jour. Optical Soc. America* **31**: 39, 1941.

tions to the $^5H^\circ$ term, it must be near 46000. The high terms u^5P°, t^5D°, u^5F° have already been assigned to $(a^6D)5p$. The corresponding septet triad should be near 50000, but has not been identified. This leaves only t^5P°, which may be assigned tentatively to the triad with b^4D as limit. The unclassified levels 10_3° and 12_5° have positions and g-values agreeing well with $^5D_3^\circ$, and $^5F_5^\circ$ and are given in parentheses; but the other levels of these terms have not been identified. The quintets are now well accounted for.

TABLE 14

ODD TRIPLET TERMS FROM 4p

Limit	Est. Level	Adopted Terms			Related Low Term
a^4F	36000	y^3D° 38175	y^3F° 36686	z^3G° 35379	a^3F
a^4D	33000	z^3P° 33947	z^3D° 31322	z^3F° 31307	$[a^5D]$
a^4P	48000	y^3S° 47556	y^3P° 46727	w^3D° 47017	b^3P
a^2G	48000	x^3F° 46889	v^3G° 49460	y^3H° 49434	a^3G
a^2P	51000	8_1° 52857?	v^3P° 52916:	u^3D° 51969	c^3P
a^2H	53000	u^3G° 51373	w^3H° 52431	y^3I° 52655	b^3H
a^2D	53000	w^3P° 50186:	t^3D° 52213:	u^3F° 56592:	a^3D
b^4P	47000	z^3S° 46601	x^3P° 48304	x^3D° 45220	a^3P
b^2P	57000	$^3S^\circ$	$^3P^\circ$	$^3D^\circ$	
a^4H	47000	y^3G° 45294	z^3H° 46982	z^3I° 45978	a^3H
b^2H	57000	$^3G^\circ$	u^3H° 56334:	x^3I° 57027	
b^4F	48000	v^3D° 49135	w^3F° 49108	x^3G° 47834	b^3F
a^2F	58000	$^3D^\circ$	$^3F^\circ$	$^3G^\circ$	
a^4G	51000	v^3F° 51304:	t^3G° 53983:	x^3H° 51023:	b^3G
b^2G	62000	$^3F^\circ$	$^3G^\circ$	$^3H^\circ$	
b^4D	57000	$^3P^\circ$	$^3D^\circ$	$^3F^\circ$	b^3D
b^2D	68000	$^3P^\circ$	$^3D^\circ$	$^3F^\circ$	
b^2F	64000	$^3D^\circ$	$^3F^\circ$	$^3G^\circ$	d^3F
a^2I	54000:	$^3H^\circ$	$^3I^\circ$	$^3K^\circ$	a^1I
c^2G	54000:	$^3F^\circ$	$^3G^\circ$	$^3H^\circ$	b^1G
a^2S	58000:		$^3P^\circ$		
c^2D	59000:	$^3P^\circ$	$^3D^\circ$	$^3F^\circ$	
c^2F	64000:	$^3D^\circ$	$^3F^\circ$	$^3G^\circ$	

(13) TRIPLETS

Table 14 is similar to the last, except that the low even terms in Fe I which are related to the triads are given. The triads already located are included. The numbers of odd triplet terms to be anticipated below 54000 and of those observed are as follows:

Term	$^3S^\circ$	$^3P^\circ$	$^3D^\circ$	$^3F^\circ$	$^3G^\circ$	$^3H^\circ$	$^3I^\circ$
Predicted	3	5	7	6	6	4	2
Observed	2	5	8	5	7	4	2

The agreement is surprisingly good in view of the roughness of the prediction. From 54000 to 62000 the comparison gives:

Term	$^3S^\circ$	$^3P^\circ$	$^3D^\circ$	$^3F^\circ$	$^3G^\circ$	$^3H^\circ$	$^3I^\circ$	$^3K^\circ$
Predicted	1	4	4	5	4	4	2	1
Observed				2	3	3	1	
Omitting faint combinations	1	2	3	3	3	2	1	

Many of the predicted terms have evidently been missed.

In assigning the observed terms, so far as practicable, to their limits, the intensities are important, since each triad has an associated low level of the same multiplicity with which its combinations should be strong. Table 15 gives that of the strongest line in each multiplet, omitting predicted lines present in the sun (§ 23). Values not in parentheses are on King's scale and fairly comparable. Those in parentheses are on a great variety of scales. Most of these lines are faint. For the combinations with a^5D, which lie far in the ultra-violet, the intensities are on a more open scale. The symbol "†" denotes that the intensities in the multiplet are seriously abnormal. The maximum separation of each term is given. When only two components are known, the separation is followed by +. Inverted separations are listed as positive, the few normal ones as negative. The separations of the odd terms average much smaller than for the even terms, doubtless because the 4p electron by itself would produce a considerable separation in the normal direction. Terms already assigned are omitted. Of the three $^3I^\circ$ terms, y^3I° combines strongly with b^3H, and belongs to its triad; and the others to a^3H. All three are near the predicted levels. Intensities and levels connect u^3G°, w^3H°, with b^3H and place y^3G°, z^3H° in the lower a^3H triad. The one related to a^3G consists of y^3H°, v^3G° and either x^3F° or w^3F°; strong combinations with a^5P connect y^3P°, w^3D° with a^4P; x^3P°, x^3D° combine well with a^3P; u^3D° is related to c^3P, and either w^3P° or v^3P° would fit, though the wide separation of the latter is favorable. The two known $^3S^\circ$ terms should belong to the lower two of these triads. The uncertain choice between them depends on the strong combination of z^3S° with a^5D. Intensities and levels suggest assigning x^3G°, w^3F°, v^3D° to b^3F (leaving x^3F° for a^3G). Beyond this, assignments are difficult; x^3H°, t^3G°, and v^3F° may be provisionally connected with b^3G and w^3P°, t^3D°, u^3F° with a^3D.

It is unprofitable to attempt specific configuration assignments for the higher terms, and some of those already made and marked with colons in the table are doubtful. The lowest missing term is the $^3S^\circ$ related to c^3P, which should be near 51000. The only unassigned odd level with $J = 1$, 8_1°, is in the expected place, and combines mainly though irregularly with the low ^3P term. The well-determined value of g for this level is 1.246—greatly perturbed from a theoretical 2.000—but the neighboring levels z^1P°, $u^3D_1^\circ$,

TABLE 15

INTENSITIES OF TRIPLET COMBINATIONS

d^6sp, d^7p		d^6s^2						d^7s									d^8
Desig		a^5D	a^3P	b^3D	b^3F	b^3G	a^3H	a^5P	a^5F	b^3P	c^3P	a^3D	a^3F	d^3F	a^3G	b^3H	c^3F
	Sep	978	1660	−52	398	555	398	377	1226	213	756	398	992	−105	534	522	892
z^3S°		(10)	1					2		6	(2)						
y^3S°		(1n)	3					2		10	3	(1)					
y^3P°	229	(3)	4	(1)				10	(2)	12	(2)						
x^3P°	211	(25)	6	(2)				4	6	6	5	(2)	(1)				1
w^3P°	−236		5	(1)	(1)			(1)		4	7	5					(1)
v^3P°	892+		10	3				(3)		7	2†	(1)†					
x^3D°	331	(15)	8		4			2	4	12	5†	(1)	12		(1)*		10
w^3D°	255	(3)	1	(1)†	1	(1)		15	(7)	12	(1)	(1)	5†	(1)†	4		40
v^3D°	163		(2)*	(1)†	3	(1)*		3	(4)	4†	(1)	(2)†		3*	3		(1)
u^3D°	543		5	(1)	4		(1)		(1)	(1)	12	2	(2)†	2	(1)		(3)
t^3D°	502		5†	(1)	6†		(1)			(1)†	(1)	8		2			(2)
s^3D°	322+	(1)			(2)				(2)		(1)	3					(1)†
x^3F°	308	(2)	(1)	(1)	4	(3)	(1)	20	(6)	5	(1)		8	(2)	10		8
w^3F°	324		(1)	(2)	8†	2	(1)	2*	(4)*		(1)		(2)	3*	12		4
v^3F°	−103			(2)	4	3	2	(1)?		(1)	2	5		3	6	3	(1)
u^3F°	266				1	(2)	(2)	(3)				8	(4)		(3)†	(3)†	
t^3F°	159				4	(2)				(1)?			(2)		(3)		
y^3G°	268	(2)	(2)		3	6	10		6				10		3	(1)	
x^3G°	−22		(1)		6	(1)	(1)		(20)				5	(1)†	3†	(1)	
w^3G°	244				2	(2)	(1)	8	(2)		(1)		3†		3	(1)	(1)
v^3G°	390				4	7	4		(1)			(1)	3		20†	(1)	4
u^3G°	452				4	(2)	6			(1)	1	(1)	(5)	4	10	10	(2)
t^3G°	617			(1)		4	(3)		(20)†			(2)	(2)	(1)	6	10	(2)
s^3G°	192				(3)	3	(2)						(4)		7		
r^3G°	438				6	(2)											
q^3G°	129				(4)	(1)					(3)†	(2)†	(2)?†			(2)†	
z^3H°	124	(1)			2	5	8	3	(6)				2		15	(4)	5
y^3H°	293				(1)	5†	(2)						2		20	3	4*
x^3H°	45+					3	3	(1)					(1)		(2)	(1)	(1)
w^3H°	337				(2)	5	15								(2)	12	(−)†
v^3H°	−60				(2)	(2)†	(4)						(1)		10	5	
u^3H°	89					(1)†	4					(3)			5	5	
t^3H°	392				(4)		(25)						(1)		(1)	10	
z^3I°	158				(1)		8		1							5	
y^3I°	244					(1)*	3									20	
x^3I°	77					(−)†	5								(2n)	10	

* Blend. † Multiplet intensities abnormal.

and $u^5P_1^\circ$ have g-values which are too great by 0.266, 0.200, and 0.133, enough to make up four-fifths of the difference. This possibility is noted in table 14.

(14) SINGLETS

The odd singlet terms to be expected from limits below 95000 are given in table 16. The estimated differences from the limits are 29000 for d^7 and 39000 for d^6s. The predicted and observed numbers of terms are given at the bottom. Some terms with large and small J have been missed. The predicted levels are closely packed, and assignment to configurations must depend mainly on the intensities. The first four triads in table 16 have related low even singlet terms; the others from d^6sp are related to low triplets and likely to give fainter combinations. The observed intensities of combinations between the known odd singlets and the relevant low terms are given in table 17. It is clear that z^1H°, z^1F° belong with a^1G; z^1I°, y^1H° with a^1H; y^1D°, z^1P° with a^1P; and w^1D° probably with a^1D. The first two triads may be filled out with z^1G° and y^1G° (though the latter is very low). These nine levels account for all the strongest singlet combinations. It is hardly practicable to assign the others. Many of them must come from d^6sp.

TABLE 16

ODD SINGLET TERMS

Limit	Est. Level	Adopted Terms			Rel. Low Term
a^2G	50000	z^1F° 50587	z^1G° 47453	z^1H° 48383	a^1G
a^2P	53000	$^1S^\circ$	z^1P° 53230	y^1D° 51708	a^1P
a^2H	55000	y^1G° 48703:	y^1H° 53722	z^1I° 53094	a^1H
a^2D	55000	$^1P^\circ$	w^1D° 55754	$^1F^\circ$	a^1D
b^2P	50000	$^1S^\circ$	$^1P^\circ$	$^1D^\circ$	a^3P
b^2H	51000	$^1G^\circ$	$^1H^\circ$	$^1I^\circ$	a^3H
a^2F	52000	$^1D^\circ$	$^1F^\circ$	$^1G^\circ$	b^3F
b^2G	55000:	$^1F^\circ$	$^1G^\circ$	$^1H^\circ$	b^3G
Type		S P	D F	G H	I
Numbers predicted		2 3	4 4	5 4	2
Numbers found		1	4 4	5 3	1

(15) UNCLASSIFIED LEVELS

The even levels called 1, 2, and 3 combine with septet terms and are at the right level for the pentad $(a^6D)5d$, while 4 may fit into the incomplete quintet pentad $(a^4D)4d$. The high odd levels fall in a region containing many anticipated terms; but 2_2°, which is undoubtedly real, lies low enough to be puzzling.

Two even levels at 40871 ($J = 3$) and 41178 ($J = 2$) present a perplexing problem. Both are confirmed by several combinations, as is shown in table 18, which gives the intensities and the residuals in 0.01 cm^{-1}. These are given in the sense o–c for transitions from lower terms, but c–o for transitions to higher terms, so that a change in these levels affects all residuals in the same sense. Predicted solar lines are here included, since the majority of them are probably real (§ 22). The intensities strongly suggest $^3D_3$, $^3D_2$; but the levels are far too low for any configuration including a 5s or 4d electron. The only ^3D term from the lower configurations which has not already been unequivocally located comes from d^7s and has d^2D for limit. The regularities shown in table 3 indicate that this term should be near 53000. It appears very improbable that so enormous a displacement can be due to perturbations, especially since those of the lower terms of the overlapping configurations d^6s^2 and d^7 are small.

There are predicted ^3P and ^3F terms of d^6s^2, related to b^3F and b^3P of Fe III, but by table 2 these should be near 48000.

The singlet $^1F_3$ from d^6s^2 should be close to 41200, and $^1D_2$ from d^8 roughly at 43000; but the combination-intensities of the observed levels are irreconcilable with their being singlets, though they do indicate that the two are related.

The writers can suggest no solution of the difficulty. In recognition of it, they have deliberately departed from the usual nomenclature to the extent of calling these two levels X_3 and X_2 rather than assigning numbers as usual.

The general result of this analysis is that the iron arc spectrum, despite its complexity, is highly regular. All the low even terms predicted by theory have been found close to their anticipated position except two which should give few and faint lines and others whose combinations should be in the infra-red. The high even terms are also well identified, though not so completely. The lower odd terms (below about 55000) are also satisfactorily accounted for except for a couple which could be detected only through highly improbable combinations, and a few of low J-value.

TABLE 17

SINGLET COMBINATIONS

	a^1P	a^1D	a^1G	a^1H	c^3P	a^3D	a^3G	b^3H	a^3P	b^3D	b^3F	b^3G	a^3H	b^1D	b^1G	a^1I
z^1P°	3	3			4	1			(2)					(2)		
z^1D°		7			(1)	(1)			(1)	(1)		(1)*				
y^1D°	7	3			2	9	(1)*			(2)*	(1)			(2)		
x^1D°	3n*	(1)			(1)	(1)	2			(1)		1*		(1)		
w^1D°	(4)	20			(1)*	3			(1)	(1)				(1)		
z^1F°		4	8		(1)	(2)				(1)	(1)	(1)		(1)	(1)	
y^1F°		2	3			(1)			(2)					m	(2)	
x^1F°		3	4		(1)	(1)*				(1)	(–)	(1)			(2)	
w^1F°		2	2			(2)*	5			(1)		2		(1)		
z^1G°			7	3			10	(1)		(2)	1	(1)			(1)	
y^1G°			15	4			10	(2)*			(1)	(1)	(1)		(2)	
x^1G°			7	(2)		(1)*	(2)	(2)			(1)		(1)		(3)*	
w^1G°			m	2			(–)	(2)*				(1)			2	
v^1G°				6				(1)							(1)	
z^1H°			20	m			3	(2)				5	(1)		(2)	4
y^1H°				10			10					2*				
x^1H°			(3)	6			(2)						(2)			2
z^1I°				15				10					(1)			(1)

* Blend. m Masked.

TABLE 18

COMBINATIONS OF X-LEVELS

	X_3	X_2		X_3	X_2
$w^1F_3^\circ$	*+04 (−3⊙)	+01 (−3⊙)	$z^5P_3^\circ$	−09 (−3⊙)	
			$z^5P_2^\circ$	−03 (−3N⊙)	
$v^1G_4^\circ$	+05 (1)				
			$z^5D_4^\circ$	−01 (−3⊙)	
$v^3P_2^\circ$	−04 (−1⊙)		$z^5D_3^\circ$	−11 (−3⊙)	+13 (−3⊙)
			$z^5D_2^\circ$	+08 (−3⊙)	+05 (−3⊙)
$z^3D_3^\circ$	+03 (20V)	+12 (−3⊙)	$z^5D_1^\circ$		+26 (−3⊙)?
$z^3D_2^\circ$	−05 (3)	+11 (10)			
$z^3D_1^\circ$		+01 (3)	$z^5F_2^\circ$		−07 (−3⊙)
			$z^5F_1^\circ$		+06 (−3N⊙)
$u^3D_3^\circ$	−06 (2)				
$t^3D_3^\circ$	0 (2)				
$t^3D_2^\circ$	+01 (−3⊙)	*−04 (−1⊙)			
$s^3D_3^\circ$	−11 (6IV?)				
$s^3D_2^\circ$		−05 (3V)			
$z^3F_4^\circ$	+10 (5)				
$u^3F_3^\circ$		−07 (1)?			

* Blend. ⊙ Predicted line present in solar spectrum.

Not many of the numerous predicted odd terms above this level have been identified. Others may be found by future intensive observations of faint lines.

(16) UNCLASSIFIED LINES

The comprehensive list of lines upon which the present analysis was based includes great numbers of faint lines which have not been classified. It is probable that many of these are not really due to Fe I. In preparing a list for this purpose, much more trouble will be caused by excluding one real iron line as doubtful than by including several impurity lines. As the analysis advances, the latter will accumulate more and more in the unclassified residuum—while, in a spectrum as complex as this, a few of them will coincide by chance with wave-lengths predicted from term values and creep into the "classified" list. In the region between 6600 and 2975A the more obviously dubious "observed" lines were rejected, and a statistical comparison of the rest was made with the solar spectrum (§ 24), with the conclusion that the majority are accounted for in the sun and are probably really due to iron.

The strongest unclassified lines of wave-length greater than 2000A are listed in table 19. All those for which King has given a temperature classification are included with the exception of four which are clearly due to Fe II. For other observers the limit was set at intensity 3 (on the narrow scale) except in the far ultra-violet, where the scale was much more open. Of these 100 lines, 20 lie between λ9700 and λ6600, 26 between λ6600 and λ2975, and 54 short of this. Of the 46 which are accessible in the sun, 42 are present or accounted for by blending or masking. It is probable, therefore, that almost all the lines in this list are really due to iron. Almost, if not quite, all the lines of wave-length less than 3000A must arise from transitions from known low even terms to unknown high odd ones (of which there are still many). The temperature class shows that the lines of longer wave-length arise from higher levels.

VI. THE SUN AS A SOURCE FOR THE IRON SPECTRUM

(17) PREDICTED IRON LINES IN THE SOLAR SPECTRUM

It has long been recognized that the arc spectrum of iron is more fully exhibited among the Fraunhofer lines of the solar spectrum than by any laboratory measures so far published. Many lines, predicted from the term values, agree so closely with unidentified solar lines that the coincidences cannot reasonably be attributed to mere chance.[28] Many accidental coincidences must, however, occur in a general comparison of predicted with solar wave-lengths, and an investigation of their probable number is in order. The present work offers ample material for statistical study.

(18) ACCEPTANCE OF PREDICTED SOLAR LINES

In the course of the analysis of the spectrum, wave numbers were computed for all combinations from low even to odd levels and for the more probable transitions from odd to high even levels, not pro-

[28] Russell and Moore, *Mount Wilson Contr.* No. 365, 1928; *Astrophys. Jour.* **68**: 151, 1928.

hibited by the inexorable inner-quantum and parity rules. This gave about 6500 predicted positions of possible lines—with the certainty that a large majority, though not theoretically forbidden, must be far too faint to be observable. The resulting wave-lengths were compared with the "Revised Rowland Table" [29] (R. R.). This line list is known to be practically complete, except near the limit in the ultraviolet set by atmospheric ozone, and in the red beyond λ6600, where Rowland's plates were greatly inferior to modern ones.

Much progress has, however, been made in the past fifteen years in the identification of solar lines then unassigned, and in the improvement of earlier identifications. A systematic study of these has been made by one of the writers (C. E. M.) in connection with the preparation of a revised and extended edition of the Multiplet Table.[30] This investigation has added about 2850 identifications to those in the R. R. Though not yet finished, it is complete enough to provide a reliable basis for this statistical investigation. Of the predicted iron lines, 1928 agreed with observed solar lines closely enough to warrant calling them coincident, blended, or masked, in the same sense in which these terms have been applied to other lines in the general study.

The examination was made multiplet by multiplet, noting first the absences and apparent coincidences. The limits of tolerance were necessarily a matter of judgment, depending on the wave-length, intensity, and diffuseness of the solar line; but they represent the product of years of experience, and are believed to be fairly consistent.

The coincidences were divided into three classes: (*a*) *Unblended*, when the solar line is not otherwise identified and the agreement in wave-length and estimated intensity is such that a line reliably observed in the laboratory would be described as present in the sun.

(*b*) *Blended*, when the solar line (especially if diffuse) disagrees slightly in wave-length, or is somewhat too strong, and is probably a blend of the line under consideration with some other line of known or unknown origin.

(*c*) *Masked*, when some neighboring strong line would conceal the presence of a line of the expected position and intensity. The decisive test was, however, based on the position of the predicted line in its multiplet. If the intensity anticipated on this basis was very small, the line was rejected, especially in the case of blends.

The predicted lines which passed all these tests, and were therefore accepted as "present," or "present blended," were graded as good, fair, or poor, with the position in the multiplet, the agreement in wave-

[29] Revision of Rowland's Preliminary Table of Solar Spectrum Wave-lengths. *Publ. Carnegie Inst. Washington*, No. 396, 1928.

[30] *A Multiplet Table of Astrophysical Interest.* Princeton, 1933.

TABLE 19

THE LEADING UNCLASSIFIED LINES OF FE I

Ref†	I A	Int	T C	Ref	I A	Int	T C
F	9666.59	2	V	W	2786.81	(3)	
F	9637.55	2	V	G	2778.842	3	III
F	9529.31	2n	V	V	2778.075	3	III
E	9430.08	3	IV*	G	2773.232	2	III
O	8145.47	4	V	G	2757.856	(3)	
O	8024.50	3n	V	V	2737.833	(3)	
E	7994.473	20	IV*	G	2698.162	(4)	
O	7808.04	6n	V	V	2695.542	(3w)	
O	7573.53	2n	V	V	2664.042	(3w)	
L	7546.177	4	IV*	G	2615.420	(3)	
R	§7376.434	3n	V	G	2608.576	(3)	
V	7254.649	2	IV*	G	2606.644	(4)	
V	6975.46	3n	V	G	2604.864	(3)	
V	6902.80	3n	V	G	2604.751	(3)	
M	6881.46	1	V	G	2603.553	(4)	
V	6838.86	3n	V	G	2600.202	(3)	
M	6793.62	1	V	V	2594.046	1	III
V	6755.609	3	IV*	G	2592.285	(3)	
D	6726.78	(3)		G	2591.252	(3)	
V	6609.56	1	V	V	§2588.010	8	III
V	6528.53	2	V	G	2582.297	6	III
V	6501.681	4	IV*	G	2578.825	(3)	
J	6042.092	2	V	C	2575.744	(4)	
R	5036.294	6		G	2553.193	(7)	
J	4552.544	(3)		C	2551.094	(8)	
J	4237.162	2	V	G	2546.864	(4)	
W	4100.17	(3)		G	2533.802	4	IV
W	3851.58	(4)		W	2527.16	(5)	
J	3739.527	3	IV	G	2525.021	(7)	
V	3681.774	1	IV	G	2523.658	(6)	
V	3680.801	1	IV	W	2523.11	(5)	
V	3656.227	3n	IV	G	2520.968	(4)	
G	3634.698	4n	IV	G	2513.328	(3)	
J	3617.317	2	IV	G	2505.627	(4)	
J	3616.572	3n	IV	G	2505.485	(5)	
G	3614.550	2n	IV	W	2460.31	(4)	
J	3587.752	3	IV	G	2436.344	(10)	
W	3506.40	(3)		G	2435.865	(3)	
V	3438.306	(3w)		C	2431.025	(20)	
V	3262.284	4	IV	G	2301.171	(6)	
V	3179.538	3	IV	C	2165.861	(20)	
V	3139.908	4n	V	C	2163.368	(10)	
W	3136.17	(3)		X	2158.49	(6)	
G	3126.175	8n	IV	N	2111.274	(20)	
W	3102.71	(4)		N	2109.861	(25)	
G	2991.632	5n	IV	N	2077.507	(20)	
G	2945.050	3	IV	N	2041.204	(25)	
W	2927.55	(3)		N	2017.090	(15)	
V	2865.191	(3)		N	2007.215	(15)	
G	2799.149	1	III	N	2006.260	(15)	

† The reference numbers in this column are the same as those used in table B—see bibliography, page 169.

§ Blend with Fe II.

length, and the intensity as criteria. This grading was decidedly severe; for example, no members of multiplets containing only predicted lines and no lines involving improbable combinations, were called "good." This whole process is substantially the same that was applied to lines of less abundant elements observed in the laboratory, except that laboratory intensities were not available as guides.

(19) SELECTION OF GROUP FOR STATISTICAL STUDY

The essential condition in this case is that the group shall be selected *impartially*, by some method which neither favors nor discriminates against the characteristics under investigation. A desirable, though not necessary, condition is that the grouping should exclude irrelevant material in which these characteristics are not present.

The complete list of predicted possible lines satisfies the first condition, but not the second, for it is overloaded with transitions so improbable that there is no hope of their giving solar lines. To escape this the statistical study has been confined to multiplets in which at least one line (observed in the laboratory or predicted) satisfies the criteria of acceptance just described. This meets the second condition efficiently; but care is required about the first. When one or more lines of the multiplet have been observed in the laboratory, there is no assignable reason why the remaining lines should be more or less likely than the average to coincide accidentally with solar lines—though the chance of real coincidences is good. When no observed line is present, *one* coincidence with a predicted line in each multiplet is forced by the method of selection, but any other coincidences should be statistically free, just as in the previous case. The exclusion of one predicted line from each multiplet of the second type leaves statistically unbiassed material. All components of each multiplet, except observed (laboratory) lines, or the one predicted line, were included in the discussion, irrespective of their anticipated intensity.

A large group of predicted lines was thus obtained, free from statistical bias, and yet likely to contain a

TABLE 20

PREDICTED LINES OF FE I IN THE SOLAR SPECTRUM

λ		2975–4000		4000–5000		5000–6000		6000–6600		All	
Accepted Unblended	Good	82		82		94		10		268	
	Fair	127		211		161		39		538	
	Poor	46		59		43		10		158	
	Total		255		352		298		59		964
Blended with unidentified lines	Good	5		5		3		1		14	
	Fair	6		10		2		4		22	
	Poor			3		2				5	
	Total		11		18		7		5		41
Blended with identified lines	Good	22		11		7				40	
	Fair	40		26		13				79	
	Poor	15		7		7				29	
	Total		77		44		27		0		148
Rejected	Unblended		42		28		9		4		83
	Blended		94		74		63		19		250
Masked			250		122		51		19		442
Absent			167		240		278		142		827
Total			896		878		733		248		2755
Lines excluded (one per multiplet	Unblended		55		77		60		9		201
	Blended unidentified				1						1
	Blended identified		9				1				10
Available accepted lines	Unblended		200		275		238		50		763
	Blended unidentified		11		17		7		5		40
	Blended identified		68		44		26		0		138
Total available lines			832		800		672		239		2543

fair proportion of lines present in the sun. It includes members of 614 multiplets, of which 402 contained observed lines, and 212 predicted lines only. Counts of these lines were made separately for four spectral regions, as summarized in table 20. When two predicted lines coincided within the tolerance with one solar line, both were counted. The numbers of lines excluded (one per multiplet, as described above) and the numbers remaining as available for statistical discussion are given at the foot of the table.

(20) THEORETICAL PROBABILITY OF ACCIDENTAL COINCIDENCES

The elementary theory of accidental coincidences between spectral lines is well known.[31] If M lines (described as Group I) are distributed at random over an interval of X units, the probability that an arbitrarily chosen wave-length will fall within a distance x from one (or more) of them is $P(x) = 1 - e^{-2mx}$ where $m = M/X$; and the probability that the interval i between two successive lines of Group I lies between i and $i + di$ is $q(i) = me^{-mi}di$.

The observed solar lines are, however, not distributed at random, for pairs or groups too close to be resolved are measured as single lines. Fortunately, this simplifies the analysis. We may treat the lines of Group I approximately as if they were sharp-edged strips of width y which will be blended if the separation of centers is less than y. Suppose that we have another set of lines, Group II, of width z, which are distributed at random. Two lines, one of each group, will merge if their centers are separated by less than the coincidence-interval $c = \frac{1}{2}(y + z)$. If the lines of Group II are the fainter, we may assume $y > z$. We may now select from the whole range X a coincidence-range C, defined as follows. From the center of every unblended line of Group I lay off an interval c in both directions, and from the outer components of every blend lay off c outward. The whole range built up of their elements, disregarding overlapping, constitutes C. The probability that an arbitrary line of Group II will "coincide" with something belonging to Group I is then $\frac{C}{X}$, and the chance that it will not, $1 - \frac{C}{X}$. If there were no overlapping, C could be very easily found. Consider any individual element of the coincidence-interval of width w and the lines of Group II which, in a large number of trials, "coincide" with it. Their centers must fall in this interval and are distributed at random. Hence if $\bar{d}$ is the average distance, regardless of sign, of one of them from the center of the element, $\bar{d} = \frac{1}{4}w$. Summing for all the separate elements, $C = \Sigma w = 4\Sigma\bar{d}$. For a single trial, there will usually be only one value for d in each element, and only a part of the elements of C will be used, but the general mean value D of $\bar{d}$ will approximate that for many trials, within the usual uncertainty arising from finite sampling. Let M' be the number of elements (which will be less than M on account of blending in Group I).

$$\text{Then} \quad D = \Sigma\bar{d}/M' \quad \text{and} \quad C = 4M'D. \qquad (1)$$

If overlapping occurs, equation (1) still holds good provided that each of the confluent portions is divided into as many sections as there were originally discrete elements, and if the residual distances d are measured from the midpoint of each section. In practice, d is the difference between the wave-length of some predicted line (of Group II) and that of some component (presumably the nearest) of Group I as blended—that is, of some solar line. For the most frequent case where each element is a single line let the separation of their centers be $2s$. The range of d will be from 0 to c on one side, and 0 to s on the other, and the numerical mean $\bar{d} = \frac{c^2 + s^2}{2(c + s)} = \frac{c + s}{4} + \frac{v^2}{4(2c - v)}$ where $v = c - s$. The first term gives the approximation just discussed, the second, a small correction. The minimum value of s is $\frac{1}{2}y$, so that v ranges from 0 to $\frac{1}{2}z$. Assuming the distribution to be uniform over this range (which is good enough for this small term), the correction is found to be $\frac{z^2}{96c} + \cdots$. The ratio of this to the leading term is $\frac{z^2}{24c(c + s)}$ which at maximum, when $z = y$, is less than 0.03. As the number of confluent elements is a rather small part of the whole, the net correction to equation (1) will be less than 1 per cent. The statistical uncertainty of $\bar{D}$ is much greater than this unless the number of lines in Group II exceeds 1000, so that equation (1) is adequate in practice and it is needless to investigate the less probable case of confluence between a wide and a narrow element.

If a second group of lines (Ib) appears in the part of the spectrum left free by the original group (Ia), the coincidence-range for these may be found in exactly the same way. If M_b' is the observed number of elements in the group and D_b the mean residual, $C_b = 4M_b'D_b$. This automatically eliminates the complications arising from masking of lines of Group Ib by those of Ia. The probabilities of coincidence of a line of Group II with elements of Group Ia, Ib, $\cdots$ will be:

$$P_a = 4M_a'D_a/X; \quad P_b = 4M_b'D_b/X \cdots. \qquad (2)$$

If there are N lines in Group II, the probable numbers of accidental coincidences will be NP_a, NP_b.

[31] *Cf.* Russell and Bowen, *Mount Wilson Contr.* No. 375, 1929; *Astrophys. Jour.* **69**: 196, 1929.

(21) PREDICTED NUMBERS OF ACCIDENTAL COINCIDENCES

The numbers M' of solar lines have been counted in a copy of the Revised Rowland in which new identifications were entered to date. As this study is not quite completed, the results are not definitive—though adequate statistically—and only a single count was made. The solar lines were divided into two groups.

Group I*a*. *Identified Lines*: All solar lines which agree, whether as unblended or blended, under the rules of acceptance already described, with lines, of whatever origin, which have been observed in the laboratory. Observed lines of Fe I, predicted lines of other elements which pass the conditions of acceptance, and all lines shown by observation to be produced in the earth's atmosphere, are included. Blends are counted only once.

Group I*b*. *Unidentified Lines*: All solar lines for which no identification as just defined is available. Predicted lines of Fe I were completely disregarded in both groupings. The numbers of these lines for the four ranges of wave-length are given at the top of table 21. These sums should equal the numbers of lines between the given limits in the R. R. which may be found by inspection. The small differences represent the errors of the single approximate count and are statistically negligible.

Next follow the mean residuals $\bar{d}$ (in Angströms) between the predicted and solar wave-lengths for the groups of lines described in table 20. They increase moderately from the "good" to the "poor" cases, and considerably more with wave-length—mainly because the widths of the solar lines increase. They are greater for the rejected lines, since poor agreement was one factor in rejection, and greater still for the masked lines, since those which mask them are strong. Finally come the probabilities [eq. (2)] of accidental coincidences for a wave-length selected at random.

(22) PHYSICAL REALITY OF PREDICTED LINES

The numbers of predicted and observed coincidences are given in table 22. The sums for the separate regions are statistically preferable to the values calculated for the whole range together, which are given in the last line. The discordances are small, and indicate that the present subdivision is adequate.

The evidence of this table is decisive. The numbers of observed coincidences with unidentified solar lines are more than three times greater, in each of the four parts of the spectrum, than can be accounted for by chance, while those with otherwise identified lines show only a small excess, and the "absences" are much fewer than for a chance distribution. This puts it beyond doubt that a large majority of the unidentified solar lines which passed the conditions of acceptance are really due to iron.

TABLE 21

PROBABILITIES OF CHANCE COINCIDENCES

λ			2975–4000	4000–5000	5000–6000	6000–6600	All
Numbers of solar lines	Identified	M_a'	5507	3131	2280	688	11606
	Unidentified	M_b'	2300	2602	1651	584	7137
	Sum		7807	5733	3931	1272	18743
	R. R.		7803	5737	3960	1280	18780
Mean residuals $\bar{d}$ between predicted lines of Fe I and identified solar lines	Blended (accepted)	G	0.011	0.014	0.017		0.0125
		F	0.008	0.013	0.010		0.0099
		P	0.011	0.006	0.016		0.0107
		All	0.0092	0.0118	0.0133		0.0108
	Blended (rejected)		0.0138	0.0158	0.0170	0.024	0.0156
	Masked		0.0225	0.0256	0.0373	0.0410	0.0260
Mean for all identified lines (D_a)			0.0180	0.0200	0.0236	0.0312	0.0202
Mean residuals for unidentified solar lines	Unblended (accepted)	G	0.007	0.009	0.012	0.006	0.0093
		F	0.008	0.010	0.013	0.016	0.0111
		P	0.009	0.013	0.015	0.029	0.0137
		All	0.0081	0.0105	0.0131	0.0162	0.0110
	Blended (accepted)		0.010	0.013	0.010	0.034	0.014
	Blended (rejected)		0.0224	0.0200	0.0267	0.0175	0.0218
Mean for all unidentified lines (D_b)			0.0101	0.0113	0.0137	0.0177	0.0120
Probability of accidental coincidence	Identified lines		0.39	0.25	0.22	0.14	0.258
	Unidentified lines		0.09	0.12	0.09	0.07	0.094
	No coincidence		0.52	0.63	0.69	0.79	0.648

TABLE 22

PREDICTED AND OBSERVED COINCIDENCES

λ	Available Lines N	Identified Lines			Unidentified Lines			No Coincidence		
		Pred	Obs	$O-P$	Pred	Obs	$O-P$	Pred	Obs	$O-P$
2975–4000	832	324	412	+ 88	75	253	+178	433	167	−266
4000–5000	800	200	240	+ 40	96	320	+224	504	240	−264
5000–6000	672	148	140	− 8	60	254	+194	464	278	−186
6000–6600	239	34	38	+ 4	17	59	+ 42	188	142	− 46
Sums	2543	706	830	+124	248	886	+638	1589	827	−762
2975–6600	2543	656	830	+174	239	886	+647	1648	827	−821

TABLE 23

NUMBERS OF PHYSICALLY SIGNIFICANT COINCIDENCES

λ	With Unidentified Lines					With Identified Lines				
	N	N′	G+F	Astrophysical	Statistical	N	N′	G+F	Astrophysical	Statistical
2975–4000	266	211	220	177	178	77	68	62	55	88
4000–5000	370	292	308	243	224	44	44	37	37	40
5000–6000	305	245	260	209	194	27	26	20	19	−8
6000–6600	64	55	54	47	42	19	19	0	0	4
Sums				676	638				111	124

The probable numbers of such real lines are given in the columns headed $O-P$. It is of much interest to compare these with the numbers of lines which had already been classified as good or fair, upon astrophysical grounds quite independent of the statistical study. In doing this, the recorded numbers of these lines should be diminished in the ratio which the number N' of coincidences in each group available for the statistical study bears to the whole number N of such coincidences (since it is unfair to impose any special characteristic upon the $N-N'$ lines which were used to pick out the multiplets). The results are shown in table 23.

The numbers of physically real coincidences derived astrophysically are $(G+F)N'/N$, while the statistical estimates are $O-P$ of table 22. The agreement of the two columns is remarkable.

It is clear that the identifications of predicted iron lines in the sun which have been classified as good or fair are physically significant. The list doubtless includes some accidental coincidences and omits about as many real ones, but should be generally trustworthy. The excess of observed coincidences with otherwise identified lines, shown in table 22, is substantially accounted for by those which showed recognizable evidence of blending.

The whole number of physically real predictions (by the statistical test), according to this table, is 762. The actual number of predicted iron lines in the solar spectrum must be considerably greater. To begin with, the 212 predicted lines excluded from the statistical study should include about the same proportion which are physically real as the 960 which were available, that is, about 170. Also, the proportion of physically real coincidences should be the same for the 706 predicted lines which coincide by accident with lines of other elements as for the other 1837. The latter give 762 coincidences; there must be about 290 among the former. If there were no masking, there should be some 1200 lines of iron which, though not yet produced in the laboratory, should be observable in the solar spectrum between λ2975 and λ6600, while nearly 1600 lines in the same multiplets are too faint to appear.

The proportion of lines present decreases toward the red, where the multiplets arise from higher energy levels, and average fainter. Despite this, many more predicted iron lines should be found in the sun when the study of its spectrum from λ6600 to λ12000 has been completed.

(23) TABLE OF PREDICTED SOLAR LINES OF FE I

Table C (p. 170) contains those predicted lines of Fe I for which the evidence of presence in the solar spectrum was adjudged to be good or fair, 1254 in all. Those in the range from 6600–2975A have just been discussed. Those of longer wave-length are taken from the "Monograph on the Red and Infra-Red Solar Spectrum," which is under preparation at Mount Wilson by H. D. Babcock and others. The identifications have been made in the manner described above, by one of the writers. The writers greatly appreciate the use of this material. The wave-lengths

TABLE 24

PRESENCE IN THE SUN OF IRON LINES OBSERVED IN THE LABORATORY

λ	2975–4000	4000–5000	5000–6000	6000–6600	All
Classified Lines					
Unblended	928	593	357	107	1985
Blended	243	95	39	15	392
Masked	25	11	10	3	49
Absent	10	6	20	13	49
Total	1206	705	426	138	2475
Unclassified Lines					
Unblended	138	69	34	11	252
Blended	67	12	10	3	92
Masked	2	1	2	2	7
Absent	39	30	104	57	230
Total	246	112	150	73	581

and intensities are from this monograph or the Revised Rowland but include some changes in Rowland's intensities.[32] The third column gives the grade g or f assigned as above. A "b" added denotes that the line is blended in the solar spectrum.

The limitations of this table should be borne in mind. First, many lines (probably almost 300 between 6600–2975A) have perforce been omitted, owing to masking in the sun. Second, some accidental coincidences are doubtless present, and some real iron lines omitted, among those graded poor and rather drastically excluded. The great majority of the tabular lines must, however, be real.

(24) OBSERVED LINES OF IRON IN THE SUN

For comparison, the solar behavior of those iron lines which have been observed in the laboratory is listed in table 24. The lines which have and have not been classified in the present analysis are listed separately.

Only 2 per cent of the former fail to appear in the sun, but 40 per cent of the latter. The unclassified lines are faint, and more of them might be absent, but not so many more. Most of the "absent" lines among these are probably due to impurities (§ 16). Most of the classified lines of Fe I which do not appear in the sun are recorded as very faint in the laboratory. Four of them (at λλ2980.532, 2981.446, 2994.50 and 3020.643) are strong lines. A letter from H. D. Babcock states that all are present in the sun with about the anticipated strength.

The remaining 47 lines of wave-length less than 6600A (two of which are blended pairs) were graded on the same system as the predicted lines. Three were graded "good," 11 "fair," and 33 "poor." For many of the last, the discrepancies in wave-length are large, and it is doubtful whether they really belong in the classified places, nor is it certain whether they are due to iron at all. Indeed, if a line attributed to iron in the laboratory is absent from the sun, this is strong presumptive evidence that it is not really due to the metal.

[32] C. E. Moore, *Atomic Lines in the Sun-Spot Spectrum*. Princeton, 1933.

(25) DESIRABILITY OF FUTURE OBSERVATIONS

It is evident that the spectrum of the iron arc is very far from being fully observed. If a laboratory source could be discovered which was as efficient in producing it as is the absorption in the sun's atmosphere, at least 1500 lines would probably be added to the list. This estimate takes account only of transitions among terms already known, and would be increased by allowance for unknown high-lying levels. No such progress is likely to be made by repeating observations with familiar sources such as the arc in air. The advantage of the solar atmosphere probably consists in the great depth of highly rarefied gas. This can only be feebly imitated in the laboratory, but experiments would be attractive.

Once the source was found, observations with pure material in the usual region (say from λ2100 to λ11000) would be easy, and measurement by known apparatus rapid. The "vacuum region" has only been reconnoitered, and must contain much of interest. The infra-red, beyond the limit of photography, appears never to have been so much as examined with modern equipment of good resolving power. Observations might lead to the identification of many new levels, and lines predicted from known levels should provide abundant standards. Much could be done here with existing sources.

All told, it is evident that a great deal of work will still be required before this "familiar" spectrum is really thoroughly known, and that it still offers attractive and remunerative problems to the observer.

(26) THE TABLES OF CLASSIFIED LINES

The long tables of classified lines of Fe I already mentioned in the text (pp. 115 and 131) conclude Part I of this paper. Table B contains all classified lines of this spectrum that have been observed in the laboratory, 3606 in all. In some cases, where the laboratory line is measured only to 0.1A, the solar wave-length is entered in the table. For such lines there is good agreement between the solar and predicted wave-lengths, and it seems reasonable to assume that the line observed in the laboratory is the one to which the assigned designation applies. Consequently, the laboratory intensity is entered instead of the solar intensity. When these poorly measured lines are absent or masked in the sun, the wave-lengths are given to 0.1A and referred to the proper source.

The letters in column 1 refer to the various sources from which the laboratory wave-length (column 2) is taken. At the end of the table a complete bibliography is given with the various letters.

The selection of the best source to use for each wave-length has not been an easy task. For the many accurately measured lines, it is difficult to select the most accurate value. Those adopted for international standards have been the first choice throughout. For less accurate measures, the sources which have appeared to be consistently satisfactory for use in many multiplets have dominated the selection. The experience gained from studying the literature for the selection of wave-lengths for all elements included in the "Revised Multiplet Table" has been the guide throughout this work. All wave-lengths have been selected by one of the writers (C. E. M.). While there is no doubt that a more homogeneous list is highly desirable, yet it is hoped that the present table will suffice for those interested in using it (compare § 3).

The intensity and temperature class are in the next two columns. These have already been described in §§ 4 and 5. Then follow the wave-number, which corresponds to the observed wave-length in column 2, and the difference between the observed wave-number and that calculated from the term values of table A. The unit of o−c is 0.01 cm^{-1}, but only one digit is given if the line has been measured only to 0.1A, as described above. Finally, the last column contains the multiplet designation, which is self-explanatory.

The symbols used in table B are described at the end of the table. An asterisk denotes that the line is a blend, i. e. that the lines designated probably all contribute sensibly to the observed intensity. For a line which is probably of observable intensity but is masked by a stronger neighbor, the designation is given in parentheses. Lines blended with Fe II are marked "§." Other special notes, made by A. S. King in the course of his work on temperature classification, are indicated by a double asterisk.

Table C contains 1254 predicted lines of Fe I that have been accepted as present in the solar spectrum and graded "good" or "fair" (see § 23). It is arranged similarly to table B, except that the solar spectrum is the source used for all wave-lengths and intensities. If the solar line is a blend, a "b" follows the grade "g" or "f" in the third column.

TABLE A

TERMS OF FE I

Config	Desig	Term	Diff	Source	g
$3d^7(a^2P)4s$	a^1P_1	27543.00		10	0.817
$3d^7(a^2D)4s$	a^1D_2	28604.61		10	1.028
$3d^64s^2$	b^1D_2	34636.82		11	
$3d^7(b^2F)4s$	a^1F_3	40534.18:		11	
$3d^7(a^2G)4s$	a^1G_4	24574.690		7	1.001
$3d^64s^2$	b^1G_4	29798.96		11	0.979
$3d^7(a^2H)4s$	a^1H_5	28819.98		10	1.000
$3d^64s^2$	a^1I_6	29313.04		11	1.014
$3d^64s^2$	a^3P_2	18378.215	1174.278	5	1.506
	a^3P_1	19552.493	485.37	5	1.500
	a^3P_0	20037.86		5	
$3d^7(a^4P)4s$	b^3P_2	22838.360	108.500	5	1.498
	b^3P_1	22946.860	104.930	5	1.489
	b^3P_0	23051.790		5	
$3d^7(a^2P)4s$	c^3P_2	24335.804	436.256	7	1.484
	c^3P_1	24772.060	319.56	7	1.466
	c^3P_0	25091.62		10	
$3d^7(a^4F)4d$	e^3P_2	54879.720	496.397	7	1.459
	e^3P_1	55376.117	350.42	7	1.459
	e^3P_0	55726.54:		10	
$3d^7(a^2D)4s$	a^3D_3	26225.03	398.70	10	1.335
	a^3D_2	26623.73	−217.24	10	1.178
	a^3D_1	26406.49		10	0.731
$3d^64s^2$	b^3D_3	29371.86	− 15.08	10	1.326
	b^3D_2	29356.78	− 36.73	10	
	b^3D_1	29320.05		11	
$3d^64s(a^4D)5s$	e^3D_3	51294.262	445.702	5	1.345
	e^3D_2	51739.964	299.975	5	1.125
	e^3D_1	52039.939		5	0.801
$3d^7(a^4F)4d$	f^3D_3	53747.547	319.274	4	1.258
	f^3D_2	54066.821	382.51	4	
	f^3D_1	54449.33		4	
$3d^7(a^4F)4s$	a^3F_4	**11976.260**	584.693	1	1.254
	a^3F_3	**12560.953**	407.620	1	1.086
	a^3F_2	**12968.573**		1	0.670
$3d^64s^2$	b^3F_4	20641.144	233.377	5	1.235
	b^3F_3	20874.521	164.500	5	1.073
	b^3F_2	21039.021		5	0.663
$3d^8$	c^3F_4	32873.68	539.10	11	1.264
	c^3F_3	33412.78	352.55	11	1.066
	c^3F_2	33765.33		11	0.677
$3d^7(b^2F)4s$	d^3F_4	37046.00	− 70.36	11	
	d^3F_3	36975.64	− 35.04	11	
	d^3F_2	36940.60		11	
$3d^7(a^4F)5s$	e^3F_4	47960.973	570.923	4	1.288
	e^3F_3	48531.896	396.527	4	1.107
	e^3F_2	48928.423		4	0.622

Config	Desig	Term	Diff	Source	g
$3d^7(a^4F)4d$	f^3F_4	54683.39	441.58	7	1.141
	f^3F_3	55124.974	253.868	7	1.071
	f^3F_2	55378.842		7	0.676
$3d^7(a^2G)4s$	a^3G_5	21715.770	283.397	6	1.197
	a^3G_4	21999.167	250.294	6	1.051
	a^3G_3	22249.461		6	0.756
$3d^64s^2$	b^3G_5	23783.654	335.200	5	1.200
	b^3G_4	24118.854	219.951	5	1.048
	b^3G_3	24338.805		5	0.761
$3d^7(a^4F)4d$	e^3G_5	53739.488	327.08	5	1.248
	e^3G_4	54066.57	312.87	5	1.096
	e^3G_3	54379.44		5	0.842
$3d^64s^2$	a^3H_6	19390.197	230.839	5	1.163
	a^3H_5	19621.036	167.244	5	1.038
	a^3H_4	19788.280		5	0.811
$3d^7(a^2H)4s$	b^3H_6	26105.95	245.14	10	1.165
	b^3H_5	26351.09	276.55	10	1.032
	b^3H_4	26627.64		10	0.811
$3d^7(a^4F)4d$	e^3H_6	53840.68:	426.08	10	1.225
	e^3H_5	54266.76:	288.69	10	1.109
	e^3H_4	54555.45:		8	0.871
$3d^64s(a^6D)4d$	e^5S_2	51148.892		7	1.952
$3d^7(a^4P)4s$	a^5P_3	**17550.210**	176.797	1	1.666
	a^5P_2	**17727.017**	200.394	1	1.820
	a^5P_1	**17927.411**		1	2.499
$3d^64s(a^6D)4d$	e^5P_3	51837.279	230.17	7	1.664
	e^5P_2	52067.45	− 47.74	11	
	e^5P_1	52019.706		7	2.432
$3d^7(a^4F)4d$	f^5P_3	53160.53	408.19	8	
	f^5P_2	53568.72	356.54	7	
	f^5P_1	53925.26		8	
$3d^64s^2$	a^5D_4	**0.000**	415.933	1	1.490
	a^5D_3	**415.933**	288.070	1	1.497
	a^5D_2	**704.003**	184.129	1	1.494
	a^5D_1	**888.132**	89.942	1	1.498
	a^5D_0	**978.074**		1	
$3d^64s(a^6D)5s$	e^5D_4	**44677.010**	384.324	1	1.502
	e^5D_3	**45061.334**	272.546	1	1.508
	e^5D_2	**45333.880**	175.275	1	1.503
	e^5D_1	**45509.155**	85.929	1	1.518
	e^5D_0	**45595.084**		1	
$3d^64s(a^6D)4d$	f^5D_4	50423.185	111.250	7	1.514
	f^5D_3	50534.435	164.231	7	1.615
	f^5D_2	50698.666	181.486	7	1.614
	f^5D_1	50880.152	100.87	7	1.662
	f^5D_0	50981.02		8	
$3d^64s(a^4D)5s$	g^5D_4	**51350.505**	420.072	3	1.487
	g^5D_3	**51770.577**	279.24	3	1.492
	g^5D_2	52049.82	164.51	3	1.57:
	g^5D_1	52214.33	43.00	3	
	g^5D_0	52257.33		3	

TABLE A—(*Continued*)

Config	Desig	Term	Diff	Source	g
$3d^7(a^4F)4d$	h^5D_4	53155.13	390.75	7	1.435
	h^5D_3	53545.882	420.838	7	
	h^5D_2	53966.720	165.76	7	
	h^5D_1	54132.48		10	
	h^5D_0				
$3d^64s(a^4D)4d$	i^5D_4	57697.59	116.38	11	1.384
	i^5D_3	57813.97	160.19	11	1.415
	i^5D_2	57974.16		11	
	i^5D_1				
	i^5D_0				
$3d^7(a^4F)4s$	a^5F_5	**6928.280**	448.495	1	1.404
	a^5F_4	**7376.775**	351.296	1	1.349
	a^5F_3	**7728.071**	257.724	1	1.248
	a^5F_2	**7985.795**	168.930	1	0.995
	a^5F_1	**8154.725**		1	−0.014
$3d^7(a^4F)5s$	e^5F_5	**47005.510**	372.457	2	1.421
	e^5F_4	**47377.967**	377.571	1	1.331
	e^5F_3	**47755.538**	281.129	1	1.236
	e^5F_2	**48036.667**	184.656	1	0.991
	e^5F_1	**48221.323**		1	0.007
$3d^64s(a^6D)4d$	f^5F_5	51103.237	358.470	7	1.384
	f^5F_4	51461.707	142.439	7	1.355:
	f^5F_3	51604.146	100.906	7	
	f^5F_2	51705.052	49.482	7	0.967
	f^5F_1	51754.534		7	
$3d^7(a^4F)4d$	g^5F_5	53061.28	332.44	7	
	g^5F_4	53393.715	437.24	7	
	g^5F_3	53830.96	426.56	7	
	g^5F_2	54257.52	128.64	7	
	g^5F_1	54386.16		8	
$3d^64s(a^6D)4d$	e^5G_6	50522.94	180.97	7	1.351
	e^5G_5	50703.912	275.715	7	1.360
	e^5G_4	50979.627	239.432	7	1.238
	e^5G_3	51219.059	151.125	7	1.294
	e^5G_2	51370.184		7	0.953
$3d^7(a^4F)4d$	f^5G_6	53169.21	112.53	7	1.323
	f^5G_5	53281.735	487.285	7	1.221
	f^5G_4	53769.020	392.162	7	
	f^5G_3	54161.182	214.537	7	1.142
	f^5G_2	54375.719		7	
$3d^64s(a^4D)4d$	g^5G_6	58001.88	269.62	11	1.40:
	g^5G_5	58271.50:	248.68	11	
	g^5G_4	58520.18:	189.91	11	
	g^5G_3	58710.09:	114.72	11	
	g^5G_2	58824.81		11	0.343
$3d^7(a^4F)4d$	e^5H_7	53275.20:	77.82	8	1.30:
	e^5H_6	53353.02:	521.28	10	1.191
	e^5H_5	53874.30:	362.90	10	1.102
	e^5H_4	54237.20	253.88	10	0.90:
	e^5H_3	54491.08		10	0.484
$3d^64s(a^6D)4d$	e^7S_3	51570.16		7	1.92:
$3d^64s(a^6D)4d$	e^7P_4	50475.32	135.98	7	1.585
	e^7P_3	50611.303	250.02	7	1.687
	e^7P_2	50861.32		7	

Config	Desig	Term	Diff	Source	g
$3d^64s(a^6D)5s$	e^7D_5	**42815.855**	347.472	2	1.585
	e^7D_4	**43163.327**	271.306	2	1.655
	e^7D_3	**43434.633**	198.902	2	1.755
	e^7D_2	**43633.535**	130.447	2	2.009
	e^7D_1	**43763.982**		2	3.002
$3d^64s(a^6D)4d$	f^7D_5	50377.92	430.13	7	1.510
	f^7D_4	50808.053	53.80	7	1.574
	f^7D_3	50861.85	136.84	8	
	f^7D_2	50998.686	49.41	7	1.844
	f^7D_1	51048.10		7	
$3d^64s(a^6D)6s$	g^7D_5	53800.90	323.72	11	1.586
	g^7D_4	54124.62	289.12	11	1.65:
	g^7D_3	54413.74:	197.98	11	
	g^7D_2	54611.72	136.02	11	
	g^7D_1	54747.74:		11	
$3d^64s(a^6D)7s$	h^7D_5	57897.17		11	
$3d^64s(a^6D)4d$	e^7F_6	50342.180	491.305	7	1.490
	e^7F_5	50833.485	358.835	7	1.505
	e^7F_4	51192.320	− 43.45	7	1.617
	e^7F_3	51148.87	182.22	7	1.499
	e^7F_2	51331.090	−123.05	7	
	e^7F_1	51208.04		7	2.490
	e^7F_0				
$3d^64s(a^6D)4d$	e^7G_7	50651.76:	316.11	8	
	e^7G_6	50967.873	260.722	7	1.415
	e^7G_5	51228.595	106.34	7	1.379
	e^7G_4	51334.94	125.59	7	1.338
	e^7G_3	51460.53	79.24	7	1.244
	e^7G_2	51539.77	27.09	7	
	e^7G_1	51566.86		10	−0.374
	X_3	40871.46	306.90	11	
	X_2	41178.36		11	
	1_5	56428.06		11	
	$2_{4,5}$	56452.04		11	
	3_4	56842.70		11	
	4_2	58213.17		11	
$3d^7(a^2P)4p$	$z^1P_1^\circ$	53229.94		10	1.266
	$z^1D_2^\circ$	49477.10		10	0.92:
$3d^7(a^2P)4p$	$y^1D_2^\circ$	51708.33		10	1.025
	$x^1D_2^\circ$	51762.12		11	0.883
$3d^7(a^2D)4p$:	$w^1D_2^\circ$	55754.29		11	0.990
$3d^7(a^2G)4p$:	$z^1F_3^\circ$	50586.89		8	1.018
	$y^1F_3^\circ$	53661.13		11	1.21:
	$x^1F_3^\circ$	53763.28		10	1.079
	$w^1F_3^\circ$	55790.72		11	0.908
$3d^7(a^2G)4p$:	$z^1G_4^\circ$	47452.770		8	1.025
$3d^7(a^2H)4p$:	$y^1G_4^\circ$	48702.57		7	1.063
	$x^1G_4^\circ$	50614.02		10	0.978
	$w^1G_4^\circ$	54810.82		11	1.001
	$v^1G_4^\circ$	56951.27		11	1.053

TABLE A—(*Continued*)

Config	Desig	Term	Diff	Source	g
$3d^7(a^2G)4p$	$z^1H_5^\circ$	48382.63		10	1.018
$3d^7(a^2H)4p$	$y^1H_5^\circ$	53722.44		10	1.03:
	$x^1H_5^\circ$	55525.58		11	1.018
$3d^7(a^2H)4p$	$z^1I_6^\circ$	53093.60		10	1.010
$3d^64s(b^4P)4p$	$z^3S_1^\circ$	46600.884		7	1.888
$3d^7(a^4P)4p$	$y^3S_1^\circ$	47555.63		5	1.884
$3d^64s(a^4D)4p$	$z^3P_2^\circ$	33946.965	415.925	1	1.493
	$z^3P_1^\circ$	34362.890	192.75	2	1.496
	$z^3P_0^\circ$	34555.64		2	
$3d^7(a^4P)4p$	$y^3P_2^\circ$	46727.137	174.755	5	1.444
	$y^3P_1^\circ$	46901.892	−229.32	5	1.600
	$y^3P_0^\circ$	46672.57		10	
$3d^64s(b^4P)4p$	$x^3P_2^\circ$	48304.707	211.44	7	1.263
	$x^3P_1^\circ$	48516.15	− 56.03	7	1.547
	$x^3P_0^\circ$	48460.12		10	
$3d^7(a^2D)4p$:	$w^3P_2^\circ$	50186.87	−143.62	7	1.469
	$w^3P_1^\circ$	50043.25	− 91.89	10	1.389
	$w^3P_0^\circ$	49951.36		10	
$3d^7(a^2P)4p$:	$v^3P_2^\circ$	52916.33	892.04	10	1.495
	$v^3P_1^\circ$	53808.37		11	1.418
	$v^3P_0^\circ$				
$3d^64s(a^4D)4p$	$z^3D_3^\circ$	**31322.639**	363.738	2	1.321
	$z^3D_2^\circ$	**31686.377**	250.973	2	1.168
	$z^3D_1^\circ$	**31937.350**		2	0.513
$3d^7(a^4F)4p$	$y^3D_3^\circ$	**38175.382**	502.685	1	1.324
	$y^3D_2^\circ$	**38678.067**	317.697	1	1.151
	$y^3D_1^\circ$	**38995.764**		1	0.493
$3d^64s(b^4P)4p$	$x^3D_3^\circ$	45220.738	61.151	7	1.352
	$x^3D_2^\circ$	45281.889	269.944	7	1.200
	$x^3D_1^\circ$	45551.833		8	0.556
$3d^7(a^4P)4p$	$w^3D_3^\circ$	47017.239	118.903	5	1.346
	$w^3D_2^\circ$	47136.142	135.953	5	1.216
	$w^3D_1^\circ$	47272.095		2	0.767
$3d^64s(b^4F)4p$	$v^3D_3^\circ$	49135.08	107.60	7	1.211
	$v^3D_2^\circ$	49242.68	54.98	8	0.954
	$v^3D_1^\circ$	49297.66		7	0.562
$3d^7(a^2P)4p$	$u^3D_3^\circ$	51969.14	327.82	10	1.306
	$u^3D_2^\circ$	52296.96	215.50	10	1.156
	$u^3D_1^\circ$	52512.46		7	0.700
$3d^7(a^2D)4p$:	$t^3D_3^\circ$	52213.29	469.64	10	1.317
	$t^3D_2^\circ$	52682.93	−502.11	10	1.145
	$t^3D_1^\circ$?	52180.82		10	0.801
	$s^3D_3^\circ$	52953.68:	321.59	11	1.231
	$s^3D_2^\circ$	53275.27		11	
	$s^3D_1^\circ$				
$3d^64s(a^4D)4p$	$z^3F_4^\circ$	**31307.272**	497.825	1	1.250
	$z^3F_3^\circ$	**31805.097**	328.917	1	1.086
	$z^3F_2^\circ$	**32134.014**		1	0.682

Config	Desig	Term	Diff	Source	g
$3d^7(a^4F)4p$	$y^3F_4^\circ$	**36686.204**	476.566	1	1.246
	$y^3F_3^\circ$	**37162.770**	358.416	1	1.086
	$y^3F_2^\circ$	**37521.186**		1	0.688
$3d^7(a^2G)4p$	$x^3F_4^\circ$	46889.207	203.569	7	1.344
	$x^3F_3^\circ$	47092.776	104.298	2	1.159
	$x^3F_2^\circ$	47197.074		2	0.743
$3d^64s(b^4F)4p$	$w^3F_4^\circ$	49108.94	134.01	7	1.181
	$w^3F_3^\circ$	49242.950	190.23	7	1.165
	$w^3F_2^\circ$	49433.18		10	0.677
$3d^64s(a^4G)4p$:	$v^3F_4^\circ$	51304.65	60.65	10	1.122
	$v^3F_3^\circ$	51365.30	−163.97	10	1.096
	$v^3F_2^\circ$	51201.33		11	0.803
$3d^7(a^2D)4p$:	$u^3F_4^\circ$	56592.76	190.57	10	1.148
	$u^3F_3^\circ$	56783.33	75.32	10	1.077
	$u^3F_2^\circ$	56858.65		10	0.687
	$t^3F_4^\circ$	57550.09	90.97	11	1.235
	$t^3F_3^\circ$	57641.06	67.70	11	
	$t^3F_2^\circ$	57708.76		11	0.698
$3d^7(a^4F)4p$	$z^3G_5^\circ$	**35379.237**	388.354	1	1.248
	$z^3G_4^\circ$	**35767.591**	311.804	1	1.100
	$z^3G_3^\circ$	**36079.395**		1	0.791
$3d^64s(a^4H)4p$	$y^3G_5^\circ$	45294.86	133.60	5	1.207
	$y^3G_4^\circ$	45428.456	134.570	5	1.053
	$y^3G_3^\circ$	45563.026		5	0.765
$3d^64s(b^4F)4p$	$x^3G_5^\circ$	47834.622	− 22.44	7	1.203
	$x^3G_4^\circ$	47812.18	22.08	7	1.061
	$x^3G_3^\circ$	47834.26		10	0.668
	$w^3G_5^\circ$	48231.33	130.59	7	1.27:
	$w^3G_4^\circ$	48361.92	113.82	7	0.934
	$w^3G_3^\circ$	48475.74		7	0.584
$3d^7(a^2G)4p$	$v^3G_5^\circ$	49460.92	167.00	7	1.163
	$v^3G_4^\circ$	49627.92	222.69	7	0.914
	$v^3G_3^\circ$	49850.61		7	0.763
$3d^7(a^2H)4p$	$u^3G_5^\circ$	51373.96	294.26	10	1.140
	$u^3G_4^\circ$	51668.22	157.58	10	1.067
	$u^3G_3^\circ$	51825.80		10	0.801
$3d^64s(a^4G)4p$:	$t^3G_5^\circ$	53983.30	254.16	8	1.234
	$t^3G_4^\circ$	54237.46	362.89	8	1.183
	$t^3G_3^\circ$	54600.35		10	0.922
	$s^3G_5^\circ$	55907.22	− 1.66	8	1.145
	$s^3G_4^\circ$	55905.56	192.29	10	
	$s^3G_3^\circ$	56097.85		10	0.857
	$r^3G_5^\circ$	59926.62:	245.44	10	1.190
	$r^3G_4^\circ$	60172.06	192.70	10	1.030
	$r^3G_3^\circ$	60364.76:		10	0.780
	$q^3G_5^\circ$	60677.23:	77.48	10	
	$q^3G_4^\circ$	60754.71:	52.01	10	
	$q^3G_3^\circ$	60806.72		10	

TABLE A—(*Continued*)

Config	Desig	Term	Diff	Source	g
$3d^64s(a^4H)4p$	$z^3H_6^\circ$	46982.383	26.045	7	1.200
	$z^3H_5^\circ$	47008.428	98.116	7	1.060
	$z^3H_4^\circ$	47106.544		7	0.880
$3d^7(a^2G)4p$	$y^3H_6^\circ$	49434.20	170.25	7	1.17:
	$y^3H_5^\circ$	49604.45	122.61	7	1.075
	$y^3H_4^\circ$	49727.058		7	0.929
$3d^64s(a^4G)4p$:	$x^3H_6^\circ$	51023.19	45.58	10	1.161
	$x^3H_5^\circ$	51068.77		8	1.038
	$x^3H_4^\circ$				
$3d^7(a^2H)4p$	$w^3H_6^\circ$	52431.47	181.61	10	1.177
	$w^3H_5^\circ$	52613.08	155.70	10	1.033
	$w^3H_4^\circ$	52768.78		10	0.810
	$v^3H_6^\circ$	55489.81	− 59.92	10	1.169
	$v^3H_5^\circ$	55429.89	16.17	10	1.057
	$v^3H_4^\circ$	55446.06		10	0.804
$3d^64s(b^2H)4p$:	$u^3H_6^\circ$	56334.01	48.68	10	1.166
	$u^3H_5^\circ$	56382.69	40.64	10	1.029
	$u^3H_4^\circ$	56423.33		10	0.859
	$t^3H_6^\circ$	60365.70:	183.48	10	1.163
	$t^3H_5^\circ$	60549.18	208.50	10	1.040
	$t^3H_4^\circ$	60757.68		10	0.805
$3d^64s(a^4H)4p$	$z^3I_7^\circ$	45978.04:	48.94	10	1.149
	$z^3I_6^\circ$	46026.98	108.94	10	1.040
	$z^3I_5^\circ$	46135.92		10	0.833
$3d^7(a^2H)4p$	$y^3I_7^\circ$	52655.04:	−141.45	10	1.147
	$y^3I_6^\circ$	52513.59	385.47	10	1.019
	$y^3I_5^\circ$	52899.06		10	0.830
$3d^64s(b^2H)4p$	$x^3I_7^\circ$	57027.56:	42.69	10	1.145
	$x^3I_6^\circ$	57070.25	34.01	10	1.028
	$x^3I_5^\circ$	57104.26		10	0.832
$3d^64s(b^4P)4p$	$z^5S_2^\circ$	**40895.022**		2	1.985
$3d^7(a^4P)4p$	$y^5S_2^\circ$	44511.86		7	1.888
$3d^64s(a^6D)4p$	$z^5P_3^\circ$	**29056.341**	412.692	1	1.657
	$z^5P_2^\circ$	**29469.033**	263.716	1	1.835
	$z^5P_1^\circ$	**29732.749**		1	2.487
$3d^64s(a^4D)4p$	$y^5P_3^\circ$	**36766.998**	390.596	1	1.661
	$y^5P_2^\circ$	**37157.594**	251.981	1	1.836
	$y^5P_1^\circ$	**37409.575**		1	2.502
$3d^64s(b^4P)4p$	$x^5P_3^\circ$	42532.76	327.07	2	1.650
	$x^5P_2^\circ$	42859.829	219.22	2	1.822
	$x^5P_1^\circ$	43079.05		2	2.464
$3d^54s^2(a^6S)4p$	$w^5P_3^\circ$	46137.14	176.47	5	1.658
	$w^5P_2^\circ$	46313.61	96.83	5	1.822
	$w^5P_1^\circ$	46410.44		5	2.436
$3d^7(a^4P)4p$	$v^5P_3^\circ$	47966.63	196.86	2	1.646
	$v^5P_2^\circ$	48163.49	126.40	2	1.740
	$v^5P_1^\circ$	48289.89		2	2.213

Config	Desig	Term	Diff	Source	g
$3d^64s(a^6D)5p$	$u^5P_3^\circ$	51691.98:	253.33	9	
	$u^5P_2^\circ$	51945.31:	165.0	9	
	$u^5P_1^\circ$	52110.3:		9	2.633
$3d^64s(b^4D)4p$	$t^5P_3^\circ$	53388.68:	723.62	9	
	$t^5P_2^\circ$	54112.30	158.81	9	1.70:
	$t^5P_1^\circ$	54271.11		9	
$3d^64s(a^6D)4p$	$z^5D_4^\circ$	**25900.002**	240.191	1	1.502
	$z^5D_3^\circ$	**26140.193**	199.515	1	1.500
	$z^5D_2^\circ$	**26339.708**	139.685	1	1.503
	$z^5D_1^\circ$	**26479.393**	71.102	1	1.495
	$z^5D_0^\circ$	**26550.495**		1	
$3d^7(a^4F)4p$	$y^5D_4^\circ$	**33095.962**	411.182	1	1.496
	$y^5D_3^\circ$	**33507.144**	294.451	1	1.492
	$y^5D_2^\circ$	**33801.595**	215.532	1	1.495
	$y^5D_1^\circ$	**34017.127**	104.496	1	1.492
	$y^5D_0^\circ$	**34121.623**		1	
$3d^64s(a^4D)4p$	$x^5D_4^\circ$	**39625.829**	344.051	1	1.489
	$x^5D_3^\circ$	**39969.880**	261.485	1	1.504
	$x^5D_2^\circ$	**40231.365**	173.179	1	1.501
	$x^5D_1^\circ$	**40404.544**	86.768	1	1.498
	$x^5D_0^\circ$	**40491.312**		1	
$3d^64s(b^4P)4p$	$w^5D_4^\circ$	43499.54	423.16	1	1.492
	$w^5D_3^\circ$	43922.70	260.94	1	1.481
	$w^5D_2^\circ$	44183.64	227.54	1	1.533
	$w^5D_1^\circ$	44411.18	47.78	10	1.315
	$w^5D_0^\circ$	44458.96		10	
$3d^64s(b^4F)4p$	$v^5D_4^\circ$	44415.13	136.31	7	1.401
	$v^5D_3^\circ$	44551.44	112.69	7	1.386
	$v^5D_2^\circ$	44664.13	96.66	7	1.378
	$v^5D_1^\circ$	44760.79	66.13	7	1.389
	$v^5D_0^\circ$	44826.92		11	
$3d^7(a^4P)4p$	$u^5D_4^\circ$	46720.85	24.18	7	1.341
	$u^5D_3^\circ$	46745.03	143.55	7	1.397
	$u^5D_2^\circ$	46888.582	288.67	7	1.260
	$u^5D_1^\circ$	47177.25	− 5.73	10	1.410
	$u^5D_0^\circ$	47171.52:		7	
$3d^64s(a^6D)5p$	$t^5D_4^\circ$	51076.68	284.78	9	1.486
	$t^5D_3^\circ$	51361.46	268.61	9	
	$t^5D_2^\circ$	51630.07:	206.80	9	
	$t^5D_1^\circ$	51836.87:	104.89	9	
	$t^5D_0^\circ$	51941.76:		9	
$3d^64s(a^6D)4p$	$z^5F_5^\circ$	**26874.562**	292.275	1	1.399
	$z^5F_4^\circ$	**27166.837**	227.866	1	1.355
	$z^5F_3^\circ$	**27394.703**	164.895	1	1.250
	$z^5F_2^\circ$	**27559.598**	106.764	1	1.004
	$z^5F_1^\circ$	**27666.362**		1	−0.012
$3d^7(a^4F)4p$	$y^5F_5^\circ$	**33695.418**	344.122	1	1.417
	$y^5F_4^\circ$	**34039.540**	289.235	1	1.344
	$y^5F_3^\circ$	**34328.775**	218.460	1	1.244
	$y^5F_2^\circ$	**34547.235**	144.937	1	0.998
	$y^5F_1^\circ$	**34692.172**		1	−0.016
$3d^64s(a^4D)4p$	$x^5F_5^\circ$	40257.367	337.10	2	1.390
	$x^5F_4^\circ$	40594.47	247.66	1	1.328
	$x^5F_3^\circ$	40842.13	165.93	1	1.254
	$x^5F_2^\circ$	41018.06	112.56	1	0.998
	$x^5F_1^\circ$	41130.62		1	−0.006

TABLE A—(*Continued*)

Config	Desig	Term	Diff	Source	g
$3d^64s(b^4F)4p$	$w^5F_5^\circ$	44243.67	−221.12	7	1.382
	$w^5F_4^\circ$	44022.55	143.69	7	1.444
	$w^5F_3^\circ$	44166.24	119.24	7	1.351
	$w^5F_2^\circ$	44285.48	92.94	1	1.117
	$w^5F_1^\circ$	44378.42:		10	0.283
$3d^64s(a^4G)4p$	$v^5F_5^\circ$	47606.10	323.94	10	1.317
	$v^5F_4^\circ$	47930.04	192.93	8	1.264
	$v^5F_3^\circ$	48122.97	115.93	10	1.236
	$v^5F_2^\circ$	48238.903	111.72	8	1.267
	$v^5F_1^\circ$	48350.62		9	0.230
$3d^64s(a^6D)5p$	$u^5F_5^\circ$	51016.72:	364.76	9	
	$u^5F_4^\circ$	51381.48	237.66	9	
	$u^5F_3^\circ$	51619.14:	208.45	9	
	$u^5F_2^\circ$	51827.59:	118.27	9	
	$u^5F_1^\circ$	51945.86:		9	
$3d^7(a^4F)4p$	$z^5G_6^\circ$	**34843.980**	− 61.532	2	1.332
	$z^5G_5^\circ$	**34782.448**	474.897	1	1.218
	$z^5G_4^\circ$	**35257.345**	354.304	1	1.103
	$z^5G_3^\circ$	**35611.649**	244.775	1	0.887
	$z^5G_2^\circ$	**35856.424**		1	0.335
$3d^64s(a^4H)4p$	$y^5G_6^\circ$	42784.387	127.531	4	1.342
	$y^5G_5^\circ$	42911:918	111.080	1	1.203
	$y^5G_4^\circ$	43022.998	114.513	2	1.024
	$y^5G_3^\circ$	43137.511	72.533	2	0.905
	$y^5G_2^\circ$	43210.044		2	0.331
$3d^64s(b^4F)4p$	$x^5G_6^\circ$	45608.35:	117.83	5	1.336
	$x^5G_5^\circ$	45726.18	107.06	5	1.269
	$x^5G_4^\circ$	45833.24	80.29	5	1.158
	$x^5G_3^\circ$	45913.53	51.45	5	0.928
	$x^5G_2^\circ$	45964.98		5	0.323
$3d^64s(a^4G)4p$	$w^5G_6^\circ$	47363.39	56.84	10	1.306
	$w^5G_5^\circ$	47420.23	169.84	10	1.305
	$w^5G_4^\circ$	47590.07	103.22	10	1.145
	$w^5G_3^\circ$	47693.289	137.91	8	0.931
	$w^5G_2^\circ$	47831.20		11	0.472

Config	Desig	Term	Diff	Source	g
$3d^64s(a^4H)4p$	$z^5H_7^\circ$				
	$z^5H_6^\circ$	43321.12:	−329.46	10	
	$z^5H_5^\circ$	42991.66	117.28	8	1.054
	$z^5H_4^\circ$	43108.944	217.04	7	0.871
	$z^5H_3^\circ$	43325.98		10	0.509
$3d^64s(a^6D)4p$	$z^7P_4^\circ$	**23711.467**	469.409	2	1.747
	$z^7P_3^\circ$	**24180.876**	326.052	2	1.908
	$z^7P_2^\circ$	**24506.928**		2	2.333
$3d^54s^2(a^6S)4p$	$y^7P_4^\circ$	40421.89	−214.77	11	1.75:
	$y^7P_3^\circ$	40207.12	−155.04	11	1.908
	$y^7P_2^\circ$	40052.08		11	2.340
$3d^64s(a^6D)4p$	$z^7D_5^\circ$	**19350.894**	211.563	2	1.597
	$z^7D_4^\circ$	**19562.457**	194.583	2	1.642
	$z^7D_3^\circ$	**19757.040**	155.471	1	1.746
	$z^7D_2^\circ$	**19912.511**	107.137	1	2.008
	$z^7D_1^\circ$	**20019.648**		1	2.999
$3d^64s(a^6D)4p$	$z^7F_6^\circ$	**22650.427**	195.453	2	1.498
	$z^7F_5^\circ$	**22845.880**	150.806	2	1.498
	$z^7F_4^\circ$	**22996.686**	114.262	2	1.493
	$z^7F_3^\circ$	**23110.948**	81.560	2	1.513
	$z^7F_2^\circ$	**23192.508**	52.339	2	1.504
	$z^7F_1^\circ$	**23244.847**	25.545	2	1.549
	$z^7F_0^\circ$	**23270.392**		2	

Desig	Level	Source	g	
1_2°	47419.72	8	1.137	
2_2°	49052.93	11		
3_3°	49227.16	11		
4_4°	51409.18	10	0.953	
5_3°	51435.90:	11		
6_5°	51630.23	10	1.061	
7_2°	51756.16	11		
8_1°	52857.84	10	1.246	$^3S_1^\circ$
9_4°	53328.87	8		
10_3°	53891.54	11	1.476	$^5D_3^\circ$?
11_3°	54004.82	11		
12_5°	54013.78	10	1.356	$^5F_5^\circ$?
13_4°	54301.36	8		

NOTE.—Term values in heavy type are those recommended by the International Astronomical Union (see § 7).

REFERENCES

1. Walters, *Jour. Optical Soc. America* **8**: 245, 1924.
2. Laporte, *Zeit. f. Physik* **23**: 135, and **26**, 1, 1924.
3. Walters (quoted by Meggers, *Astrophys. Jour.* **60**: 60, 1924).
4. Laporte, *Proc. Nat. Acad. Sci.* **12**: 496, 1926.
5. Walters, unpublished (Moore and Russell, *Astrophys. Jour.* **68**: 151, 1928).
6. Russell, *Astrophys. Jour.* **68**: 151, 1928.
7. Burns and Walters, *Publ. Allegheny Observ.* **6**: 159, 1929.
8. Catalán, *Anales Soc. Española Fisica y Quimica* **28**: 1239, 1930.
9. Green, *Phys. Rev.* **55**: 1209, 1939.
10. Catalán, private communication, unpublished.
11. Present work, unpublished.

TABLE B

CLASSIFIED LINES OF FE I

Ref	λ I A	Int	T C	Wave Number Observed	Wave Number o−c	Desig
D	11973.01	8		8349.84	+05	$a^5P_3 - z^5D_4^\circ$
D	11884.12	3		8412.29	−01	$a^5P_1 - z^5D_2^\circ$
D	11882.80	7		8413.22	+04	$a^5P_2 - z^5D_3^\circ$
D	11783.28	6		8484.28	00	$b^3P_2 - z^3D_3^\circ$
D	11689.98	8		8551.99	+01	$a^5P_1 - z^5D_1^\circ$
D	11638.25	7		8590.00	+02	$a^5P_3 - z^5D_3^\circ$
D	11607.57	12		8612.70	+01	$a^5P_2 - z^5D_2^\circ$
D	11593.55	5		8623.13	+05	$a^5P_1 - z^5D_0^\circ$
D	11439.06	15		8739.59	+07	$b^3P_1 - z^3D_2^\circ$
D	11422.30	6		8752.41	+03	$a^5P_2 - z^5D_1^\circ$
D	11374.02	3		8789.56	+06	$a^5P_3 - z^5D_2^\circ$
D	11355.97	1		8803.53	+01	$b^3D_3 - y^3D_3^\circ$
D	11298.83	3		8848.05	+03	$b^3P_2 - z^3D_2^\circ$
D	11251.09	3		8885.60	+04	$b^3P_0 - z^3D_1^\circ$
D	11149.34	2		8966.69	−05	$b^3P_2 - z^3F_3^\circ$
D	11119.80	10		8990.51	+02	$b^3P_1 - z^3D_1^\circ$
D	11013.27	1		9077.47	00	$y^3D_2^\circ - e^5F_3$
D	10925.80	1		9150.14	+09	$w^5F_5^\circ - g^5F_4$
D	10896.30	3		9174.92	+02	$c^3P_1 - z^3P_2^\circ$
D	10884.30	3		9185.03	−05	$z^3D_2^\circ - X_3$
D	10881.65	1		9187.27	+12	$b^3P_1 - z^3F_2^\circ$
D	10863.60	5		9202.53	−05	$y^3D_3^\circ - e^5F_4$
D	10849.68	2		9214.34	+05	$e^7G_4 - t^3H_5^\circ$?
D	10818.36	3		9241.02	+01	$z^3D_1^\circ - X_2$
D	10783.09	3		9271.24	−03	$c^3P_0 - z^3P_1^\circ$
D	10752.99	3		9297.20	+05	$e^7G_3 - t^3H_4^\circ$?
D	10532.21	10		9492.09	+11	$z^3D_2^\circ - X_2$
D	10469.59	20	V	9548.85	+03	$z^3D_3^\circ - X_3$
D	10452.70	5		9564.29	+10	$z^3F_4^\circ - X_3$
D	10435.36	0N		9580.19	+03	$y^3D_3^\circ - e^5F_3$
D	10423.65	3		9590.95	+12	$c^3P_1 - z^3P_1^\circ$
D	10422.99	0		9591.55	+05	$a^3G_5 - z^3F_4^\circ$
F	10395.75	8		9616.68	+05	$a^5P_3 - z^5F_4^\circ$
SS	10353.82	(2n)		9655.63	+04	$w^5D_4^\circ - h^5D_4$
F	10348.16	4n		9660.91	−08	$w^5D_4^\circ - f^5P_3$
F	10340.77	4		9667.81	+12	$a^5P_2 - z^5F_3^\circ$
F	10218.36	3		9783.63	+05	$c^3P_1 - z^3P_0^\circ$
E	10216.351	100	V	9785.55	−04	$y^3D_3^\circ - e^3F_4$
F	10195.11	2		9805.94	+01	$a^3G_4 - z^3F_3^\circ$
F	10167.4	1		9832.64	+06	$a^5P_2 - z^5F_2^\circ$
E	10145.601	80	V	9853.79	−04	$y^3D_2^\circ - e^3F_3$
F	10142.82	2		9856.49	−05	$x^5F_3^\circ - f^5D_2$
F	10113.86	2		9884.71	+16	$a^3G_3 - z^3F_2^\circ$
E	10065.080	60	V	9932.62	−04	$y^3D_1^\circ - e^3F_2$
F	10057.64	3	V	9939.97	00	$x^5F_4^\circ - f^5D_3$
F	9980.55	2n		10016.74	−09	$x^5F_4^\circ - e^7P_3$
F	9977.52	1		10019.78	+06	$x^5F_3^\circ - f^7D_3$
F	9944.13	3n		10053.43	00	$y^7P_4^\circ - e^7P_4$
F	9917.93	2	IV*	10079.99	+15	$a^1F_3 - x^1G_4^\circ$
E	9889.082	40	V	10109.39	−05	$x^5F_4^\circ - e^5G_5$
SS	9881.54	(1)		10117.11	−03	$d^3F_3 - x^3F_3^\circ$
F	*9868.09	3	V	10130.90	+07	$x^5F_2^\circ - e^5S_2$
					+09	$x^5F_2^\circ - e^7F_3$
E	9861.793	30	V	10137.37	−13	$x^5F_3^\circ - e^5G_4$
F	9839.38	1		10160.46	−04	$d^3F_3 - w^3D_2^\circ$
F	9834.04	3n	V	10165.97	+15	$x^5F_5^\circ - f^5D_4$
F	9811.36	2		10189.47	+06	$y^7P_4^\circ - e^7P_3$
E	9800.335	20	V	10200.94	−06	$x^5F_2^\circ - e^5G_3$
F	9786.62	2		10215.24	+04	$y^3F_3^\circ - e^5F_4$
F	9783.96	3	V	10218.01	+06	$x^5F_5^\circ - e^7P_4$
E	9763.913	15	V	10238.99	−02	$x^5F_4^\circ - e^7F_5$
E	9763.450	15	V	10239.48	−08	$x^5F_1^\circ - e^5G_2$

Ref	λ I A	Int	T C	Wave Number Observed	Wave Number o−c	Desig
E	9753.129	10	V	10250.31	−05	$y^3D_2^\circ - e^3F_2$
F	9747.24	2		10256.50	+03	$d^3F_2 - x^3F_2^\circ$
E	9738.624	200	V	10265.58	+01	$x^5F_5^\circ - e^5G_6$
F	*9699.70	6n	V	10306.77	+01	$x^5F_3^\circ - e^5S_2$
					+03	$x^5F_3^\circ - e^7F_3$
F	9693.69	1		10313.16	+13	$x^5F_2^\circ - e^7F_2$
F	9683.57	1		10323.94	−10	$z^5S_2^\circ - e^5G_3$
F	9676.42	1		10331.57	+15	$w^5D_4^\circ - g^5F_3$
F	9673.16	1n		10335.05	−06	$b^3H_5 - y^3F_4^\circ$
F	9658.94	3		10350.26	+07	$x^5F_3^\circ - e^7F_4$
F	9657.30	4	V	10352.03	−09	$x^5F_2^\circ - e^5G_2$
E	9653.143	20	V	10356.48	−03	$y^3D_3^\circ - e^3F_3$
Y	9636.69	(1)		10374.17	−06	$d^3F_4 - w^5G_5^\circ$
F	9634.22	5	V	10376.82	−11	$x^5F_3^\circ - e^5G_3$
E	9626.562	30n	V	10385.08	−08	$x^5F_4^\circ - e^5G_4$
F	9602.07	2		10411.57	−02	$y^7P_4^\circ - e^7F_5$
E	9569.960	40n	V	10446.50	−04	$x^5F_5^\circ - e^5G_5$
F	9556.56	1		10461.14	−03	$a^3D_3 - y^3F_4^\circ$
F	9550.90	2		10467.35	+05	$x^5D_2^\circ - f^5D_2$
F	9527.7	1		10492.8	0	$x^5F_3^\circ - e^7G_4$
E	9513.24	10n	V	10508.78	+01	$x^5F_4^\circ - f^5F_5$
F	9462.97	2	V	10564.61	+05	$x^5D_3^\circ - f^5D_3$
F	9454.24	4n	V	10574.37	−06	$x^5F_1^\circ - f^5F_2$
F	*9452.45	2	V	10576.37	+25	$x^5F_5^\circ - e^7F_5$
					−11	$x^5D_1^\circ - f^5D_0$
F	9443.98	10n	V	10585.85	−22	$x^5F_2^\circ - f^5F_3$
F	9437.91	2		10592.66	−11	$y^3F_3^\circ - e^5F_3$
F	9414.14	20n	V	10619.41	−17	$x^5F_3^\circ - f^5F_4$
F	9410.1	1n		10624.0	+ 1	$x^5F_1^\circ - f^5F_1$
F	9401.09	10n	V	10634.15	+02	$x^5F_4^\circ - e^7G_5$
F	9394.71	3n	V	10641.37	−05	$x^5D_3^\circ - e^7P_3$
F	9388.28	3n		10648.66	−13	$x^5D_2^\circ - f^5D_1$
F	9382.83	3n		10654.84	+11	$y^7P_3^\circ - f^7D_3$
E	9372.900	6	IV*	10666.13	00	$b^3F_4 - z^3F_4^\circ$
E	9362.370	4	IV*	10678.13	00	$a^3P_2 - z^5P_3^\circ$
E	9359.420	3	IV*	10681.50	00	$b^3F_4 - z^3D_3^\circ$
E	9350.46	10	V	10691.73	−03	$y^3F_4^\circ - e^5F_4$
F	9343.40	3	V	10699.80	+01	$x^5F_4^\circ - e^3D_3$
F	9333.94	2		10710.65	+14	$x^5F_5^\circ - e^7G_6$
O	9324.07	(1)		10721.99	+09	$x^5F_2^\circ - e^3D_2$
F	9318.13	3	V	10728.82	+03	$x^5D_3^\circ - f^5D_2$
F	9307.94	2		10740.57	+10	$x^5F_4^\circ - e^7G_4$
F	9294.66	2	V	10755.92	−11	$x^5F_4^\circ - g^5D_4$
F	9259.05	15	V	10797.29	−07	$x^5D_4^\circ - f^5D_4$
E	9258.30	20	V	10798.16	−04	$y^3F_3^\circ - e^3F_4$
F	9246.54	2	IV*	10811.89	+03	$b^3F_3 - z^3D_2^\circ$
F	9242.32	2		10816.82	+08	$x^5D_2^\circ - f^7D_1$
SS	9233.18	(1)		10827.54	−03	$y^5G_5^\circ - e^3G_5$
O	9225.55	(1)		10836.49	−05	$d^3F_3 - x^3G_4^\circ$
F	9217.54	5n	V	10845.91	+04	$x^5F_5^\circ - f^5F_5$
F	9214.45	6	V	10849.54	+05	$x^5D_4^\circ - e^7P_4$
E	9210.030	6	IV*	10854.75	+01	$b^3P_1 - y^5D_2^\circ$
F	9199.52	2n		10867.15	−09	$x^5F_4^\circ - f^5F_4$
F	9178.57	1n		10891.96	−01	$x^5D_3^\circ - f^7D_3$
U	9173.83	(1)		10897.58	+12	$a^3D_2 - y^3F_2^\circ$
F	9173.46	4nd		10898.02	+19	$x^5F_3^\circ - e^3D_2$
U	9164.51	(1)		10908.67	+06	$x^5D_4^\circ - f^5D_3$
SS	*9156.94	(2)		10917.68	+15	$x^5D_2^\circ - e^5S_2$
					+17	$x^5D_2^\circ - e^7F_3$
F	9155.9	1		10918.9	−3	$x^5F_1^\circ - g^5D_2$
E	9147.800	5n	V	10928.59	+14	$x^5F_3^\circ - g^5D_3$
F	9146.11	3	IV*	10930.61	+03	$b^3F_3 - z^3F_3^\circ$

TABLE B—(*Continued*)

Ref	λ I A	Int	T C	Wave Number Observed	o−c	Desig
E	9118.888	20	IV*	10963.24	00	$b^3P_2 - y^5D_2^\circ$
F	9117.10	2		10965.39	+05	$b^3P_0 - y^5D_1^\circ$
F	9103.64	1		10981.60	+01	$y^5F_5^\circ - e^5D_4$
F	9100.50	5n	V	10985.40	−07	$x^5D_4^\circ - e^7P_3$
E	9089.413	30	IV*	10998.80	+01	$b^3G_5 - z^5G_5^\circ$
E	9088.326	50	IV*	11000.11	+01	$b^3P_1 - z^3P_2^\circ$
SS	9084.22	(1)		11005.09	−01	$y^5F_3^\circ - e^5D_2$
F	*9080.48	3n		11009.62	−13	$x^5D_3^\circ - e^5G_4$
					−06	$x^5F_4^\circ - f^5F_3$
E	9079.599	8	V	11010.68	−03	$y^3F_2^\circ - e^3F_3$
F	*9070.42	2		11021.82	+03	$y^5F_4^\circ - e^5D_3$
					−06	$x^5F_2^\circ - e^3D_1$
F	9062.24	2		11031.77	+01	$x^5F_2^\circ - g^5D_2$
F	9052.6	1		11043.5	−1	$y^5G_4^\circ - e^3G_4$
SS	9036.72	(1)		11062.93	+03	$x^5D_2^\circ - e^3D_3$
F	9030.67	1		11070.34	+07	$b^3P_1 - y^5D_1^\circ$
F	9024.47	15	V	11077.94	−14	$x^5D_4^\circ - e^5G_5$
F	9019.84	2		11083.63	−08	$x^5F_1^\circ - g^5D_1$
F	9013.90	1		11090.93	+11	$a^3P_2 - z^5P_2^\circ$
E	9012.098	30	V	11093.15	+01	$x^5F_5^\circ - g^5D_4$
F	9010.55	2	IV*	11095.06	+07	$b^3F_2 - z^3F_2^\circ$
F	9008.37	2		11097.74	+06	$X_3 - u^3D_3^\circ$
F	9006.72	1		11099.78	+06	$x^5D_2^\circ - e^7F_2$
E	8999.561	100	III	11108.61	+01	$b^3P_2 - z^3P_2^\circ$
F	8984.87	3	V	11126.77	+06	$x^5F_1^\circ - g^5D_0$
E	8975.408	10	IV*	11138.50	+01	$b^3G_4 - z^5G_4^\circ$
F	8946.25	1		11174.80	+04	$b^3P_1 - y^5D_0^\circ$
E	8945.204	20	V	11176.11	00	$x^5F_4^\circ - g^5D_3$
F	8943.00	3	IV*	11178.86	+09	$b^3P_2 - y^5D_1^\circ$
F	8929.04	5	V	11196.34	+07	$x^5F_2^\circ - g^5D_1$
F	8919.95	10	V	11207.75	+06	$x^5F_3^\circ - g^5D_2$
					+09	$(x^5D_4^\circ - e^7F_5)$
F	8916.26	1	IIA	11212.39	+09	$a^3F_2 - z^7P_3^\circ$
F	8876.13	2	IV*	11263.08	−14	$x^5D_0^\circ - f^5F_1$
F	8868.42	3	IV*	11272.87	+03	$b^3G_3 - z^5G_3^\circ$
E	8866.961	150	V	11274.73	−04	$y^3F_4^\circ - e^3F_4$
F	8863.64	1p?		11278.95	−06	$y^7P_2^\circ - e^7F_2$
F	8846.82	5	V	11300.40	−11	$x^5D_1^\circ - f^5F_2$
E	8838.433	30	IV*	11311.12	+02	$b^3P_0 - z^3P_1^\circ$
E	8824.227	200	II	11329.33	+01	$a^5P_2 - z^5P_3^\circ$
F	8814.5	2		11341.8	0	$X_3 - t^3D_3^\circ$
SS	8808.173	4n		11349.98	−01	$x^5D_1^\circ - f^5F_1$
E	8804.624	10	IV*	11354.55	+02	$a^3P_2 - z^5P_1^\circ$
F	8796.42	2		11365.14	+08	$x^5D_3^\circ - e^7G_4$
E	8793.376	120	V	11369.08	−05	$y^3F_3^\circ - e^3F_3$
F	8790.62	10n	V	11372.64	−14	$x^5D_2^\circ - f^5F_3$
F	8784.44	5	V	11380.64	+02	$x^5D_3^\circ - g^5D_4$
E	8764.000	100	V	11407.19	−05	$y^3F_2^\circ - e^3F_2$
E	8757.192	50	IV	11416.05	+02	$b^3P_1 - z^3P_1^\circ$
F	8747.32	2		11428.94	+15	$b^3G_3 - z^3G_4^\circ$
SS	8729.171	(2)		11452.70	−06	$a^1P_1 - y^3D_1^\circ$
F	*8713.19	10	V	11473.70	+01	$b^3G_5 - z^5G_4^\circ$
					+01	$x^5D_2^\circ - f^5F_2$
F	8710.29	20n	V	11477.53	+12	$x^5D_4^\circ - f^5F_5$
O	8699.43	4n	V	11491.85	+02	$x^5D_3^\circ - f^5F_4$
E	8688.633	1500	II	11506.13	00	$a^5P_3 - z^5P_3^\circ$
E	8674.751	60	III	11524.55	+02	$b^3P_2 - z^3P_1^\circ$
E	8661.908	600	II	11541.63	+01	$a^5P_1 - z^5P_2^\circ$
E	8621.612	10	IV*	11595.58	00	$b^3G_5 - z^3G_5^\circ$
E	8611.807	40	III	11608.78	00	$b^3P_1 - z^3P_0^\circ$
O	8598.79	4	V	11626.35	+08	$z^3G_5^\circ - e^5F_5$
O	8592.97	2n	V	11634.23	−04	$x^5D_3^\circ - f^5F_3$
E	8582.267	15	IV*	11648.74	00	$b^3G_4 - z^3G_4^\circ$
W	8559.98	(1)		11679.06	−05	$a^1F_3 - t^3D_3^\circ$
E	8526.685	8	V	11724.67	−01	$x^5D_4^\circ - g^5D_4$

Ref	λ I A	Int	T C	Wave Number Observed	o−c	Desig
O	8515.08	20	IV*	11740.65	+06	$b^3G_3 - z^3G_3^\circ$
E	8514.075	150	II	11742.04	+02	$a^5P_2 - z^5P_2^\circ$
O	8497.00	8	V	11765.63	−02	$y^3F_3^\circ - e^3F_2$
O	8471.75	2	V	11800.70	00	$x^5D_3^\circ - g^5D_3$
E	8468.413	300	II	11805.35	+01	$a^5P_1 - z^5P_1^\circ$
E	8439.603	20	V	11845.65	−04	$y^3F_4^\circ - e^3F_3$
O	8424.14	2n	V	11867.39	−01	$x^5D_3^\circ - e^5P_3$
O	8422.95	2	V	11869.07	−04	$c^3F_3 - x^3D_2^\circ$
O	8401.42	2	IV*	11899.49	00	$a^3P_0 - z^3D_1^\circ$
E	8387.781	1200	II	11918.83	+01	$a^5P_3 - z^5P_2^\circ$
E	8365.642	25	IV?	11950.38	+03	$a^3D_3 - y^3D_3^\circ$
E	8360.822	8	V	11957.26	−01	$z^3G_3^\circ - e^5F_2$
R	8342.95	(−)		11982.88	−09	$x^5D_2^\circ - g^5D_1$
E	8339.431	80	V	11987.94	−01	$z^3G_4^\circ - e^5F_3$
E	8331.941	200	V	11998.71	−02	$z^3G_5^\circ - e^5F_4$
E	8327.063	1200	II	12005.74	+01	$a^5P_2 - z^5P_1^\circ$
E	8293.527	20	V	12054.29	−05	$a^3D_2 - y^3D_2^\circ$
O	8275.91	4n	V	12079.95	+01	$x^5D_3^\circ - g^5D_2$
O	8274.28	6	IV?	12082.33	+11	$X_3 - s^3D_3^\circ$
M	8264.27	3	V	12096.96	+05	$X_2 - s^3D_2^\circ$
E	8248.151	30	V	12120.60	−02	$z^5G_4^\circ - e^5F_4$
E	8239.130	8	IV*	12133.87	−01	$a^3P_1 - z^3D_2^\circ$
E	8232.347	50	V	12143.87	−02	$z^5G_3^\circ - e^5F_3$
E	8220.406	1500	V	12161.51	−02	$z^5G_6^\circ - e^5F_5$
E	8207.767	40	V	12180.24	00	$z^5G_2^\circ - e^5F_2$
E	8198.951	80	V	12193.34	−04	$z^3G_4^\circ - e^3F_4$
O	8186.80	10nd?	V	12211.43	−02	$x^5D_4^\circ - e^5P_3$
O	8179.03	(1)	IV*?	12223.03	−03	$z^5G_5^\circ - e^5F_5$
O	*8149.59	3	V	12267.18	+14	$d^3F_3 - v^3D_2^\circ$
					−13	$d^3F_3 - w^3F_3^\circ$
E	8096.874	10	IV*	12347.06	00	$c^3F_4 - x^3D_3^\circ$
E	8085.200	500	V	12364.88	−02	$z^5G_2^\circ - e^5F_1$
E	8080.668	10nd?	V	12371.82	−21	$a^3D_2 - y^3D_1^\circ$
O	8075.13	4	II	12380.30	+04	$a^5F_4 - z^7D_3^\circ$
O	8047.60	15	II	12422.65	+04	$a^5F_5 - z^7D_5^\circ$
E	8046.073	600	V	12425.01	−01	$z^5G_3^\circ - e^5F_2$
E	8028.341	50	V	12452.45	−05	$z^3G_3^\circ - e^3F_3$
E	7998.972	700	V	12498.17	−02	$z^5G_4^\circ - e^5F_3$
O	7959.21	(1)		12560.61	−05	$x^5F_4^\circ - h^5D_4$
O	7955.81	(1)		12565.98	−08	$x^5F_4^\circ - f^5P_3$
E	7945.878	600	V	12581.68	−06	$z^3G_5^\circ - e^3F_4$
					+16	$(a^3P_1 - z^3F_2^\circ)$
O	7941.09	10	IV*	12589.27	00	$a^3D_1 - y^3D_1^\circ$
E	7937.166	700	V	12595.50	−02	$z^5G_5^\circ - e^5F_4$
E	7912.866	6	IIA	12634.18	00	$a^5F_5 - z^7D_4^\circ$
O	7879.84	1	V	12687.13	−14	$x^5F_4^\circ - f^5G_5$
O	7869.65	4	V	12703.55	−08	$z^5G_4^\circ - e^3F_4$
O	7855.48	4n	V	12726.47	−12	$x^5F_3^\circ - f^5P_2$
O	7844.66	2	V	12744.02	−18	$y^3D_1^\circ - e^3D_2$
E	7832.224	400	V	12764.26	−04	$z^3G_4^\circ - e^3F_3$
E	7780.586	300	V	12848.97	−06	$z^3G_3^\circ - e^3F_2$
O	7751.18	5n	V	12897.72	−04	$x^5F_5^\circ - h^5D_4$
E	7748.281	125	IV	12902.54	−01	$b^3G_5 - y^3F_4^\circ$
O	7742.71	4n	V	12911.83	−01	$x^5F_5^\circ - f^5G_6$
O	7723.20	4	IV*	12944.44	+02	$a^3P_2 - z^3D_3^\circ$
E	7710.390	25	V	12965.95	−02	$y^5F_4^\circ - e^5F_5$
E	7664.302	80	IV	13043.92	00	$b^3G_4 - y^3F_3^\circ$
E	7661.223	30	V	13049.16	−03	$y^5F_3^\circ - e^5F_4$
L	7653.783	6	V	13061.84	−06	$y^3D_2^\circ - e^3D_2$
E	7620.538	25	V	13118.83	−05	$y^3D_3^\circ - e^3D_3$
O	7605.32	2n	V	13145.08	+06	$x^5F_4^\circ - e^3G_5$
E	7586.044	150	V	13178.48	−04	$z^5G_5^\circ - e^3F_4$
E	7583.796	50	IV*	13182.38	00	$b^3G_3 - y^3F_2^\circ$
E	7568.925	30	V	13208.28	−02	$y^5F_2^\circ - e^5F_3$
O	7563.03	1n	V	13218.58	+01	$y^3D_1^\circ - g^5D_1$

TABLE B—(*Continued*)

Ref	λ I A	Int	T C	Wave Number Observed	o−c	Desig
O	7559.68	1n	V	13224.44	00	$x^5F_3^\circ - e^3G_4$
U	7541.61	(1)		13256.14	−10	$z^3F_3^\circ - e^5D_3$?
E	7531.171	60	V	13274.50	−05	$z^5G_4^\circ - e^3F_3$
E	7511.045	800	V	13310.07	−02	$y^5F_5^\circ - e^5F_5$
L	7507.300	8	V	13316.69	−08	$z^5G_3^\circ - e^3F_2$
V	7498.56	1	IV*	13332.23	−02	$c^3F_3 - u^5D_3^\circ$
E	7495.088	400	V	13338.40	−03	$y^5F_4^\circ - e^5F_4$
L	7491.678	12	V	13344.48	−02	$y^5F_1^\circ - e^5F_2$
SS	7481.934	(1)		13361.85	−02	$y^3D_2^\circ - e^3D_1$
SS	7476.376	(1)		13371.79	+04	$y^3D_2^\circ - g^5D_2$
O	7473.56	(1)		13376.82	−02	$y^5P_2^\circ - f^5D_3$
V	7461.534	(1)		13398.40	00	$b^3F_4 - y^5F_4^\circ$
V	7454.02	(1)		13411.89	−03	$c^3F_2 - u^5D_1^\circ$
V	7447.43	1	V	13423.76	−08	$x^5D_3^\circ - g^5F_4$
E	7445.776	200	V	13426.74	−02 −14	$y^5F_3^\circ - e^5F_3$ $(a^3P_2 - z^3F_3^\circ)$
L	7443.031	2	IV*	13431.69	−05	$c^3F_2 - x^3F_2^\circ$
V	7440.98	2n	V	13435.40	−05	$x^5D_4^\circ - g^5F_5$
M	7430.90	1−	IV*	13453.62	−09	$y^5P_2^\circ - e^7P_3$
O	7430.73	(1)		13453.93	+01	$w^5F_5^\circ - i^5D_4$
M	7430.58	1−	IV*	13454.20	−05	$b^3F_3 - y^5F_3^\circ$
O	7421.60	1−	V	13470.48	−10	$y^5P_1^\circ - f^5D_1$
E	7418.674	5	IV*	13475.79	−01	$c^3F_3 - u^5D_2^\circ$
E	7411.178	100	V	13489.42	−01	$y^5F_2^\circ - e^5F_2$
E	7401.689	4	IV*	13506.72	−05	$c^3F_2 - w^3D_1^\circ$
E	7389.425	80	V	13529.13	−02	$y^5F_1^\circ - e^5F_1$
L	7386.394	8n	V	13534.68	−02	$x^5D_4^\circ - f^5P_3$
V	7382.99	1n	V	13540.92	−15	$y^5P_2^\circ - f^5D_2$
O	7370.16	1	V	13564.50	−08	$y^3D_3^\circ - e^3D_2$
O	7366.37	1	V	13571.48	+04	$y^5P_1^\circ - f^5D_0$
O	7363.96	1n	V	13575.92	−08	$x^5D_3^\circ - h^5D_3$
V	7353.528	1	V	13595.17	−03	$y^3D_3^\circ - g^5D_3$
V	7351.56	4	V	13598.81	−03	$x^5D_3^\circ - f^5P_2$
V	7351.160	2n	V	13599.55	−05	$x^5D_2^\circ - g^5F_3$
V	7333.62	1n	V	13632.08	−12	$y^5F_3^\circ - e^3F_4$
L	§*7320.694	5n	V	13656.15	−04 +24	$y^5P_3^\circ - f^5D_4$ $x^5D_4^\circ - f^5G_5$
I	7311.101	12	V	13674.07	−02	$y^5F_2^\circ - e^5F_1$
L	§7307.938	8	IV*	13679.99	−01	$c^3F_3 - x^3F_3^\circ$
V	7306.61	3	V	13682.47	−08	$y^5F_5^\circ - e^5F_4$
O	7300.47	1n	V	13693.98	+08	$x^5D_2^\circ - f^5P_1$
V	7295.00	1−	V	13704.25	−01	$y^5P_2^\circ - f^7D_3$
I	7293.068	15	V	13707.88	−01	$y^5F_3^\circ - e^5F_2$
V	7292.856	3n	V	13708.28	−04	$y^5P_3^\circ - e^7P_4$
I	7288.760	10	V	13716.02	+02	$y^5F_4^\circ - e^5F_3$
V	7285.286	1	V	13722.52	−04	$y^5P_2^\circ - f^5D_1$
L	7284.843	4	IV*	13723.35	−01	$c^3F_3 - w^3D_2^\circ$
V	7282.39	1n	V	13727.98	+04	$x^5D_1^\circ - h^5D_1$
V	7261.54	3n	V	13767.40	−04	$y^5P_3^\circ - f^5D_3$
SS	7256.142	1−	V	13777.64	−03	$x^5D_3^\circ - f^3D_3$
V	7244.86	2n	V	13799.09	−05	$x^5D_3^\circ - f^5G_4$
I	7239.885	6	V	13808.58	+01	$z^3P_2^\circ - e^5F_3$
O	7228.69	1	IV*	13829.96	+03	$a^3G_3 - z^3G_3^\circ$
I	7223.668	12	IV*	13839.58	00	$c^3P_2 - y^3D_3^\circ$
V	*7222.88	(1)		13841.08	−18 −01	$x^5F_3^\circ - f^3F_4$ $y^5P_2^\circ - f^7D_2$
V	7221.22	2n	V	13844.27	−03	$y^5P_3^\circ - e^7P_3$
I	7219.686	5	IV*	13847.21	+04	$c^3F_4 - u^5D_4^\circ$
V	7212.47	1n	V	13861.06	−02	$x^5D_3^\circ - g^5F_3$
E	7207.406	500	V	13870.80	−02	$y^5D_3^\circ - e^5F_4$
V	7207.123	6	IV*	13871.34	−01	$c^3F_4 - u^5D_3^\circ$
O	7194.92	1	V	13894.87	+02	$x^5D_0^\circ - g^5F_1$
O	7191.66	(1)		13901.17	+05	$x^5D_2^\circ - h^5D_1$
V	7189.17	3	IV*	13905.98	−03	$c^3P_1 - y^3D_2^\circ$
E	7187.341	800	V	13909.52	−03	$y^5D_4^\circ - e^5F_5$

Ref	λ I A	Int	T C	Wave Number Observed	o−c	Desig
V	7181.93	1n	V	13920.00	−05	$x^5D_4^\circ - h^5D_3$
L	7181.222	10	V	13921.38	−05	$y^5F_4^\circ - e^3F_4$
V	7180.020	1	IV*	13923.71	−03	$a^3F_4 - z^5D_4^\circ$
V	7176.886	2n	V	13929.78	−04	$x^5D_2^\circ - f^5G_3$
V	7175.937	3	V	13931.64	−03	$y^5P_3^\circ - f^5D_2$
E	7164.469	250	V	13953.93	−01	$y^5D_2^\circ - e^5F_3$
V	7158.502	1	V	13965.56	−04	$z^5P_2^\circ - e^7D_3$
V	7155.64	3n	V	13971.14	−04	$x^5D_1^\circ - f^5G_2$
V	7151.495	1	IV*	13979.24	−03	$a^3P_0 - y^5D_1^\circ$
R	*7148.69	(−)		13984.73	+07 +03	$y^5F_2^\circ - e^3F_3$ $z^5S_2^\circ - e^3P_2$
V	*7145.317	5	V	13991.33	+05 +03	$y^5P_2^\circ - e^7F_3$ $y^5P_2^\circ - e^5S_2$
V	7142.522	4n	V	13996.80	−04	$x^5D_3^\circ - h^5D_2$
I	7132.989	8	IV*	14015.51	−02	$c^3F_4 - x^3F_4^\circ$
I	7130.942	150	V	14019.52	−02	$y^5D_1^\circ - e^5F_2$
I	7112.176	3	IV*	14056.52	−01	$b^3G_4 - y^3D_3^\circ$
I	7107.461	4	IV*	14065.84	−03	$c^3F_2 - w^5G_2^\circ$
I	7095.425	3	V	14089.73	+03	$z^3P_2^\circ - e^5F_2$
SS	7091.942	(1)		14096.68	−01	$x^5D_3^\circ - e^3G_4$
O	7091.83	(1)		14096.86	−08	$x^5D_3^\circ - f^3D_2$
I	7090.404	40	V	14099.69	−01	$y^5D_0^\circ - e^5F_1$
V	*7086.76	2	V	14106.94	+03 −05	$x^5F_2^\circ - f^3F_3$ $z^5P_3^\circ - e^7D_4$
V	7083.396	1n	V	14113.64	−02	$x^5D_4^\circ - e^3G_5$
V	7071.88	1	V	14136.62	−05	$y^5P_2^\circ - e^3D_3$
SS	7069.54	1−	IV*	14141.30	00	$b^3F_4 - z^5G_5^\circ$
I	7068.415	40	IV*	14143.54	−02	$c^3F_4 - w^3D_3^\circ$
O	7044.60	(1)		14191.36	+06	$x^5D_3^\circ - f^5G_3$
V	7038.818	2	V	14203.02	−10	$y^5F_3^\circ - e^3F_3$
I	7038.251	40	V	14204.16	−04	$y^5D_1^\circ - e^5F_1$
V	7027.60	(1)		14225.69	00	$d^3F_3 - v^3F_2^\circ$
V	7024.649	10n	V	14231.66	−03	$y^5P_3^\circ - f^7D_2$
V	7024.084	5	IV*	14232.81	−05	$c^3F_4 - z^3H_4^\circ$
L	7022.976	50	V	14235.06	−01	$y^5D_2^\circ - e^5F_2$
V	7016.436	60	V	14248.41	+02	$y^5D_3^\circ - e^5F_3$
V	7016.075	20	IV*	14249.06	−04	$a^3P_1 - y^5D_2^\circ$
O	7014.99	(1)		14251.26	00	$a^3H_4 - y^5F_4^\circ$
V	7011.364	3	IV*	14258.64	−01	$d^3F_4 - v^3F_4^\circ$
V	7010.362	2	IV*	14260.67	−06	$d^3F_2 - v^3F_2^\circ$
V	7008.014	5	V	14265.45	−11	$y^5F_5^\circ - e^3F_4$
V	7000.633	3	IV*	14280.49	−02	$c^3F_3 - w^5G_3^\circ$
I	6999.902	30	V	14281.98	−02	$y^5D_4^\circ - e^5F_4$
I	6988.530	5	IV*	14305.23	+01	$a^3H_6 - y^5F_5^\circ$
I	6978.855	100	III	14325.06	+03	$a^3P_0 - z^3P_1^\circ$
V	6977.445	4	IV*	14327.95	−01	$d^3F_4 - u^3G_5^\circ$
V	6976.934	3	IV*	14329.00	−01	$d^3F_3 - v^3F_4^\circ$
V	6976.306	1	V	14330.29	−10	$y^5P_1^\circ - e^3D_2$
V	6971.95	1	IV*	14339.24	−02	$b^3G_3 - y^3D_2^\circ$
U	6960.343	2	V	14363.15	−03	$d^3F_4 - 4_4^\circ$
SS	6951.656	1−	V	14381.10	−09	$y^5F_2^\circ - e^3F_2$
I	*6951.261	25	V	14381.92	+03 +05	$y^5P_3^\circ - e^5S_2$ $y^5P_3^\circ - e^7F_3$
V	*6947.501	3	V	14389.70	+04 −20	$d^3F_3 - v^3F_3^\circ$ $d^3F_4 - 5^\circ$
I	6945.208	150	III	14394.45	−02	$a^3P_1 - z^3P_2^\circ$
L	*6933.628	6	IV*	14418.49	−01 +07	$a^3H_5 - y^5F_4^\circ$ $c^3F_3 - w^5G_2^\circ$
U	6933.04	1	V	14419.72	−01	$y^5D_2^\circ - e^5F_1$
V	6930.64	1	V	14424.71	+01	$d^3F_2 - v^3F_3^\circ$
I	6916.702	60	V	14453.78	−05	$y^5D_3^\circ - e^3F_4$
V	6911.52	1	IV*	14464.61	−02	$a^3P_1 - y^5D_1^\circ$
V	6898.31	3	V	14492.31	−05	$y^5F_4^\circ - e^3F_3$
L	6885.772	20	V	14518.70	−05	$y^3F_2^\circ - e^3D_1$
M	6881.74	1	V	14527.21	−05	$y^5P_3^\circ - e^3D_3$

TABLE B—(*Continued*)

Ref	λ I A	Int	T C	Wave Number Observed	Wave Number o−c	Desig
V	6880.65	2	V	14529.51	−01	$y^5D_3^\circ - e^5F_2$
V	6875.98	1	IV*	14539.38	00	$c^3F_2 - x^3P_2^\circ$
V	6875.45	1	IV*	14540.50	00	$a^3H_4 - y^5F_3^\circ$
V	6862.481	4n	V	14567.98	+04	$y^5P_3^\circ - e^7G_4$
V	6861.93	2	IV*	14569.15	+02	$a^3P_1 - y^5D_0^\circ$
SS	6860.953	1−	IV*	14571.22	00	$b^3P_2 - y^5P_1^\circ$
V	6860.29	1	IV*	14572.63	00	$b^3F_2 - z^5G_3^\circ$
K	6858.164	40	V	14577.15	−04	$y^3F_3^\circ - e^3D_2$
V	6857.25	4	IV*	14579.09	00	$c^3F_4 - z^1G_4^\circ$
V	6855.74	2	V	14582.30	−07	$y^5P_2^\circ - e^3D_2$
I	6855.176	150	V	14583.50	−01	$y^5P_3^\circ - g^5D_4$
V	6854.82	2	IV*	14584.26	+03	$d^3F_4 - 6_5^\circ$
SS	6851.652	1−	IV*	14591.00	−02	$a^3F_2 - z^5F_2^\circ$
SS	6847.603	1−	V	14599.63	−02	$y^5F_3^\circ - e^3F_2$
SS	6844.683	1−	IIIA?	14605.86	−02 +02	$a^3F_3 - z^5F_4^\circ$ $(b^1D_2 - v^3D_2^\circ)$
I	6843.671	60	V	14608.02	−04	$y^3F_4^\circ - e^3D_3$
V	6842.668	6n	V	14610.16	+03	$y^5P_1^\circ - e^5P_1$
I	6841.349	80	V	14612.97	−01	$y^5P_2^\circ - g^5D_3$
V	6839.828	4	IV*	14616.22	+02	$b^3F_4 - z^5G_4^\circ$
O	6837.00	3	IV*	14622.27	+05	$d^3F_4 - u^3G_4^\circ$
V	6833.24	1	V	14630.32	−04	$y^5P_1^\circ - e^3D_1$
I	6828.610	50	V	14640.24	00	$y^5P_1^\circ - g^5D_2$
SS	*6822.042	1−	V	14654.33	+02 −10	$a^3P_0 - y^5F_1^\circ$ $d^3F_3 - t^5D_2^\circ$
O	6820.43	8n	V	14657.80	−07	$y^5P_1^\circ - e^5P_2$
SS	6819.595	(1)	V	14659.59	+01	$y^5D_4^\circ - e^5F_3$
V	6810.28	20n	V	14679.64	−04	$y^5P_2^\circ - e^5P_3$
L	6806.851	10	IV	14687.03	−01	$a^3G_4 - y^3F_4^\circ$
V	6804.27	3	IV*	14692.61	+03	$d^3F_3 - u^3G_4^\circ$
V	6804.020	5	V	14693.15	+01	$y^3F_2^\circ - g^5D_1$
V	6796.11	2	V	14710.25	+06	$c^3F_3 - v^5F_3^\circ$
V	6793.26	2	V	14716.42	+03	$c^3F_4 - w^5G_4^\circ$
V	6786.88	5	V	14730.25	−05	$y^5D_2^\circ - e^3F_3$
V	6783.71	2	IV*	14737.14	+01	$b^3F_3 - z^5G_3^\circ$
V	*6777.44	1	V	14750.77	−05 +06	$c^3F_2 - x^3P_1^\circ$ $c^3F_3 - v^5P_2^\circ$
L	6752.724	10	V	14804.76	+01	$y^5P_1^\circ - g^5D_1$
I	6750.152	100	III	14810.41	+01	$a^3P_1 - z^3P_1^\circ$
V	6745.11	1	IV*	14821.47	−05	$d^3F_2 - x^1D_2^\circ$
V	6739.54	1	IIIA	14833.72	−03	$a^3F_3 - z^5F_3^\circ$
U	6738.02	4nl	V	14837.07	−08	$y^5P_3^\circ - f^5F_3$
L	6733.164	6	V	14847.77	+02	$y^5P_1^\circ - g^5D_0$
V	6732.06	1	IV*	14850.20	+04	$d^3F_3 - u^3G_3^\circ$
V	6726.668	20n	V	14862.10	−01	$y^5P_2^\circ - e^5P_1$
V	6725.39	2	V	14864.93	−08	$y^5D_4^\circ - e^3F_4$
V	6717.556	3	V	14882.27	−07	$y^5P_2^\circ - e^3D_1$
V	6716.24	3	IV*	14885.18	−02	$d^3F_2 - u^3G_3^\circ$
V	6715.410	5	V	14887.02	−03	$y^3F_3^\circ - g^5D_2$
V	6713.76	3n	V	14890.68	+03	$y^3D_2^\circ - f^5P_2$
V	*6713.14	6d	V	14892.06	+13 −17	$c^3F_3 - x^3P_2^\circ$ $y^5P_2^\circ - g^5D_2$
V	6710.31	2	III?	14898.34	+04	$a^3F_4 - z^5F_5^\circ$
I	6705.117	15n	V	14909.88	+02	$y^5P_2^\circ - e^5P_2$
SS	6704.500	1	V	14911.25	−05	$y^5D_1^\circ - e^3F_2$
L	6703.573	10	IV	14913.31	00	$a^3G_3 - y^3F_3^\circ$
V	6699.14	2	V	14923.18	+04	$d^3F_4 - u^3D_3^\circ$
R	6692.5	(1)		14937.98	−07	$y^5P_3^\circ - f^5F_2$
B	6677.993	600	III	14970.43	00	$a^3G_5 - y^3F_4^\circ$
W	6671.36	(2)		14985.32	+07	$y^5G_5^\circ - h^7D_5$
V	6667.73	(1)		14993.48	−02	$d^3F_3 - u^3D_3^\circ$
SS	6667.455	1−	IV*	14994.09	−08	$a^3H_4 - z^5G_5^\circ$
U	6665.48	(−)		14998.54	−10	$a^3F_3 - z^5F_2^\circ$
B	6663.446	80	III	15003.12	−03	$a^3P_1 - z^3P_0^\circ$
V	6663.26	(1)		15003.53	−05	$y^5P_3^\circ - g^5D_3$
V	6653.88	(1)		15024.68	−07	$y^5D_3^\circ - e^3F_3$
SS	6648.121	1−	IIA	15037.70	−08	$a^5F_1 - z^7F_2^\circ$
V	6646.98	(1)		15040.28	−09 −18	$b^3F_2 - z^3G_3^\circ$ $(z^3G_4^\circ - f^7D_4)$
SS	6639.897	2	V	15056.32	−04	$c^3F_4 - v^5F_4^\circ$
SS	6639.717	4	V	15056.73	−01	$y^5P_2^\circ - g^5D_1$
SS	6634.123	4n	V	15069.41	−07	$y^3D_2^\circ - f^3D_3$
K	6633.764	50	V	15070.24	−04	$y^5P_3^\circ - e^5P_3$
M	6633.44	4n	V	15070.98	−08	$y^3D_1^\circ - f^3D_2$
L	6627.558	5	V	15084.36	−01	$y^3F_4^\circ - g^5D_3$
V	6625.04	1	IIA	15090.09	−03	$a^5F_1 - z^7F_1^\circ$
SS	6613.808	1−	IIIA	15115.72	+05	$a^5F_1 - z^7F_0^\circ$
I	6609.116	30	III	15126.45	00	$b^3F_4 - z^3G_4^\circ$
V	6608.03	2	IV*	15128.93	00	$a^3P_2 - y^5D_3^\circ$
V	6604.67	(1)		15136.63	−09	$y^3D_1^\circ - h^5D_1$
V	6597.607	15n	V	15152.83	−06	$y^3D_2^\circ - g^5F_3$
I	6593.878	60	III	15161.40	−01	$a^3H_5 - z^5G_5^\circ$
B	6592.919	300	III	15163.61	+01	$a^3G_4 - y^3F_3^\circ$
V	6591.32	2	V	15167.29	00	$d^3F_4 - t^3D_3^\circ$
V	6581.22	2	III?	15190.56	−02	$a^3F_4 - z^5F_4^\circ$
I	6575.022	30	IV	15204.88	+01	$b^3F_3 - z^3G_3^\circ$
V	6574.238	3	IIA	15206.70	−01	$a^5F_2 - z^7F_2^\circ$
U	6571.22	(1)		15213.68	−11	$b^1D_2 - v^3G_3^\circ$
I	6569.231	50n	V	15218.29	−04	$y^3D_3^\circ - g^5F_4$
U	6556.79	(1)		15247.16	−03	$y^3D_2^\circ - f^5P_1$
W	6552.77	(2)		15256.51	−03	$a^1F_3 - w^1F_3^\circ$
B	6546.245	200	III	15271.72	00	$a^3G_3 - y^3F_2^\circ$
U	6543.98	(1)		15277.01	−08	$z^5G_4^\circ - f^5D_3$
W	6539.72	(2)		15286.96	−06	$b^3G_3 - x^5D_4^\circ$?
V	6533.97	8n	V	15300.41	−04	$y^5P_3^\circ - e^5P_2$
I	6518.376	20	IV	15337.02	00	$b^3P_2 - y^3D_3^\circ$
V	6509.56	(1)		15357.79	+14	$c^3F_4 - w^3G_5^\circ$
V	6498.950	5	IIA	15382.86	−02	$a^5F_3 - z^7F_3^\circ$
K	6496.456	20n	V	15388.77	+02	$y^3D_2^\circ - f^3D_2$
U	6495.779	3	V	15390.37	−03	$y^3D_1^\circ - g^5F_1$
B	6494.985	1000	II	15392.25	00	$a^3H_6 - z^5G_5^\circ$
I	6481.878	20	III	15423.37	−01	$a^3P_2 - y^5D_2^\circ$
I	6475.632	12	IV	15438.25	00	$b^3F_4 - z^3G_3^\circ$
V	6474.61	(1)		15440.69	−05	$b^3D_1 - v^5D_1^\circ$
I	6469.214	15n	V	15453.57	00 −21	$y^3D_1^\circ - f^3D_1$ $(a^3H_6 - z^5G_6^\circ)$
I	6462.731	30	II	15469.07	+01 −03	$a^3H_4 - z^5G_4^\circ$ $(a^5F_4 - z^7F_5^\circ)$
U	6451.587	(2)		15495.79	−11	$b^1G_4 - y^3G_5^\circ$
V	6450.99	(1)		15497.22	+04	$y^5G_4^\circ - g^5G_4$
U	6438.775	(1)		15526.62	−05	$z^3G_4^\circ - e^3D_3$
V	6436.43	(1)		15532.28	−05	$c^3F_2 - v^3D_1^\circ$
B	6430.851	300	II	15545.75	00	$a^5P_3 - y^5D_4^\circ$
U	6428.793	(1)		15550.73	+02	$z^5G_4^\circ - f^7D_4$
B	6421.355	200	II	15568.74	−01	$a^3P_2 - z^3P_2^\circ$
K	6419.982	30n	V	15572.07	−09	$y^3D_3^\circ - f^3D_3$
I	6411.658	400	IV	15592.29	−01	$z^5P_2^\circ - e^5D_3$
SS	6411.125	1n	V	15593.59	−05	$y^3D_3^\circ - f^5G_4$
I	6408.031	60	V	15601.11	−02	$z^5P_1^\circ - e^5D_2$
W	6406.42	(1)		15605.04	+07	$X_2 - u^3F_3^\circ$?
W	6402.4	(1)		15614.8	0	$y^5G_2^\circ - g^5G_2$
U	6400.318	(50)	IA	15619.91	00	$a^5F_4 - z^7F_4^\circ$
I	6400.010	800	IV	15620.67	00	$z^5P_3^\circ - e^5D_4$
B	6393.605	400	II	15636.31	00	$a^3H_5 - z^5G_4^\circ$
U	6392.547	(1)		15638.90	−01	$a^3P_2 - y^5D_1^\circ$
I	6380.748	3	V	15667.82	−03	$c^3F_2 - w^3F_2^\circ$
V	6364.717	(1)		15707.28	−01	$d^3F_3 - t^3D_2^\circ$
V	6364.384	(1)		15708.11	+02	$y^3D_2^\circ - g^5F_1$
V	6362.889	(2)		15711.80	+03	$c^3F_2 - z^1D_2^\circ$
I	6358.692	3	IA	15722.17	+02	$a^5F_5 - z^7F_6^\circ$

TABLE B—(*Continued*)

Ref	λ I A	Int	T C	Wave Number Observed	o−c	Desig
U	6356.293	(1)		15728.10	+12	$b^3F_2 - y^5P_3^\circ$
I	6355.038	4	III	15731.21	00	$b^3P_1 - y^3D_2^\circ$
I	6344.154	2	III	15758.19	−01	$a^3H_5 - z^3G_5^\circ$
V	6338.896	(1n)		15771.27	+01	$y^3D_2^\circ - f^3D_1$
K	6336.835	12	V	15776.40	−01	$z^5P_1^\circ - e^5D_1$
B	6335.335	10	III	15780.13	00	$a^5P_2 - y^5D_3^\circ$
U	6330.856	(1n)		15791.29	−05	$y^3D_3^\circ - h^5D_2$
I	6322.693	5	III	15811.74	+06	$b^3F_3 - y^3F_4^\circ$
B	6318.022	10	III	15823.37	00	$a^3H_4 - z^5G_3^\circ$
V	6315.814	(2)		15828.90	+01	$c^3F_4 - y^1G_4^\circ$
J	6315.316	(3)		15830.15	−02	$c^3F_3 - w^3F_3^\circ$
V	6311.506	(1)		15839.71	00	$b^3P_2 - y^3D_2^\circ$
U	6310.543	(1)		15842.13	−05	$b^3G_5 - x^5D_4^\circ$
V	6303.46	(1n)		15859.93	00	$z^5G_6^\circ - e^5G_5$
K	6302.507	6	V	15862.32	−02	$z^5P_1^\circ - e^5D_0$
K	6301.515	15	IV	15864.82	−03	$z^5P_2^\circ - e^5D_2$
I	6297.800	5	III	15874.18	00	$a^5P_1 - y^5D_2^\circ$
I	6290.968	3n	V	15891.42	−02	$y^3D_3^\circ - f^3D_2$
I	6280.625	2	IA	15917.59	−01	$a^5F_5 - z^7F_5^\circ$
U	6271.289	(1)		15941.29	00	$z^5F_5^\circ - e^7D_5$
J	6270.238	(2)		15943.96	−01	$b^3P_0 - y^3D_1^\circ$
U	6267.845	(1)		15950.05	−02	$b^1D_2 - z^1F_3^\circ$
B	6265.140	6	III	15956.93	00	$a^5P_3 - y^5D_3^\circ$
I	6256.370	4	III	15979.30	−01	$a^3H_4 - z^3G_4^\circ$
I	6254.262	6	III	15984.69	+01	$a^3P_2 - z^3P_1^\circ$
B	6252.561	20	III	15989.03	−01	$a^3H_6 - z^3G_5^\circ$
K	6246.334	15	V	16004.97	−02	$z^5P_3^\circ - e^5D_3$
V	6245.84	(1)		16006.24	+07	$y^7P_4^\circ - 1_5$
I	6240.656	(2)		16019.53	−02	$a^5P_1 - z^3P_2^\circ$
U	6240.266	(1)		16020.54	+14	$c^3F_3 - w^3F_2^\circ$
Q	6232.735	(−)		16039.89	−04	$z^5F_3^\circ - e^7D_3$
K	6232.661	5	V	16040.08	−04	$z^5P_2^\circ - e^5D_1$
B	6230.728	25	III	16045.06	00	$b^3F_4 - y^3F_4^\circ$
V	6229.234	(1)		16048.91	+01	$b^3P_1 - y^3D_1^\circ$
U	6226.756	(1)		16055.29	−04	$z^3D_3^\circ - e^5F_4$
U	6221.661	(−)		16068.44	+03	$a^5F_5 - z^7F_4^\circ$
U	6221.405	(1)		16069.10	−06	$z^3D_2^\circ - e^5F_3$
U	6220.774	(1)		16070.73	+03	$z^3F_4^\circ - e^5F_4$
I	6219.290	6	III	16074.57	−01	$a^5P_2 - y^5D_2^\circ$
U	6217.283	(1)		16079.76	−05	$X_3 - v^1G_4^\circ$
J	6215.152	(2)		16085.27	−01	$c^3F_2 - v^3G_3^\circ$
I	6213.438	5	III	16089.71	−01	$a^5P_1 - y^5D_1^\circ$
U	6212.045	(1)		16093.32	+16	$z^5G_4^\circ - g^5D_4$
J	6200.323	4	IV	16123.74	−01	$b^3F_2 - y^3F_3^\circ$
U	6199.475	(1)		16125.95	+10	$b^3F_4 - y^5P_3^\circ$
B	6191.562	20	II	16146.56	00	$a^3H_5 - z^3G_4^\circ$
V	6188.037	(2ld)		16155.75	−13	$z^3F_3^\circ - e^3F_4$
U	6180.212	(2)		16176.21	−01	$a^3G_4 - y^3D_3^\circ$
J	6173.343	3	III	16194.21	00	$a^5P_1 - y^5D_0^\circ$
K	6170.492	4n	V	16201.69	+04	$y^3D_2^\circ - e^3P_2$
J	6165.366	(2)		16215.16	+02	$c^3F_3 - v^3G_4^\circ$
U	6163.544	(1)		16219.95	00	$a^5P_2 - z^3P_2^\circ$
U	6159.409	(1n)		16230.84	−10	$y^3F_3^\circ - g^5F_4$
J	6157.734	4	V	16235.26	00	$c^3F_4 - w^3F_4^\circ$
L	6151.624	(2)		16251.38	00	$a^5P_3 - y^5D_2^\circ$
V	6147.85	(−)		16261.36	−04	$c^3F_4 - v^3D_3^\circ$
K	6141.734	4	V	16277.54	00	$z^5P_3^\circ - e^5D_2$
B	6137.696	18	III	16288.26	+01	$b^3F_3 - y^3F_3^\circ$
J	6136.999	(2)		16290.11	00	$a^5P_2 - y^5D_1^\circ$
B	6136.620	20	III	16291.12	00	$a^3H_4 - z^3G_3^\circ$
SS	6130.358	(1)		16307.75	+02	$a^3D_3 - x^5P_3^\circ$
J	*6127.913	(2)		16314.26	−02 +18	$c^3F_3 - y^3H_4^\circ$ $y^5F_2^\circ - e^7P_2$
U	6109.308	(1)		16363.95	−07	$b^3H_4 - z^5H_5^\circ$

Ref	λ I A	Int	T C	Wave Number Observed	o−c	Desig
SS	6107.104	(1)		16369.85	−04	$y^5F_3^\circ - f^5D_2$
K	6103.190	3	V	16380.35	−01	$y^3D_1^\circ - e^3P_1$
K	6102.178	5	V	16383.07	−01	$y^3D_1^\circ - f^3F_2$
SS	*6100.284	(1)		16388.15	+02 −14	$y^5P_3^\circ - h^5D_4$ $y^5P_2^\circ - h^5D_3$
V	6096.689	(1)		16397.82	−06	$z^3F_2^\circ - e^3F_3$
U	6094.419	(1)		16403.92	−15	$y^3F_2^\circ - f^5P_1$
V	6093.66	(1)		16405.97	+02	$y^3F_3^\circ - f^5P_2$
L	6089.566	(1)		16416.99	−10	$a^1F_3 - v^1G_4^\circ$
V	6085.267	(1)		16428.59	−02	$a^3G_3 - y^3D_2^\circ$
U	6082.709	(1)		16435.50	+02	$a^5P_1 - z^3P_1^\circ$
V	6079.02	(1)		16445.48	−05	$y^3F_2^\circ - h^5D_2$
K	6078.496	4n	V	16446.90	−01	$y^3D_2^\circ - f^3F_3$
B	6065.487	15	III	16482.17	+01	$b^3F_2 - y^3F_2^\circ$
V	6062.89	(1)		16489.23	−10	$a^5P_3 - y^5F_4^\circ$
K	6055.987	4	V	16508.02	+01	$y^3D_3^\circ - f^3F_4$
U	6054.100	(2)		16513.17	−06	$z^5G_4^\circ - g^5D_3$
U	6043.738	(1)		16541.48	−19	$b^3D_3 - x^5G_3^\circ$?
SS	6034.057	(2)		16568.02	−04	$z^5G_5^\circ - g^5D_4$
V	6032.67	(1)		16571.83	+07	$y^5F_4^\circ - e^7P_3$
B	6027.057	4	V	16587.26	+02	$c^3F_4 - v^3G_5^\circ$
K	6024.066	15	V	16595.50	−03	$y^3F_4^\circ - f^5G_5$
W	*6021.82	(2n)		16601.69	+06 +03 −07	$y^5F_2^\circ - e^7F_3$ $y^5F_2^\circ - e^5S_2$ $a^5P_2 - y^5F_3^\circ$
K	6020.173	10n	V	16606.23	−02	$y^3F_3^\circ - f^5G_4$
W	6016.66	(2)		16615.93	−20	$a^1D_2 - x^3D_3^\circ$
K	6008.577	9	V	16638.28	−05	$z^3D_3^\circ - e^3F_4$
K	6007.961	(3n)		16639.98	−02	$y^3F_2^\circ - f^5G_3$
V	*6005.53	(1)		16646.72	+06 −04	$b^3F_3 - y^3F_2^\circ$ $y^5F_5^\circ - e^7F_6$
K	6003.033	8	V	16653.64	−06	$z^3F_4^\circ - e^3F_4$
U	5997.805	(1)		16668.16	−03	$y^3F_3^\circ - g^5F_3$
K	5987.057	6	V	16698.08	+03	$y^3D_2^\circ - e^3P_1$
K	5984.805	8	IV	16704.37	+03	$y^3D_3^\circ - e^3P_2$
K	5983.704	6	V	16707.44	−07	$y^3F_4^\circ - g^5F_4$
K	5976.799	5	V	16726.74	−06	$z^3F_3^\circ - e^3F_3$
J	*5975.355	4	V	16730.78	00 +01	$y^3D_1^\circ - e^3P_0$ $c^3F_4 - y^3H_5^\circ$
U	5969.554	(2)		16747.04	+01	$y^5F_2^\circ - e^3D_3$
V	5963.25	(1)		16764.74	−02	$a^5P_1 - y^5F_1^\circ$
U	5959.878	(1)		16774.23	+14	$c^3F_3 - w^3P_2^\circ$
SS	*5958.246	(2)		16778.82	−04 −06	$a^5F_3 - z^7P_2^\circ$ $y^5P_3^\circ - h^5D_3$
J	5956.702	(3)		16783.17	−02	$a^5F_5 - z^7P_4^\circ$
U	5955.682	(1)		16786.05	+05	$z^3P_1^\circ - e^5S_2$
V	5952.749	3	V	16794.32	−09	$z^3F_2^\circ - e^3F_2$
V	*5949.35	(2)		16803.91	−19 −04	$a^5F_4 - z^7P_3^\circ$ $y^3F_3^\circ - h^5D_2$
SS	5947.517	(1)		16809.09	−04	$y^5P_2^\circ - h^5D_2$
V	5940.972	(2)		16827.61	+09	$y^5F_5^\circ - e^5G_6$
K	5934.658	5	V	16845.51	−01	$z^3D_2^\circ - e^3F_3$
K	5930.173	8	V	16858.25	00	$y^3F_2^\circ - e^3G_3$
U	5929.700	(1)		16859.60	−08	$y^3F_4^\circ - h^5D_3$
V	5927.798	(2wd)		16865.01	+04	$y^3F_2^\circ - g^5F_1$
V	5920.520	(2)		16885.74	+03	$b^3H_6 - z^5H_5^\circ$
U	5919.024	(1)		16890.01	−27	$y^5F_3^\circ - e^5G_3$?
V	5916.250	(3)		16897.93	+01	$a^3H_4 - y^3F_4^\circ$
V	*5914.16	8	V	16903.90	+10 −15	$y^3F_3^\circ - e^3G_4$ $y^3F_3^\circ - f^3D_3$
U	5909.986	(3)		16915.84	−01	$z^5D_4^\circ - e^7D_5$
U	5908.252	(2)		16920.80	−15	$z^7D_1^\circ - d^3F_2$
K	5905.673	3n	V	16928.19	+05	$y^3F_2^\circ - f^3D_1$
U	5902.527	(1)		16937.21	−09	$d^3F_4 - t^3G_5^\circ$
U	5898.212	(1)		16949.60	+01	$y^3D_3^\circ - f^3F_3$

TABLE B—(*Continued*)

Ref	λ I A	Int	T C	Wave Number Observed	o−c	Desig
U	5895.007	(1)		16958.82	00	$d^3F_4 - 11_3^\circ$
W	5892.71	(2)		16965.43	−06	$y^5F_3^\circ - e^3D_3$
SS	5891.896	(1)		16967.77	−01	$d^3F_4 - 12_5^\circ$
U	5891.12	(1)		16970.01	−02	$b^3H_5 - z^5H_6^\circ$
K	5883.838	4	V	16991.01	−06	$z^3D_1^\circ - e^3F_2$
V	5880.00	(2wd)		17002.10	+08	$y^5P_3^\circ - f^5G_4$
U	5877.770	(1)		17008.55	+06	$y^5F_5^\circ - e^5G_5$
U	5873.219	(2)		17021.73	00	$y^5F_3^\circ - g^5D_4$
U	5871.289	(1)		17027.33	+04	$y^5D_3^\circ - f^5D_3$
V	5871.04	(1)		17028.05	−04	$z^7D_2^\circ - d^3F_2$
SS	5864.252	(1)		17047.76	−03	$y^5F_1^\circ - e^3D_2$
K	5862.357	8	V	17053.27	−01	$y^3F_4^\circ - e^3G_5$
K	5859.608	5	V	17061.27	−07	$y^3F_4^\circ - f^3D_3$
U	5859.197	(1)		17062.47	+11	$y^5F_1^\circ - f^5F_1$
U	5856.081	(2)		17071.55	+04	$b^1D_2 - y^1D_2^\circ$
U	5855.130	(1)		17074.32	−11	$y^3F_3^\circ - e^5H_4$
U	5853.195	(1)		17079.96	−12	$a^3F_4 - z^5P_3^\circ$
W	5852.19	(2n)		17082.90	+08	$y^3F_4^\circ - f^5G_4$
W	*5848.09	(2n)		17094.87	−05	$z^5D_2^\circ - e^7D_3$
					+12	$y^3F_3^\circ - g^5F_2$
U	5844.879	(1)		17104.26	+10	$y^5D_3^\circ - e^7P_3$
U	5838.418	(1)		17123.19	−14	$z^3F_3^\circ - e^3F_2$
U	5837.703	(1)		17125.29	−01	$b^1D_2 - x^1D_2^\circ$
V	5816.36	(3d)		17188.13	+03	$y^3F_4^\circ - e^5H_5$
V	5815.16	(1)		17191.68	+16	$y^5D_3^\circ - f^5D_2$?
U	5814.816	(1)		17192.69	−03	$y^5F_2^\circ - e^3D_2$
U	5811.936	(1)		17201.21	−03	$c^3F_3 - x^1G_4^\circ$
U	5809.245	(2)		17209.18	−08	$z^3D_3^\circ - e^3F_3$
V	5806.727	(2)		17216.64	−03	$y^3F_3^\circ - e^3G_3$
SS	5805.774	(1)		17219.47	−03	$x^5F_4^\circ - i^5D_3$
U	5804.478	(1)		17223.31	−03	$y^5F_2^\circ - g^5D_3$
U	5804.072	(1)		17224.52	−10	$z^3F_4^\circ - e^3F_3$
V	5798.194	(2)		17241.98	−07	$z^3D_2^\circ - e^3F_2$
V	5793.932	(2)		17254.66	−06	$y^5F_4^\circ - e^3D_3$
V	5791.044	(2)		17263.27	−05	$z^5D_4^\circ - e^7D_4$
V	5784.69	(1)		17282.23	−08	$z^5F_3^\circ - e^5D_4$
V	*5780.83	(1)		17293.77	−05	$z^5D_2^\circ - e^7D_2$
					+08	$z^3G_4^\circ - g^5F_5$
					−05	$b^1G_4 - x^3F_3^\circ$
V	5780.621	(2)		17294.40	−04	$z^5D_3^\circ - e^7D_3$
V	5778.47	(1)		17300.83	−03	$b^3F_3 - y^3D_3^\circ$
J	5775.090	(5)		17310.96	00	$y^5F_4^\circ - g^5D_4$
SS	5769.336	(1)		17328.22	−09	$y^3F_3^\circ - e^5H_3$
K	5762.992	10	V	17347.30	00	$z^3P_2^\circ - e^3D_3$
V	5762.434	(1)		17348.98	−01	$b^3D_3 - u^5D_4^\circ$
U	5761.246	(1)		17352.56	+04	$b^3D_1 - y^3P_0^\circ$
V	5760.351	(1)		17355.25	−03	$b^3D_3 - y^3P_2^\circ$
SS	*5759.550	(2)		17357.67	+02	$y^5F_1^\circ - g^5D_2$
					+05	$y^5P_3^\circ - g^7D_4$
U	5759.270	(1)		17358.51	−02	$y^3F_2^\circ - e^3P_2$
V	5754.41	(1)		17373.17	00	$b^3D_3 - u^5D_3^\circ$
J	5753.136	5	V	17377.02	−05	$z^3P_1^\circ - e^3D_2$
J	5752.043	(2)		17380.32	−05	$y^3F_4^\circ - e^3G_4$
U	5747.959	(1)		17392.67	−01	$y^3F_3^\circ - e^3H_4$
SS	5742.972	(1)		17407.77	−05	$y^5F_5^\circ - f^5F_5$
V	5741.861	(2)		17411.14	−05	$y^5F_2^\circ - e^3D_2$
J	5731.771	(3)		17441.79	−01	$y^5F_3^\circ - g^5D_3$
U	5727.75	(1)		17454.03	−10	$y^5P_2^\circ - g^7D_2$
U	5724.445	(1)		17464.11	+04	$z^3P_0^\circ - e^5P_1$
SS	5723.673	(1)		17466.47	−02	$z^3G_3^\circ - h^5D_3$
W	5720.8	(1n)		17475.2	−1	$y^7P_4^\circ - h^7D_5$
L	5717.845	(3)		17484.27	−03	$z^3P_0^\circ - e^3D_1$
V	*5715.107	(1)		17492.64	−06	$y^5F_2^\circ - e^3D_1$
					−03	$y^5D_2^\circ - e^3D_3$
U	5712.145	(2)		17501.71	−03	$z^5F_2^\circ - e^5D_3$

Ref	λ I A	Int	T C	Wave Number Observed	o−c	Desig
V	5711.867	(2)		17502.57	−01	$y^5F_2^\circ - g^5D_2$
K	5709.378	10	IV	17510.20	+03	$z^5F_4^\circ - e^5D_4$
V	5708.109	(1)		17514.09	−05	$z^3G_4^\circ - f^5G_5$
U	5707.055	(1)		17517.33	−02	$b^3D_3 - x^3F_4^\circ$
U	5705.992	(2)		17520.59	−03	$y^3F_3^\circ - f^3F_4$
U	5705.475	(1)		17522.18	+02	$y^5F_1^\circ - g^5D_1$
U	5702.434	(1)		17531.52	−28	$b^3D_2 - u^5D_2^\circ$?
J	5701.553	7	III?	17534.23	−01	$b^3F_4 - y^3D_3^\circ$
W	5698.37	(2)		17544.02	+02	$b^1D_2 - t^3D_1^\circ$
W	5698.05	(1)		17545.01	−10	$b^3D_2 - y^3P_1^\circ$
V	5691.509	(1)		17565.17	+01	$y^5F_1^\circ - g^5D_0$
V	5686.532	(3)		17580.54	−02	$y^3F_4^\circ - e^3H_5$
W	5680.26	(1)		17599.96	−01	$c^3F_2 - v^3F_3^\circ$
V	5679.023	(2)		17603.79	00	$y^3F_2^\circ - f^3F_3$
SS	5672.273	(1)		17624.74	+03	$d^3F_3 - t^3G_3^\circ$
U	*5666.837	(1)		17641.64	−11	$y^5D_3^\circ - e^5S_2$
					−09	$y^5D_3^\circ - e^7F_3$
U	5662.938	(1)		17653.79	−02	$b^1G_4 - z^1G_4^\circ$
B	5662.525	6	V	17655.08	−01	$y^5F_5^\circ - g^5D_4$
W	5661.36	(1)		17658.71	+02	$z^3P_0^\circ - g^5D_1$
W	5660.79	(1)		17660.49	+03	$b^3D_2 - w^3D_3^\circ$
B	5658.826	10	IV	17666.62	−01	$z^5F_3^\circ - e^5D_3$
U	5658.537	(1)		17667.52	00	$z^5F_1^\circ - e^5D_2$
V	*5655.506	4	V	17676.99	−06	$z^3P_1^\circ - e^3D_1$
					−04	$x^5F_4^\circ - g^5G_5$
V	5655.179	(2)		17678.01	−04	$x^5F_3^\circ - g^5G_4$
U	5653.889	(1w)		17682.05	+01	$z^3G_5^\circ - g^5F_5$
U	5652.317	(1)		17686.96	+03	$z^3P_1^\circ - g^5D_2$
U	5650.721	(1)		17691.96	−07	$x^5F_2^\circ - g^5G_3$
V	5650.01	(1)		17694.18	−01	$x^5F_1^\circ - g^5G_2$
V	5649.66	(1)		17695.28	−11	$a^1I_6 - z^3H_5^\circ$
U	5641.453	(2)		17721.02	−02	$y^5F_3^\circ - g^5D_2$
W	5640.46	(1n)		17724.14	+06	$y^5P_3^\circ - e^5H_3$
I	5638.266	3	V	17731.04	00	$y^5F_4^\circ - g^5D_3$
U	5636.693	(1)		17735.99	−01	$b^3D_2 - x^3F_3^\circ$
U	5635.845	(1)		17738.66	−01	$y^5F_3^\circ - e^5P_2$
V	5633.970	(2)		17744.56	+05	$x^5F_5^\circ - g^5G_6$
U	5631.72	(2)		17751.65	+09	$z^3G_3^\circ - g^5F_3$
B	5624.549	10	IV	17774.28	00	$z^5F_2^\circ - e^5D_2$
V	5624.056	(1)		17775.84	−05	$z^3G_5^\circ - h^5D_4$
V	5620.527	(1)		17787.00	−12	$y^5D_3^\circ - e^3D_3$
W	*5620.04	(1)		17788.54	+09	$y^5P_3^\circ - e^3H_4$
					−01	$c^3F_3 - v^3F_2^\circ$
V	5619.60	(1)		17789.94	−03	$z^3G_5^\circ - f^5G_6$
U	5618.633	(1)		17793.00	00	$z^3P_2^\circ - e^3D_2$
W	5617.22	(1)		17797.47	−05	$a^3D_3 - w^5F_4^\circ$
B	5615.652	50	IV	17802.44	−01	$z^5F_5^\circ - e^5D_4$
U	5615.301	(2)		17803.55	00	$b^3F_3 - y^3D_2^\circ$
J	5602.955	10	IV	17842.78	−01	$z^5F_1^\circ - e^5D_1$
U	5602.770	(2)		17843.37	+01	$y^5D_3^\circ - g^5D_4$
V	*5600.242	(1)		17851.43	−01	$z^3P_1^\circ - g^5D_1$
					−04	$b^3D_1 - u^5D_0^\circ$
J	5598.303	4	IV?	17857.61	−05	$y^3F_2^\circ - f^3F_2$
U	5594.670	(2)		17869.21	−04	$y^3F_4^\circ - e^3H_4$
U	5587.576	(1)		17891.89	+02	$c^3F_3 - v^3F_4^\circ$
B	5586.763	40	IV	17894.50	00	$z^5F_4^\circ - e^5D_3$
U	5584.766	(1)		17900.90	+03	$a^1H_5 - u^5D_4^\circ$
J	5576.097	10	IV	17928.73	+01	$z^5F_1^\circ - e^5D_0$
U	5573.105	(1)		17938.35	−02	$y^5D_2^\circ - e^3D_2$
B	5572.849	30	IV	17939.17	−01	$z^5F_3^\circ - e^5D_2$
B	5569.625	20	IV	17949.56	00	$z^5F_2^\circ - e^5D_1$
U	5568.81	(1)		17952.18	+13	$b^3D_1 - w^3D_1^\circ$?
U	5567.403	(2)		17956.72	−02	$b^3F_2 - y^3D_1^\circ$
I	5565.708	4	V	17962.19	−01	$y^3F_3^\circ - f^3F_3$
I	5563.604	3	V	17968.98	00	$y^5D_2^\circ - g^5D_3$

TABLE B—(*Continued*)

Ref	λ I A	Int	T C	Wave Number Observed	o−c	Desig
V	*5562.712	(2)		17971.86	−04 −07	$z^3G_4^\circ - e^3G_5$ $a^3D_1 - w^5F_1^\circ$
V	5560.230	(1)		17979.89	−07	$z^3G_4^\circ - f^3D_3$
U	5557.962	(1)		17987.22	+05	$z^3G_3^\circ - e^3G_4$
I	5554.895	4	V	17997.16	−03	$y^3F_4^\circ - f^3F_4$
V	5553.586	(1)		18001.40	−03	$z^3G_4^\circ - f^5G_4$
U	5549.94	(2)		18013.22	00	$b^1G_4 - x^3G_4^\circ$
W	5547.00	(2)		18022.77	−04	$y^5D_1^\circ - e^3D_1$
U	5546.486	(1)		18024.44	+05	$z^5G_4^\circ - f^5G_5$
J	5543.930	(2)		18032.75	+06	$y^5D_1^\circ - g^5D_2$
V	5543.184	(2)		18035.18	−12	$b^1G_4 - x^3G_3^\circ$
U	5539.831	(1)		18046.09	−02	$b^1D_2 - t^3D_2^\circ$
U	5539.27	(1)		18047.92	+06	$b^3D_3 - 1_2^\circ$
V	*5538.54	(1)		18050.30	−02 −05	$y^5D_1^\circ - e^5P_2$ $a^1I_6 - w^5G_6^\circ$
W	5737.71?	(1)		18053.00	+09	$y^5D_4^\circ - e^7F_3$
J	*5535.419	(2)		18060.48	+03 +01	$a^3D_3 - w^5F_2^\circ$ $c^3F_2 - u^3G_3^\circ$
U	*5534.64	(1)		18063.02	00 +08	$y^5D_3^\circ - e^7S_3$ $b^3D_2 - 1_2^\circ$
U	5532.742	(1)		18069.21	−02	$a^1H_5 - x^3F_4^\circ$
U	5531.949	(1)		18071.80	+04	$x^5D_4^\circ - i^5D_4$
U	5529.13	(2)		18081.02	+11	$b^3D_3 - z^1G_4^\circ$
V	5525.552	(3)		18092.73	+02 −24	$y^5D_0^\circ - g^5D_1$ $(z^3P_2^\circ - e^3D_1)$
SS	5524.273	(1)		18096.92	−08	$y^5D_3^\circ - f^5F_3$
V	5522.46	(2)		18102.86	00	$z^3P_2^\circ - g^5D_2$
SS	5521.141	(1)		18107.18	−01	$a^1I_6 - w^5G_5^\circ$
W	5517.08	(1n)		18120.51	+02	$z^3P_2^\circ - e^5P_2$
V	5512.277	(1)		18136.30	−07	$z^5G_4^\circ - g^5F_4$
B	5506.782	18	IB	18154.40	00	$a^5F_2 - z^5D_3^\circ$
T	5505.893	(−)		18157.33	−04	$z^5G_3^\circ - f^5G_4$
B	5501.469	12	IB	18171.93	00	$a^5F_3 - z^5D_4^\circ$
B	5497.519	15	IB	18184.98	00	$a^5F_1 - z^5D_2^\circ$
U	5494.462	(1)		18195.10	+01	$c^3F_4 - x^3H_5^\circ$
T	*5493.850	(0)		18197.13	−07 +17	$y^5D_1^\circ - g^5D_1$ $c^3P_2 - x^5P_3^\circ$
U	5493.511	(1)		18198.25	−06	$y^5D_4^\circ - e^3D_3$
W	5491.84	(2)		18203.79	−02	$c^3F_2 - u^3D_3^\circ$
K	5487.747	(8)		18217.37	+08	$c^3F_3 - t^5D_2^\circ$
U	5487.144	(1)		18219.37	+06	$z^5G_3^\circ - g^5F_3$
U	5483.116	(1)		18232.75	−07	$y^5D_3^\circ - e^3D_2$
T	5481.451	(3)		18238.29	−05	$y^5D_2^\circ - e^3D_1$
U	5481.256	(2)		18238.94	−04	$y^5D_4^\circ - e^7G_4$
U	5480.873	(2)		18240.21	+01	$y^5D_1^\circ - g^5D_0$
U	5478.463	(1)		18248.24	+02	$y^5D_2^\circ - g^5D_2$
J	5476.571	10	IV	18254.54	00	$y^5D_4^\circ - g^5D_4$
J	5476.298	(2)		18255.45	+01	$c^3F_3 - u^3G_4^\circ$
J	5473.908	(3)		18263.42	−01	$y^5D_3^\circ - g^5D_3$
U	5472.729	(1)		18267.36	−01	$z^3P_2^\circ - g^5D_1$
W	5470.17	(1)		18275.90	−16	$z^5G_2^\circ - h^5D_1$
V	*5466.993	(1)		18286.52	+02 −04	$z^5P_2^\circ - e^5F_3$ $a^1H_5 - z^3H_4^\circ$
J	5466.404	(3)		18288.49	−05	$z^5G_4^\circ - h^5D_3$
W	5465.1	(1)		18292.9	−2	$a^1I_6 - v^5F_5^\circ$
V	5464.286	(1)		18295.58	+03	$c^3F_3 - y^1D_2^\circ$
J	5463.282	10n	V	18298.94	−04	$z^3G_4^\circ - e^3G_4$
J	5462.970	(2)		18299.99	−05	$z^3G_3^\circ - e^3G_3$
U	5461.553	(1n)		18304.73	−03	$z^5G_2^\circ - f^5G_3$
U	5460.909	(1)		18306.90	−09	$c^3P_1 - x^5P_1^\circ$
U	5456.468	(1)		18321.80	+17	$z^5P_3^\circ - e^5F_4$
B	5455.613	40	IB	18324.67	00	$a^5F_1 - z^5D_1^\circ$
K	5455.433	(5)		18325.27	+04	$z^5G_6^\circ - f^5G_6$
U	5452.119	(1)		18336.41	−10	$b^3D_2 - w^5G_3^\circ$
B	5446.920	40	IB	18353.91	00 −16	$a^5F_2 - z^5D_2^\circ$ $(a^3F_2 - z^3D_3^\circ)$
J	5445.045	15n	V	18360.23	−02	$z^3G_5^\circ - e^3G_5$
U	5441.321	(1)		18372.80	+12	$z^5G_5^\circ - h^5D_4$
V	5436.594	(2)		18388.77	−01	$a^3P_2 - y^5P_3^\circ$
U	5436.299	(1)		18389.77	−01	$z^3G_5^\circ - f^5G_4$
B	5434.527	30	IB	18395.77	00	$a^5F_1 - z^5D_0^\circ$
U	5432.950	(2n)		18401.11	+01	$z^5G_2^\circ - g^5F_2$
B	5429.699	40	IB	18412.12	00	$a^5F_3 - z^5D_3^\circ$
I	5424.072	45n	V	18431.22	00	$z^5G_6^\circ - e^5H_7$
U	5417.045	(1)		18455.13	−04	$z^5G_3^\circ - f^3D_2$
I	5415.201	35n	V	18461.42	−02	$z^3G_5^\circ - e^3H_6$
I	5410.913	15n	V	18476.05	00	$z^3G_3^\circ - e^3H_4$
V	5409.125	(1)		18482.15	+01	$z^5G_4^\circ - e^3G_5$
B	5405.778	40	IB	18493.60	00	$a^5F_2 - z^5D_1^\circ$
I	5404.144	30n	V	18499.19	+02 −10	$z^3G_4^\circ - e^3H_5$ $(z^5G_5^\circ - f^5G_5)$
U	5403.819	(1)		18500.30	+02	$c^3F_4 - u^3G_5^\circ$
J	5400.509	(5)		18511.64	−04	$z^5G_4^\circ - f^5G_4$
U	5398.280	(1)		18519.28	−02	$z^5G_2^\circ - f^5G_2$
U	5397.616	(1)		18521.56	−02	$a^1I_6 - x^3G_5^\circ$
B	5397.131	40	IB	18523.23	00	$a^5F_4 - z^5D_4^\circ$
W	5395.25	(1n)		18529.69	−05	$z^5G_2^\circ - g^5F_1$
T	5394.682	(−)		18531.63	00	$c^3F_2 - u^3D_2^\circ$
I	5393.174	10	IV	18536.82	00	$z^5D_3^\circ - e^5D_4$
U	5391.470	(1)		18542.67	−01	$y^5D_3^\circ - g^5D_2$
K	5389.461	(5)		18549.59	+06	$z^5G_3^\circ - f^5G_3$
R	5387.51	3		18556.30	−06	$c^3F_3 - u^3D_3^\circ$
T	5386.958	(1)		18558.20	+02	$b^3D_3 - v^5F_4^\circ$
V	5386.341	(1)		18560.33	+02	$y^5D_3^\circ - e^5P_2$
I	5383.374	35n	V	18570.56	−01	$z^5G_5^\circ - e^5H_6$
T	5382.750	(−)		18572.72	+08	$a^1D_2 - u^5D_1^\circ$
J	5379.580	(2)		18583.66	−01	$b^1G_4 - z^1H_5^\circ$
U	5376.849	(2)		18593.10	−02	$b^1D_2 - z^1P_1^\circ$
V	5373.704	(1)		18603.98	−01	$z^3G_3^\circ - f^3F_4$
B	5371.493	50	IB	18611.64	00 −21	$a^5F_3 - z^5D_2^\circ$ $(z^3G_4^\circ - e^3G_3)$
I	5369.965	25n	V	18616.93	−03	$z^5G_4^\circ - e^5H_5$
I	5367.470	20n	V	18625.59	+04	$z^5G_3^\circ - e^5H_4$
J	5365.403	3	V	18632.76	−03	$a^1H_5 - z^1G_4^\circ$
I	5364.874	15n	V	18634.60	−06	$z^5G_2^\circ - e^5H_3$
U	5361.637	(1)		18645.85	−02	$z^5G_3^\circ - g^5F_2$
U	5353.389	(2)		18674.58	−04	$y^5D_4^\circ - g^5D_3$
T	5349.742	(3)		18687.31	−02	$z^3G_5^\circ - e^3G_4$
B	5341.026	20	II	18717.80	00	$a^3F_2 - z^3D_2^\circ$
I	5339.935	12	V	18721.63	00	$z^5D_2^\circ - e^5D_3$
J	5332.903	4	IB?	18746.31	−01	$a^3F_3 - z^3F_4^\circ$
U	5332.681	(1)		18747.09	−04	$c^3F_2 - u^3D_1^\circ$
J	5329.994	(2)		18756.54	−01	$c^3F_4 - 6_5^\circ$
B	5328.534	15	II	18761.68	−01	$a^3F_3 - z^3D_3^\circ$
I	5328.042	50	IB	18763.42	00	$a^5F_4 - z^5D_3^\circ$
T	5326.793	(−)		18767.82	+03	$z^5G_3^\circ - e^3G_3$
V	*5326.154	(1)		18770.07	−02 −07	$a^1H_5 - w^5G_4^\circ$ $b^3G_3 - z^5H_4^\circ$?
I	5324.185	30	IV	18777.01	00	$z^5D_4^\circ - e^5D_4$
SS	5323.510	(1)		18779.39	+01	$a^3P_2 - y^5P_2^\circ$
U	5322.054	(2)		18784.53	−03	$a^3P_2 - y^3F_3^\circ$
V	5321.106	(1)		18787.87	+01	$z^3G_4^\circ - e^3H_4$
U	5320.046	(1)		18791.62	−01	$b^3D_3 - v^5P_2^\circ$
V	5317.394	(1)		18800.99	+17	$b^3H_4 - y^3G_4^\circ$
U	5315.080	(1)		18809.17	−06	$z^5G_4^\circ - e^3G_4$
T	5313.839	(−)		18813.57	−12	$d^3F_2 - w^1D_2^\circ$
B	5307.365	2	III?	18836.52	00	$a^3F_2 - z^3F_3^\circ$
W	5304.1	(1)		18848.1	0	$z^3D_2^\circ - f^5D_3$
I	5302.307	10	V	18854.48	−01	$z^5D_1^\circ - e^5D_2$

TABLE B—(*Continued*)

Ref	λ I A	Int	T C	Wave Number		Desig
				Observed	o−c	
U	5298.779	(1)		18867.04	00	$b^3D_3 - v^5F_2^\circ$
U	5295.316	(1)		18879.38	+05	$z^5G_3^\circ - e^5H_3$
T	5294.555	(–)		18882.09	−03	$b^3D_2 - v^5F_2^\circ$
U	5293.965	(1)		18884.19	+01	$c^3F_3 - u^3D_2^\circ$
U	5288.537	(2)		18903.58	−03	$b^1G_4 - y^1G_4^\circ$
W	5285.6	(1)		18914.1	0	$z^3F_2^\circ - f^7D_1$
T	5284.416	(–)		18918.32	+03	$a^1I_6 - w^3G_5^\circ$
I	5283.628	18	IV	18921.14	00	$z^5D_3^\circ - e^5D_3$
I	5281.796	10	IV	18927.70	00	$z^7P_2^\circ - e^7D_3$
V	5280.364	(1)		18932.83	−02	$b^3D_3 - x^3P_2^\circ$
W	5277.6	(1)		18942.7	−1	$z^3D_1^\circ - f^5D_1$
U	5275.021	(1n)		18952.01	−11	$c^3F_4 - u^3G_3^\circ$
J	5273.379	4	IV	18957.91	+01	$a^3P_0 - y^3D_1^\circ$
K	5273.176	(5)		18958.64	−02	$z^5D_0^\circ - e^5D_1$
B	5270.360	30	II	18968.77	−01	$a^3F_2 - z^3D_1^\circ$
I	5269.541	60	IB	18971.72	00	$a^5F_5 - z^5D_4^\circ$
I	5266.562	30	IV	18982.45	00	$z^7P_3^\circ - e^7D_4$
U	5263.870	(1)		18992.16	−04	$a^1H_5 - x^3G_4^\circ$
J	5263.314	8	V	18994.16	−01	$z^5D_2^\circ - e^5D_2$
V	5254.956	1	IA	19024.37	−01 +04	$a^5D_1 - z^7D_2^\circ$ $(b^1D_2 - y^1F_3^\circ)$
V	5253.479	(2)		19029.72	−04	$z^5D_1^\circ - e^5D_1$
B	5250.650	6	IV	19039.98	00	$a^5P_2 - y^5P_3^\circ$
U	5250.211	1	IA	19041.57	−01	$a^5D_0 - z^7D_1^\circ$
U	5249.099	(1n)		19045.60	+02	$z^3G_3^\circ - f^3F_3$
U	5247.065	1	IA	19052.98	−06	$a^5D_2 - z^7D_3^\circ$
U	5243.789	(1)		19064.89	−05	$y^5F_3^\circ - g^5F_4$
B	5242.495	4	IV	19069.59	00	$a^1I_6 - z^1H_5^\circ$
SS	5241.931	(1)		19071.65	−09	$z^5G_3^\circ - f^3F_4$
U	5236.204	(1)		19092.50	−01	$c^3F_2 - 8_1^\circ$
V	*5235.392	(2)		19095.47	+11 +01	$b^3F_3 - x^5D_3^\circ$ $c^3F_4 - u^3D_3^\circ$
I	5232.946	40	III	19104.39	00	$z^7P_4^\circ - e^7D_5$
U	5231.41	(1)		19110.00	−06	$a^1H_5 - v^5F_4^\circ$
J	*5229.857	5n	V	19115.67	−02 +08	$z^5D_1^\circ - e^5D_0$ $y^5F_4^\circ - h^5D_4$
U	5228.391	(1n)		19121.03	+04	$y^5F_4^\circ - f^5P_3$
B	5227.192	40	II	19125.42	00 −15	$a^3F_3 - z^3D_2^\circ$ $(a^3P_1 - y^3D_2^\circ)$
J	5226.868	15	IV	19126.61	00	$z^7P_2^\circ - e^7D_2$
SS	5226.063	(1)		19129.55	−02	$a^1P_1 - y^3P_0^\circ$
U	5225.531	1	IA	19131.50	−02	$a^5D_1 - z^7D_1^\circ$
U	5223.193	(1)		19140.06	−01	$b^3D_1 - x^3P_0^\circ$
W	5221.8	(1)		19145.2	−1	$a^3D_1 - x^3D_1^\circ$?
T	5217.927	(2)		19159.38	+01	$b^3D_2 - x^3P_1^\circ$
J	5217.395	5	V	19161.33	00	$z^5D_4^\circ - e^5D_3$
B	5216.278	10	II	19165.44	00	$a^3F_2 - z^3F_2^\circ$
J	5215.185	6	IV	19169.45	00	$z^5D_2^\circ - e^5D_1$
J	5208.601	7	IV	19193.68	−01	$z^5D_3^\circ - e^5D_2$
SS	5207.937	(1)		19196.13	+03	$b^3D_1 - x^3P_1^\circ$
J	**5204.582	2	IA	19208.51	00	$a^5D_2 - z^7D_2^\circ$
B	5202.339	8	IV	19216.79	00	$a^5P_3 - y^5P_3^\circ$
W	5202.27?	(1)		19217.04	−07	$y^5F_3^\circ - h^5D_3$
V	5198.843	(1)		19229.71	+06	$a^1D_2 - x^3G_3^\circ$
B	5198.714	4	IV	19230.19	+01	$a^5P_1 - y^5P_2^\circ$
V	5196.100	(2w)		19239.86	−08	$y^5F_3^\circ - f^5P_2$
K	5195.471	(8)		19242.19	−01	$y^5F_4^\circ - f^5G_5$
I	5194.943	10	IB	19244.15	+01	$a^3F_3 - z^3F_3^\circ$
I	5192.350	30	IV	19253.76	00	$z^7P_3^\circ - e^7D_3$
J	5191.460	20	IV	19257.06	+01	$z^7P_2^\circ - e^7D_1$
U	5187.924	(2)		19270.18	+03	$c^3F_3 - t^3D_2^\circ$
U	5184.292	(3n)		19283.68	−04	$y^5F_2^\circ - g^5F_3$
T	5180.065	(–)		19299.42	−03	$z^3G_3^\circ - f^3F_2$
U	5178.798	(1n)		19304.14	−01	$z^3G_5^\circ - f^3F_4$
T	5177.230	(–)		19309.99	+01	$b^1G_4 - w^3F_4^\circ$

Ref	λ I A	Int	T C	Wave Number		Desig
				Observed	o−c	
B	5171.599	20	II	19331.01	00	$a^3F_4 - z^3F_4^\circ$
B	5168.901	4	IA	19341.10	−01	$a^5D_3 - z^7D_3^\circ$
B	5167.491	40	II	19346.38	00	$a^3F_4 - z^3D_3^\circ$
J	5166.286	4	IA	19350.89	00	$a^5D_4 - z^7D_5^\circ$
J	5165.422	(4)		19354.13	−05	$y^5F_4^\circ - g^5F_4$
S	5164.922	(–)		19356.00	00	$c^3F_3 - w^3H_4^\circ$
W	5164.56	(1)		19357.36	00	$z^3G_4^\circ - f^3F_3$
J	5162.288	10n	IV?	19365.88	+02 +36	$y^5F_5^\circ - g^5F_5$ $(b^3F_2 - x^5D_1^\circ)$
V	5159.066	(2w)		19377.97	−05	$y^5F_2^\circ - f^5P_1$
J	5151.915	4	IB	19404.87	00	$a^5F_1 - z^5F_2^\circ$
B	5150.843	6	IB	19408.91	00	$a^5F_2 - z^5F_3^\circ$
U	5148.234	(3)		19418.74	−03	$y^5F_3^\circ - f^3D_3$
V	5148.061	(3)		19419.40	−08	$y^5F_2^\circ - h^5D_2$
T	5145.105	(–)		19430.55	−03	$a^5P_2 - y^5P_2^\circ$
J	5142.932	6	IB	19438.76	−01	$a^5F_3 - z^5F_4^\circ$
J	*5142.541	(3w)		19440.24	00 −07	$y^5F_3^\circ - f^5G_4$ $y^5F_1^\circ - h^5D_1$
U	5141.747	(2)		19443.24	−03	$a^3P_1 - y^3D_1^\circ$
J	5139.468	20	IV	19451.86	00	$z^7P_4^\circ - e^7D_4$
J	5139.260	10	IV	19452.65	−01	$z^7P_3^\circ - e^7D_2$
J	5137.388	6n	V	19459.74	+03	$y^5F_5^\circ - h^5D_4$
W	5136.09	(1)		19464.66	+05	$c^3F_2 - z^1P_1^\circ$
J	5133.692	20n	V	19473.75	−04	$y^5F_5^\circ - f^5G_6$
J	5131.475	(2)		19482.16	00	$a^5P_1 - y^5P_1^\circ$
T	5129.658	(1)		19489.06	−10	$z^3F_3^\circ - e^3D_3$
B	5127.363	5	IB	19497.79	00	$a^5F_4 - z^5F_5^\circ$
U	5126.598	(1)		19500.70	−08	$z^3F_4^\circ - f^7D_4$
T	5126.218	(1)		19502.14	−04	$y^5F_3^\circ - g^5F_3$
J	5125.130	6n	V	19506.28	−06	$y^5F_4^\circ - h^5D_3$
W	5124.1	(1)		19510.2	+3	$c^3F_2 - s^3D_2^\circ$
B	5123.723	6	IB	19511.64	00	$a^5F_1 - z^5F_1^\circ$
U	5121.636	(2n)		19519.59	00	$y^5F_2^\circ - f^3D_2$
T	5115.788	(1)		19541.90	−04	$a^1H_5 - w^3G_4^\circ$
B	5110.414	10	IB	19562.45	−01 −20	$a^5D_4 - z^7D_4^\circ$ $(a^1H_5 - z^1H_5^\circ)$
U	5109.646	(2)		19565.39	+04	$y^5F_1^\circ - g^5F_2$
J	5107.645	8	II	19573.06	00	$a^3F_3 - z^3F_2^\circ$
J	5107.452	6	IB	19573.79	−01	$a^5F_2 - z^5F_2^\circ$
SS	5104.441	(1)		19585.34	+10	$y^5F_2^\circ - h^5D_1$
U	5104.21	(1)		19586.23	−09	$y^5F_5^\circ - f^5G_5$
T	5104.038	(–)		19586.89	−01	$c^3P_2 - w^5D_3^\circ$
T	5099.091	(1)		19605.89	−06	$z^3F_2^\circ - e^3D_2$
J	5098.703	8	IV	19607.38	00	$a^5P_3 - y^5P_2^\circ$
K	5098.594	(3)		19607.80	−08	$z^3D_2^\circ - e^3D_3$
J	5096.998	(6)		19613.94	−01	$y^5F_2^\circ - f^5G_3$
K	5090.787	(6n)		19637.87	−07	$y^5F_3^\circ - h^5D_2$
SS	5088.159	(1)		19648.02	+03	$y^5D_3^\circ - h^5D_4$
B	5083.342	7	IB	19666.63	00	$a^5F_3 - z^5F_3^\circ$
J	5079.742	4	IB	19680.57	00	$a^5F_2 - z^5F_1^\circ$
J	5079.226	6	IV	19682.57	+01	$a^5P_2 - y^5P_1^\circ$
U	5078.983	(1n)		19683.51	−04	$y^5F_1^\circ - f^5G_2$
T	5076.288	(2)		19693.96	−03	$y^5F_1^\circ - g^5F_1$
J	5074.757	10n	V	19699.90	−05	$y^5F_4^\circ - e^3G_5$
T	5072.690	(1)		19707.93	−08	$y^5F_4^\circ - f^3D_3$
K	5072.077	(1)		19710.31	+03	$y^5F_2^\circ - g^5F_2$
J	5068.774	10	V	19723.15	−02	$z^7P_4^\circ - e^7D_3$
V	5067.162	(1)		19729.43	−05	$y^5F_4^\circ - f^5G_4$
U	5065.213	(2)		19737.02	−06	$b^3D_3 - w^3F_4^\circ$
J	5065.020	6n	V	19737.77	−02	$y^5F_3^\circ - e^3G_4$
T	5063.296	(–)		19744.49	+20	$y^5D_2^\circ - h^5D_3$
T	*5060.079	(1)		19757.05	+01 −11	$a^5D_4 - z^7D_3^\circ$ $y^5F_1^\circ - f^3D_1$
U	5058.507	(1)		19763.18	−04	$b^3D_3 - v^3D_3^\circ$
W	5058.00	(1)		19765.16	+10	$z^3F_3^\circ - e^7S_3$

TABLE B—(*Continued*)

Ref	λ I A	Int	T C	Wave Number Observed	o−c	Desig
W	*5057.49	(1)		19767.16	+04 −03	$y^5D_2^\circ - f^5P_2$ $z^5G_3^\circ - f^3F_2$
U	5056.856	(1)		19769.64	+05	$z^3P_1^\circ - h^5D_1$
U	5056.023	(1)		19772.89	−11	$z^5G_5^\circ - e^3H_4$
T	5054.647	1		19778.28	−02	$b^3D_2 - v^3D_3^\circ$
B	5051.636	10	IB	19790.07	+01	$a^5F_4 - z^5F_4^\circ$
B	5049.825	15	III	19797.16	−01	$a^3P_2 - y^3D_3^\circ$
U	5048.457	(2)		19802.53	−08	$z^3D_1^\circ - e^3D_2$
T	5044.221	(2)		19819.16	−01	$z^7F_4^\circ - e^7D_5$
B	5041.759	10	III	19828.83	−01	$a^3F_4 - z^3F_3^\circ$
J	5041.074	7	IB	19831.53	00	$a^5F_3 - z^5F_2^\circ$
V	*5040.902	(2)		19832.20	00 −21	$y^5F_2^\circ - e^3G_3$ $y^5F_3^\circ - f^5G_3$
U	5039.261	(2)		19838.66	−01	$z^5F_4^\circ - e^5F_5$
R	§5036.931	2		19847.84	00	$c^3P_2 - w^5D_2^\circ$
R	5035.025	3		19855.35	+05	$b^3D_3 - 3_3^\circ$
R	5031.901	8		19867.68	+05	$z^5G_4^\circ - f^3F_3$
R	*5031.030	2		19871.12	−01 +03	$a^1D_2 - w^3G_3^\circ$ $b^3D_3 - w^3F_3^\circ$
R	§5030.784	5		19872.09	00	$b^3H_6 - z^3I_7^\circ$
V	5029.623	(1)		19876.68	−04	$a^1P_1 - 1_2^\circ$
J	5028.129	4	V	19882.58	−01	$a^1H_5 - y^1G_4^\circ$
T	5027.785	(−)		19883.95	−05	$z^3P_2^\circ - g^5F_3$
V	5027.212	(1)		19886.21	+04	$b^3D_2 - w^3F_3^\circ$
J	5027.136	5n	V	19886.51	−06	$y^5D_3^\circ - g^5F_4$
T	5023.476	(−)		19901.00	+06	$z^5G_5^\circ - f^3F_4$
T	5023.226	(−)		19901.99	−10	$y^5F_2^\circ - f^3D_1$
J	5022.244	6	V	19905.88	−04	$z^3F_2^\circ - e^3D_1$
V	5021.894	(1)		19907.27	+15	$a^3D_1 - w^5P_2^\circ$
U	5020.819	(1)		19911.53	−01	$a^1D_2 - x^3P_1^\circ$
J	5014.950	10	V	19934.83	−04	$z^3F_3^\circ - e^3D_2$
B	5012.071	12	IB	19946.28	00	$a^5F_5 - z^5F_5^\circ$
J	*5007.289	(3n)		19965.33	−15 +01	$z^3F_3^\circ - g^5D_3$ $y^5D_4^\circ - g^5F_5$
I	5006.126	20	III	19969.97	−01	$z^7F_5^\circ - e^7D_5$
J	5005.720	10	V	19971.59	−03	$z^3D_3^\circ - e^3D_3$
T	5004.034	(1)		19978.32	+02	$z^3P_2^\circ - f^5P_1$
J	5002.800	(6)		19983.25	−01	$z^5F_3^\circ - e^5F_4$
B	5001.871	12	V	19986.96	−03	$z^3F_4^\circ - e^3D_3$
T	4999.114	(1)		19997.98	+03	$c^3F_2 - x^1F_3^\circ$
B	4994.133	8	IB	20017.93	00	$a^5F_4 - z^5F_3^\circ$
U	4993.687	(1)		20019.72	−04	$z^3P_2^\circ - h^5D_2$
J	4991.277	(3)		20029.38	+02	$y^5D_2^\circ - g^5F_3$
J	4988.963	(6)		20038.67	−07	$y^5D_3^\circ - h^5D_3$
U	4986.223	(1)		20049.68	−01	$y^5D_1^\circ - f^3D_2$
J	4985.553	7	V	20052.38	00	$z^7F_3^\circ - e^7D_4$
J	4985.261	7	V	20053.55	−04	$z^3D_2^\circ - e^3D_2$
J	4983.855	6n	V	20059.21	+04	$y^5D_4^\circ - h^5D_4$
J	4983.258	5n	V	20061.61	+03	$y^5D_3^\circ - f^5P_2$
J	4982.507	8n	V	20064.64	+07	$y^5D_4^\circ - f^5P_3$
U	4979.586	(1)		20076.41	+01	$b^3D_2 - w^3F_2^\circ$
J	4978.606	2	V	20080.36	+04	$z^3F_2^\circ - g^5D_1$
U	4977.653	(1)		20084.20	00	$z^3D_2^\circ - g^5D_3$
U	4975.415	(1)		20093.24	+03	$b^3H_4 - u^5D_4^\circ$
J	4973.108	3	V	20102.56	−03	$z^3D_1^\circ - e^3D_1$
SS	4972.398	(1)		20105.43	−05	$y^5F_5^\circ - g^7D_5$
U	4970.493	(2)		20113.13	00	$b^3D_1 - w^3F_2^\circ$
J	4969.927	(3)		20115.42	+07	$y^5D_1^\circ - h^5D_1$
U	4968.702	(1)		20120.38	+06	$b^3D_2 - z^1D_2^\circ$
J	4967.899	(3)		20123.64	−02	$y^5D_2^\circ - f^5P_1$
B	4966.096	8	V	20130.94	−01	$z^5F_5^\circ - e^5F_5$
U	4962.564	(1)		20145.27	+01	$y^5F_5^\circ - e^3H_6$
U	4961.908	(1)		20147.93	+05	$a^1I_6 - v^3G_5^\circ$
I	4957.603	60	III	20165.43	00	$z^7F_6^\circ - e^7D_5$
J	4957.302	20	III	20166.65	+01	$z^7F_4^\circ - e^7D_4$

Ref	λ I A	Int	T C	Wave Number Observed	o−c	Desig
V	*4952.646	(1n)		20185.61	−16 +09	$y^5D_4^\circ - f^5G_5$ $z^3P_2^\circ - h^5D_1$
J	4950.112	(2)		20195.94	00	$z^5F_2^\circ - e^5F_3$
J	4946.394	4	IV	20211.12	−01	$z^5F_4^\circ - e^5F_4$
U	4945.63	(1)		20214.25	+03	$z^3P_2^\circ - f^5G_3$
B	4939.690	4	IB	20238.55	−01	$a^5F_5 - z^5F_4^\circ$
J	*4939.244	(2)		20240.38	−02 −01	$y^5D_3^\circ - f^3D_3$ $y^5D_1^\circ - g^5F_2$
J	4938.820	10	IV	20242.12	00	$z^7F_2^\circ - e^7D_3$
K	4938.183	(2)		20244.73	+01	$z^3F_3^\circ - g^5D_2$
K	4934.023	(2n)		20261.80	−08	$y^5D_3^\circ - f^5G_4$
Q	4933.878	(1)		20262.39	+04	$z^3F_3^\circ - e^5P_2$
K	4933.348	(2n)		20264.57	+03	$y^5D_0^\circ - g^5F_1$
K	4930.331	(2)		20276.97	−01	$z^3D_1^\circ - g^5D_1$
U	4927.447	(1)		20288.84	−12	$a^1H_5 - w^3F_4^\circ$
U	4925.293	(1)		20297.71	−04	$y^5D_4^\circ - g^5F_4$
B	4924.776	3	V	20299.84	−01	$a^3P_2 - y^3D_2^\circ$
I	4920.509	60	III	20317.45	00	$z^7F_5^\circ - e^7D_4$
B	4918.999	30	III	20323.68	00	$z^7F_3^\circ - e^7D_3$
U	4918.023	(1)		20327.72	+01	$y^5D_0^\circ - f^3D_1$
U	4917.242	(1)		20330.95	+07	$y^5D_2^\circ - h^5D_1$
U	4911.786	(1)		20353.53	−03	$z^3D_2^\circ - e^3D_1$
J	4910.570	(1w)		20358.57	−02	$y^5D_1^\circ - f^5G_2$
J	4910.328	(1w)		20359.57	−02	$y^5D_2^\circ - f^5G_3$
J	4910.027	(2)		20360.82	−02	$z^5F_3^\circ - e^5F_3$
J	4909.387	(1)		20363.47	+03	$z^3D_2^\circ - g^5D_2$
K	4907.743	(1)		20370.29	−01	$z^5F_1^\circ - e^5F_2$
W	4905.15	(1)		20381.06	−01	$z^3D_2^\circ - e^5P_2$
B	4903.317	12	III	20388.68	−01	$z^7F_1^\circ - e^7D_2$
U	4896.437	(1)		20417.33	+01	$z^3D_3^\circ - e^3D_2$
U	4892.866	(1)		20432.23	+03	$y^5D_1^\circ - f^3D_1$
I	4891.496	50	III	20437.95	00	$z^7F_4^\circ - e^7D_3$
J	4890.762	25	III	20441.02	−01	$z^7F_2^\circ - e^7D_2$
U	4889.113	(2)		20447.91	−03	$z^3D_3^\circ - g^5D_3$
U	*4889.009	(1)		20448.35	−01 +03	$a^5P_2 - y^3D_3^\circ$ $a^1D_2 - 2_2^\circ$
V	4888.651	(1)		20449.85	−08	$y^5D_4^\circ - h^5D_3$
K	4887.189	(−)		20455.96	+04	$y^5D_2^\circ - g^5F_2$
J	4886.335	(1)		20459.54	−04	$y^5D_3^\circ - h^5D_2$
J	4885.435	2	V	20463.31	+01	$z^3F_4^\circ - g^5D_3$
J	4882.151	(2)		20477.07	00	$z^5F_2^\circ - e^5F_2$
J	*4881.726	(2)		20478.85	−05 +09	$b^3H_4 - z^3H_4^\circ$ $c^3F_3 - 10_3^\circ$
B	4878.218	12	III	20493.58	−01	$z^7F_0^\circ - e^7D_1$
U	4875.897	(1)		20503.34	−06	$z^5F_5^\circ - e^5F_4$
I	4872.144	20	III	20519.13	−01	$z^7F_1^\circ - e^7D_1$
I	4871.323	25	III	20522.59	00	$z^7F_3^\circ - e^7D_2$
J	4863.653	(2)		20554.95	−01	$z^5F_1^\circ - e^5F_1$
SS	4860.994	(1)		20566.20	−07	$z^5F_3^\circ - e^3F_4$
B	4859.748	15	III	20571.47	00	$z^7F_2^\circ - e^7D_1$
U	4859.142	(1)		20574.04	−08	$y^5D_2^\circ - f^5G_2$
J	4855.683	(3)		20588.69	−01	$z^5F_4^\circ - e^5F_3$
U	4854.888	(1n)		20592.06	+02	$c^3F_3 - 11_3^\circ$
U	4848.885	(1)		20617.55	00	$a^3P_2 - y^3D_1^\circ$
V	*4845.656	(2)		20631.29	00 −02	$b^3H_5 - z^3H_6^\circ$ $b^3D_1 - w^3P_0^\circ$
U	4844.004	(2)		20638.33	−01	$a^1D_2 - w^3F_3^\circ$
J	4843.155	(3)		20641.95	−01	$z^5F_3^\circ - e^5F_2$
U	4842.788	(1)		20643.51	−02	$y^5D_4^\circ - e^3G_5$
W	4841.80	(1)		20647.72	−01	$y^5D_2^\circ - f^3D_1$
U	4840.319	(1n)		20654.04	00	$y^5D_3^\circ - f^5G_3$
J	4839.549	(3)		20657.34	00	$b^3H_5 - z^3H_5^\circ$
J	4838.519	(2n)		20661.72	00	$z^5F_2^\circ - e^5F_1$
K	4835.862	(3)		20673.08	+02	$y^5D_4^\circ - f^5G_4$
V	4834.511	(1)		20678.86	−01	$a^3P_1 - x^5D_2^\circ$

TABLE B—(*Continued*)

Ref	λ I A	Int	T C	Wave Number Observed	o−c	Desig
J	*4832.734	(2)		20686.46	−01 −21	$b^3D_2 - w^3P_1^\circ$ $y^5F_1^\circ - f^3F_2$
U	4824.165	(1)		20723.20	00	$b^3D_1 - w^3P_1^\circ$
U	4817.773	(1)		20750.70	+04	$a^5P_1 - y^3D_2^\circ$
U	4813.115	(1)		20770.78	+02	$a^3D_1 - u^5D_1^\circ$
V	4811.04	(1)		20779.74	−03	$c^3P_1 - x^3D_1^\circ$
U	4809.950	(1)		20784.45	−02	$a^1H_5 - y^3H_5^\circ$
W	4809.3	(1)		20787.3	−1	$c^3F_4 - y^1F_3^\circ$
U	4809.154	(1)		20787.89	−04	$b^1G_4 - z^1F_3^\circ$
U	4808.159	(1)		20792.19	−02	$a^3D_3 - w^3D_3^\circ$
K	4807.725	(2)		20794.06	−08	$z^5F_4^\circ - e^3F_4$
S	*4807.243	(−)		20796.15	−05 +16	$y^5F_3^\circ - f^3F_3$ $a^3D_2 - 1_2^\circ$
W	4804.6	(1)		20807.6	0	$a^1P_1 - v^5F_1^\circ$
U	4804.531	(1)		20807.89	−05	$a^1H_5 - v^3G_4^\circ$
J	*4802.883	(3)		20815.03	+02 −03	$b^3D_3 - w^3P_2^\circ$ $b^1G_4 - x^1G_4^\circ$
J	4800.652	(2)		20824.70	+02	$c^3F_3 - t^3G_4^\circ$
U	4800.137	(1)		20826.94	−01	$z^7P_2^\circ - e^5D_2$
U	4799.414	(1)		20830.07	−02	$b^3D_2 - w^3P_2^\circ$
U	4798.735	(1)		20833.02	00	$a^3F_2 - y^5D_2^\circ$
U	4798.271	(1)		20835.03	+01	$c^3F_2 - t^3G_3^\circ$
W	4794.0	(1)		20853.6	−2	$a^1G_4 - y^3G_4^\circ$
U	4791.248	(1)		20865.57	−04	$a^3D_1 - w^3D_1^\circ$
B	4789.654	7	V	20872.52	+03	$a^1D_2 - z^1D_2^\circ$
J	4788.757	(4)		20876.43	00	$b^3H_6 - z^3H_6^\circ$
U	4787.839	(1)		20880.43	−03	$z^7P_3^\circ - e^5D_3$
B	4786.810	5	IV?	20884.92	−01	$c^3P_2 - x^3D_3^\circ$
U	4785.959	(1)		20888.63	+05	$c^3F_3 - 13_4^\circ$
J	4779.444	(1)		20917.10	−02	$a^1P_1 - x^3P_0^\circ$
V	4776.34	(1n)		20930.70	+11	$y^5P_3^\circ - i^5D_4$
U	4776.074	(1)		20931.86	−04	$a^3D_2 - y^3S_1^\circ$
B	*4772.817	3	III	20946.15	+07 −04	$c^3P_2 - x^3D_2^\circ$ $a^3F_3 - y^5D_3^\circ$
J	4771.702	(1)		20951.04	−01	$a^5P_2 - y^3D_2^\circ$
V	4768.397	3n	V	20965.56	+02	$z^7P_4^\circ - e^5D_4$
V	4768.334	(1)		20965.84	−08	$z^5P_1^\circ - f^5D_2$
U	4765.482	(1)		20978.39	00	$a^3F_2 - z^3P_2^\circ$
J	*4757.582	(2)		21013.22	−01 −01	$z^3P_1^\circ - e^3P_1$ $a^3D_1 - 1_2^\circ$
V	4749.93	(1)		21047.07	+10	$y^5P_3^\circ - i^5D_3$
B	*4745.806	3n	V	21065.36	−04 +14	$z^5P_2^\circ - f^5D_3$ $y^5D_4^\circ - f^5G_3$
U	4745.129	(1)		21068.37	+02	$a^5P_1 - y^3D_1^\circ$
B	4741.533	3	V	21084.35	+01	$b^3P_2 - w^5D_3^\circ$
J	4741.081	(1)		21086.36	−05	$z^5F_5^\circ - e^3F_4$
J	4740.343	(1)		21089.64	−01	$b^3G_3 - y^3G_4^\circ$
J	4737.633	(1)		21101.70	+02	$b^3H_5 - z^1G_4^\circ$
I	4736.780	12	II?	21105.50	−01	$z^5D_4^\circ - e^5F_5$
J	4735.846	(2)		21109.67	+05	$c^3F_4 - t^3G_5^\circ$
J	4734.100	(1)		21117.45	−02	$b^1D_2 - w^1D_2^\circ$
B	4733.596	4	IB?	21119.70	00	$a^3F_4 - y^5D_4^\circ$
V	4729.699	(1)		21137.10	−09	$z^5F_3^\circ - e^3F_3$
J	4729.028	(1)		21140.10	00	$c^3F_4 - 12_5^\circ$
J	4728.555	3n	IV	21142.21	−06	$z^5P_2^\circ - e^7P_3$
J	4727.405	3n	IV	21147.36	−04	$z^5P_1^\circ - f^5D_1$
U	4726.160	(1)		21152.93	−07	$z^7P_3^\circ - e^5D_2$
U	4725.945	(1n)		21153.89	−01	$b^1D_2 - w^1F_3^\circ$
J	*4720.997	(1)		21176.06	+05 −19	$b^3G_4 - y^3G_5^\circ$ $y^5D_3^\circ - f^3F_4$
V	4714.182	(1n)		21206.67	+05	$b^3H_4 - x^3G_3^\circ$
V	4714.074	(1n)		21207.16	00	$y^5P_3^\circ - i^5D_2$
V	4712.104	(1)		21216.03	00	$c^3P_2 - x^3D_1^\circ$
B	4710.286	5	IV	21224.21	−01	$b^3G_3 - y^3G_2^\circ$
J	4709.092	(3)		21229.59	−04	$z^5P_2^\circ - f^5D_2$

Ref	λ I A	Int	T C	Wave Number Observed	o−c	Desig
V	4708.972	(1)		21230.14	+03	$b^3D_2 - z^1F_3^\circ$
J	4707.487	(2)		21236.83	+05	$b^3P_1 - w^5D_2^\circ$
B	4707.281	8	IV	21237.76	−01	$z^5D_3^\circ - e^5F_4$
J	4705.464	(1)		21245.96	−04	$a^1D_2 - v^3G_3^\circ$
J	4704.958	(5)		21248.25	−02	$z^5P_1^\circ - f^5D_0$
J	4701.052	(1)		21265.90	−04	$z^5P_1^\circ - f^7D_2$
J	4700.171	(2n)		21269.89	+08	$b^1G_4 - x^3H_5^\circ$
B	4691.414	6	IV	21309.59	−01	$b^3G_4 - y^3G_4^\circ$
J	4690.146	(3)		21315.35	00	$z^5P_1^\circ - f^7D_1$
J	4687.387	(1)		21327.90	+02	$b^3P_2 - w^5F_3^\circ$
J	4683.565	(2)		21345.30	+02	$b^3P_2 - w^5D_2^\circ$
U	4682.583	(1)		21349.78	−09	$z^7P_4^\circ - e^5D_3$
V	4680.475	(1)		21359.39	00	$b^3P_0 - w^5D_1^\circ$
J	4680.297	(2)		21360.21	+01	$a^3F_2 - y^5F_3^\circ$
V	4679.229	(1)		21365.08	+02	$z^5F_4^\circ - e^3F_3$
B	4678.852	7	V	21366.80	−04	$z^5P_3^\circ - f^5D_4$
J	4673.169	(4)		21392.78	−04	$z^5P_2^\circ - f^7D_3$
J	4669.174	(4)		21411.09	−03	$z^5P_2^\circ - f^5D_1$
J	4668.142	6	IV	21415.82	−01	$z^5D_2^\circ - e^5F_3$
B	4667.459	6	V	21418.96	−02	$z^5P_3^\circ - e^7P_4$
J	4663.183	(1)		21438.60	−04	$a^1D_2 - w^3P_1^\circ$
J	4661.975	(2)		21444.15	−02	$b^3G_4 - y^3G_3^\circ$
J	4661.538	(2n)		21446.16	−01	$y^5P_3^\circ - 4_2$
U	4658.29	(1)		21461.12	+03	$b^3H_5 - x^3G_4^\circ$
U	4657.596	(1)		21464.31	−01	$b^3P_1 - w^5D_1^\circ$
J	*4654.628	5	V	21478.00	+04 −09	$z^5D_1^\circ - e^5F_4$ $z^5P_3^\circ - f^5D_3$
J	4654.501	5	II?	21478.59	00	$a^3F_3 - y^5F_4^\circ$
U	4649.828	(1)		21500.17	+02	$b^3H_6 - v^5F_5^\circ$
B	4647.437	6	IV	21511.23	+02	$b^3G_5 - y^3G_5^\circ$
J	4643.468	(2)		21529.62	−03	$z^5P_2^\circ - f^7D_2$
J	4638.016	3	IV	21554.92	−04	$z^5P_3^\circ - e^7P_3$
J	4637.512	3	IV	21557.27	00	$z^5D_1^\circ - e^5F_2$
J	4635.846	(1)		21565.01	+01	$b^3P_1 - y^5S_2^\circ$
U	4633.764	(1)		21574.71	−02	$b^3G_3 - x^5G_3^\circ$
J	4632.915	2	III?	21578.65	−01	$a^3F_2 - y^5F_2^\circ$
U	4631.501	(1)		21585.25	−11	$z^5G_4^\circ - 3$
U	4630.785	(1)		21588.58	−04	$z^3F_3^\circ - g^5F_4$
J	4630.125	(2)		21591.66	00	$a^3P_2 - x^5D_3^\circ$
S	4626.758	(−)		21607.37	+04	$b^3G_4 - x^5G_5^\circ$
I	4625.052	3	IV	21615.35	+01	$z^5D_3^\circ - e^5F_3$
J	4619.294	3n	IV	21642.29	−03	$z^5P_3^\circ - f^5D_2$
J	4618.765	(2)		21644.77	−03	$b^3G_5 - y^3G_4^\circ$
V	4618.568	(2w)		21645.69	+08	$z^5G_5^\circ - 1$
J	4614.216	(1)		21666.10	−06	$a^3D_2 - v^5P_1^\circ$
J	4613.210	2n	V	21670.83	00	$z^5D_0^\circ - e^5F_1$
I	4611.285	5n	III	21679.88	+02 +31 +04	$z^5P_2^\circ - e^5S_2$ $(a^5F_4 - z^5P_3^\circ)$ $(z^5P_2^\circ - e^7F_3)$
J	*4607.655	3n	V	21696.95	−01 00	$z^5D_2^\circ - e^5F_2$ $z^3F_2^\circ - g^5F_3$
V	4603.956	(1)		21714.39	00	$b^3G_4 - x^5G_4^\circ$
B	4602.944	9	IB?	21719.16	00	$a^3F_4 - y^5F_5^\circ$
J	4602.005	(2)		21723.59	−01	$a^3F_2 - y^5F_1^\circ$
J	4600.937	(1)		21728.64	−03	$b^3H_6 - x^3G_5^\circ$
J	4598.122	(2n)		21741.94	+01	$z^5D_1^\circ - e^5F_1$
U	4596.433	(1)		21749.93	−10	$z^5P_2^\circ - e^5G_3$
K	4596.059	(2n)		21751.70	−01	$z^5P_3^\circ - f^7D_4$
J	4595.363	(2)		21754.99	00	$b^3H_4 - z^1H_5^\circ$
U	4594.957	(2)		21756.91	−09	$a^3D_1 - v^5P_2^\circ$
U	4593.544	(1)		21763.61	−01	$z^3F_3^\circ - f^5P_2$
B	4592.655	5	IB	21767.82	00	$a^3F_3 - y^5F_3^\circ$
J	4587.132	(2)		21794.03	−01	$a^1H_5 - x^1G_4^\circ$
K	4584.824	(2)		21805.00	+02	$z^5P_3^\circ - e^7P_2$
U	4584.723	(1)		21805.48	−03	$z^5P_3^\circ - f^7D_3$

TABLE B—(*Continued*)

Ref	λ I A	Int	T C	Wave Number Observed	Wave Number o−c	Desig
U	4582.941	(1)		21813.96	+03	$b^3P_1 - v^5D_1^\circ$
J	4581.517	(2)		21820.74	−04	$z^5D_3^\circ - e^3F_4$
K	4580.600	(2)		21825.10	−13	$z^5P_2^\circ - e^3D_3$
V	4579.825	(1)		21828.80	−02	$c^3P_1 - z^3S_1^\circ$
V	*4579.344	(1)		21831.09	−04	$z^7F_5^\circ - e^5D_4$
					−18	$b^1G_4 - 6_5^\circ$
U	*4575.80	(1)		21848.00	−10	$b^3H_4 - w^3G_3^\circ$
					+14	$z^3F_4^\circ - h^5D_4$
J	4574.724	(2)		21853.14	−01	$a^3P_2 - x^5D_2^\circ$
V	4574.240	(1)		21855.45	−09	$z^5D_4^\circ - e^5F_3$
W	4572.9	(1)		21861.9	−2	$z^5P_2^\circ - e^7F_2$
U	4568.840	(1)		21881.28	00	$b^3D_1 - v^3F_2^\circ$
U	4568.787	(1)		21881.53	−09	$z^5D_2^\circ - e^5F_1$
U	4566.988	(1)		21890.15	−03	$a^1P_1 - w^3F_2^\circ$
J	4566.520	(2)		21892.40	−02	$a^3D_2 - x^3P_1^\circ$
U	4565.667	(2)		21896.49	+02	$z^5D_3^\circ - e^5F_2$
V	4565.324	(2n)		21898.13	−09	$a^3D_1 - x^3P_2^\circ$
V	4564.832	(1)		21900.49	−02	$c^3P_1 - y^3P_0^\circ$
U	4564.713	(1)		21901.06	−09	$z^5P_2^\circ - e^5G_2$
J	4560.096	(2)		21923.24	−05	$z^5P_3^\circ - e^5G_4$
J	*4558.108	(1)		21932.80	+01	$b^3D_3 - v^3F_4^\circ$
					−01	$z^3F_2^\circ - f^3D_2$
J	4556.939	(1)		21938.43	−03	$a^3D_3 - v^5P_2^\circ$
J	*4556.129	4n	V	21942.32	−02	$z^5P_3^\circ - f^7D_2$
					−13	$z^3F_3^\circ - f^3D_3$
					−21	$b^3G_5 - x^5G_5^\circ$
U	4554.465	(1)		21950.34	−05	$z^7F_3^\circ - e^5D_3$
U	4551.667	(1)		21963.83	−09	$z^3F_3^\circ - f^5G_4$
B	4547.851	4	V	21982.26	−02	$a^1D_2 - z^1F_3^\circ$
J	4547.022	(2)		21986.27	−01	$a^3F_3 - y^5F_2^\circ$
V	4542.720	(1)		22007.09	−13	$z^5P_1^\circ - e^3D_2$
U	4542.420	(2)		22008.55	+03	$b^3D_2 - v^3F_3^\circ$
U	4541.953	(1)		22010.81	−02	$b^3H_5 - w^3G_4^\circ$?
W	4538.84	(2)		22025.91	+05	$z^3F_3^\circ - g^5F_3$
V	4538.764	(1)		22026.28	−05	$a^3P_2 - x^5D_1^\circ$
J	4537.677	(1n)		22031.55	+01	$b^3H_5 - z^1H_5^\circ$
U	4536.509	(1)		22037.22	−10	$b^3D_3 - 4_4^\circ$
V	4533.143	(1n)		22053.59	−04	$a^3D_1 - x^3P_0^\circ$
J	*4531.633	(2)		22060.93	+01	$a^1I_6 - u^3G_5^\circ$
					−04	$z^5D_4^\circ - e^3F_4$
					−24	$z^3D_2^\circ - f^3D_3$
B	4531.152	8	II	22063.28	00	$a^3F_4 - y^5F_4^\circ$
V	4529.562	(1)		22071.02	−06	$z^3D_3^\circ - g^5F_4$
B	4528.619	18	II	22075.62	00	$a^5P_3 - x^5D_4^\circ$
W	4527.9	(1)		22079.1	0	$b^3D_2 - 5^\circ$
U	4527.784	(1)		22079.69	+01	$a^3D_3 - x^3P_2^\circ$
J	4526.563	(2)		22085.64	+01	$c^3P_0 - u^5D_1^\circ$
U	4525.868	(1)		22089.04	+01	$z^7F_1^\circ - e^5D_2$
I	4525.142	5n	IV	22092.58	+06	$z^5P_3^\circ - e^5S_2$
					+05	$(z^5P_3^\circ - e^7F_3)$
J	4523.403	(2)		22101.07	−06	$z^5P_2^\circ - e^7S_3$
W	§4520.3	(1)		22116.2	−3	$c^3P_1 - u^5D_2^\circ$
U	4518.45	(1)		22125.30	−08	$b^3H_6 - w^3G_5^\circ$
B	4517.530	(2)		22129.80	−03	$c^3P_1 - y^3P_1^\circ$
SS	4515.178	(1)		22141.33	−04	$z^7F_2^\circ - e^5D_2$
J	4514.189	(2)		22146.18	+02	$a^1G_4 - u^5D_4^\circ$
U	*4509.306	(1)		22170.16	−02	$b^1G_4 - u^3D_3^\circ$
					−18	$a^1G_4 - u^5D_3^\circ$
J	4504.838	(2)		22192.15	−04	$z^5D_2^\circ - e^3F_3$
U	4502.590	(1)		22203.23	+02	$a^1H_5 - x^3H_6^\circ$
J	4495.966	(1)		22235.95	−07	$z^5P_2^\circ - f^5F_2$
J	4495.566	(1)		22237.92	00	$z^5P_3^\circ - e^3D_3$
V	*4495.386	(1)		22238.81	+05	$z^7F_0^\circ - e^5D_1$
					+20	$z^3F_4^\circ - h^5D_3$
B	4494.568	12	III	22242.86	00	$a^5P_2 - x^5D_3^\circ$

Ref	λ I A	Int	T C	Wave Number Observed	Wave Number o−c	Desig
W	4493.3	(1)		22249.1	+3	$a^1H_5 - x^3H_5^\circ$?
V	4492.693	(1n)		22252.14	−01	$z^3F_2^\circ - g^5F_1$
J	*4490.773	(2n)		22261.66	+19	$z^3F_3^\circ - e^3G_4$
					−06	$z^3F_3^\circ - f^3D_2$
J	4490.084	(2)	IV	22265.07	−01	$c^3P_2 - z^3S_1^\circ$
B	4489.741	3	IA	22266.77	00	$a^5D_0 - z^7F_1^\circ$
J	*4488.917	(2)	IV	22270.86	+09	$b^3F_4 - y^5G_5^\circ$
					−07	$z^5P_2^\circ - e^3D_2$
J	4488.140	(2n)		22274.71	−04	$z^5P_3^\circ - e^7F_2$
J	4485.679	(2)	IV	22286.93	−03	$z^5P_1^\circ - e^5P_1$
I	4484.227	4	IV	22294.15	−01	$z^5P_3^\circ - g^5D_4$
V	4482.750	(2)		22301.50	−04	$z^5P_2^\circ - g^5D_3$
J	4482.257	6	I	22303.95	00	$a^5P_1 - x^5D_2^\circ$
J	4482.171	4		22304.38	00	$a^5D_1 - z^7F_2^\circ$
J	4481.621	(2)		22307.12	−07	$z^5P_1^\circ - e^3D_1$
J	4480.142	(3)	IV	22314.48	−04	$a^1G_4 - x^3F_4^\circ$
J	*4479.612	(3)	IV	22317.12	+05	$z^5P_1^\circ - g^5D_2$
					−07	$a^1I_6 - 6_5^\circ$
U	4478.040	(1)		22324.95	−11	$a^5P_2 - y^7P_2^\circ$
Q	4476.082	(4)		22334.72	+02	$z^5P_1^\circ - e^5P_2$
I	4476.021	10	III	22335.03	00	$b^3P_1 - x^3D_2^\circ$
J	*4472.721	(2)		22351.50	+02	$b^3H_5 - y^1G_4^\circ$
					−05	$b^3D_2 - y^1D_2^\circ$
SS	4471.810	(1)		22356.06	−02	$z^3F_3^\circ - f^5G_3$
SS	4471.685	(1)		22356.68	−04	$a^5D_1 - z^7F_1^\circ$
I	4469.381	5n	IV	22368.21	−04	$z^5P_2^\circ - e^5P_3$
U	4467.446	(1)		22377.89	−05	$c^3F_3 - w^1F_3^\circ$
V	4466.939	(2)		22380.43	−01	$z^3D_2^\circ - f^3D_2$
B	4466.554	12	II	22382.36	−02	$b^3P_2 - x^3D_3^\circ$
					+10	$(a^5D_1 - z^7F_0^\circ)$
U	4466.181	(1)		22384.23	−07	$b^3D_3 - 7_2^\circ$
W	4465.3	(1)		22388.6	+1	$y^5F_4^\circ - 1_5$
U	4464.766	(2)	IV	22391.33	00	$c^3P_2 - y^3P_2^\circ$
J	*4461.989	(4)	IV	22405.26	−11	$z^5P_3^\circ - f^5F_4$
					+07	$c^3P_1 - u^5D_1^\circ$
					−08	$b^3D_2 - x^1D_2^\circ$
B	4461.654	8	I	22406.94	00	$a^5D_2 - z^7F_3^\circ$
U	4461.373	(1)		22408.36	00	$a^1P_1 - w^3P_0^\circ$
V	4461.205	(2)		22409.20	−03	$c^3P_2 - u^5D_3^\circ$
B	4459.121	10	III	22419.67	00	$a^5P_3 - x^5D_3^\circ$
V	4458.101	(3)		22424.80	−11	$z^3D_3^\circ - f^3D_3$
J	4456.331	(1)		22433.71	−03	$a^1G_4 - z^3H_5^\circ$
J	4455.032	(2)		22440.25	−03	$z^3F_4^\circ - f^3D_3$
J	4454.655	(1)		22442.15	+08	$b^3D_1 - x^1D_2^\circ$
B	4454.383	5	III	22443.52	−01	$b^3P_2 - x^3D_2^\circ$
J	4450.320	(3)		22464.01	00	$c^3P_0 - y^3S_1^\circ$
B	4447.722	9	III	22477.13	00	$a^5P_1 - x^5D_1^\circ$
J	4447.134	(2)Mn?	IV	22480.10	00	$a^5P_2 - y^7P_3^\circ$
J	4446.842	(2)		22481.58	00	$z^5P_1^\circ - g^5D_1$
U	4445.48	(1)	IA	22488.47	−03	$a^5D_2 - z^7F_2^\circ$
B	4443.197	7	III	22500.02	−02	$b^3P_0 - x^3D_1^\circ$
J	4442.835	(2)	IV	22501.85	−02	$a^5P_3 - y^7P_2^\circ$
B	4442.343	12	III	22504.35	00	$a^5P_2 - x^5D_2^\circ$
V	4440.972	(2)		22511.29	−06	$a^3D_2 - v^3D_3^\circ$
U	4440.838	(1)		22511.97	−01	$z^3D_1^\circ - f^3D_1$
V	4440.479	(1)		22513.79	−03	$z^5P_3^\circ - e^7S_3$
J	4439.883	(2)	IV	22516.81	00	$a^3P_2 - z^5S_2^\circ$
V	4439.643	(1)		22518.03	−06	$a^1G_4 - x^3F_3^\circ$
K	4438.353	(2)		22524.58	00	$z^5P_1^\circ - g^5D_0$
V	4436.931	(2)		22531.80	−05	$a^1G_4 - z^3H_4^\circ$
J	4435.151	2	IIA	22540.84	00	$a^5D_2 - z^7F_1^\circ$
J	4433.793	(3n)		22547.74	−06	$z^5P_3^\circ - f^5F_3$
J	4433.223	3n	IV	22550.64	−03	$z^5P_2^\circ - e^5P_1$
J	4432.572	(3)		22553.95	−03	$a^1H_5 - u^3G_5^\circ$
B	4430.618	6	III	22563.90	00	$a^5P_1 - x^5D_0^\circ$

TABLE B—(*Continued*)

Ref	λ I A	Int	T C	Wave Number Observed	o−c	Desig
V	4430.197	(2)	IV	22566.04	−05	$c^3P_2 - y^3P_1^\circ$
U	4429.32	(1)		22570.51	−11	$z^3F_3^\circ - f^5G_2$
B	4427.312	10	I	22580.75	00	$a^5D_3 - z^7F_4^\circ$
					−04	$(z^5P_2^\circ - g^5D_2)$
U	4425.660	(1)		22589.18	−02	$a^1H_5 - 4_4^\circ$
U	4424.192	(1)		22596.67	−05	$a^1D_2 - v^3F_2^\circ$
V	4423.858	(2l)		22598.38	−04	$z^5P_2^\circ - e^5P_2$
U	4423.142	(1)		22602.04	+04	$b^3G_4 - u^5D_4^\circ$
U	4422.884	(1n)		22603.36	−07	$a^3D_2 - 3_3^\circ$
B	4422.570	6	III	22604.96	−01	$b^3P_1 - x^3D_1^\circ$
U	4418.429	(1)		22626.15	−03	$b^3G_4 - u^5D_3^\circ$
B	4415.125	20	II	22643.08	00	$a^3F_2 - z^5G_3^\circ$
SS	4414.464	(1)		22646.47	+03	$a^3D_1 - 2_2^\circ$
J	4409.123	(2l)		22673.90	−03	$a^3D_2 - v^3D_1^\circ$
B	4408.419	6	III?	22677.52	−01	$a^5P_2 - x^5D_1^\circ$
J	4407.714	5	III?	22681.15	−01	$a^5P_3 - x^5D_2^\circ$
B	4404.752	30	II	22696.40	+01	$a^3F_3 - z^5G_4^\circ$
U	4401.450	(2)		22713.43	−04	$b^3P_2 - x^3D_1^\circ$
J	4401.293	(5)		22714.24	00	$z^5P_3^\circ - g^5D_3$
V	*4395.514	(1w)		22744.10	+17	$z^3D_3^\circ - e^3G_4$
					−08	$z^3D_3^\circ - f^3D_2$
U	4395.286	(2)		22745.28	−02	$z^5P_2^\circ - g^5D_1$
U	4392.58	(1)		22759.29	−01	$z^3F_4^\circ - e^3G_4$
B	4390.954	4	IV	22767.72	−02	$b^3G_3 - z^3H_4^\circ$
U	4390.458	(1)		22770.29	−06	$b^3G_4 - x^3F_4^\circ$
J	4389.244	2	IIA	22776.59	+01	$a^5D_3 - z^7F_2^\circ$
J	4388.412	4n	IV	22780.91	−03	$z^5P_3^\circ - e^5P_3$
J	4387.897	3	IV	22783.58	+01	$c^3P_1 - y^3S_1^\circ$
W	4386.6	(1n)		22790.3	0	$b^3D_1 - u^5P_1^\circ$
U	4385.258	(1)		22797.29	−05	$b^3G_3 - w^3D_2^\circ$
V	4384.682	(1)		22800.29	−05	$c^3P_2 - w^3D_2^\circ$
B	4383.547	45r	II	22806.19	00	$a^3F_4 - z^5G_5^\circ$
U	4382.773	(2)		22810.22	−03	$a^1H_5 - 6_5^\circ$
U	4377.793	(1)		22836.16	−03	$a^3D_1 - v^3D_2^\circ$
U	4377.330	(1)		22838.58	+04	$z^3D_3^\circ - f^5G_3$
V	*4376.782	(1)		22841.44	−01	$c^3P_2 - u^5D_1^\circ$
					+01	$b^3D_3 - t^3D_3^\circ$
B	4375.932	9	I	22845.88	00	$a^5D_4 - z^7F_5^\circ$
U	4374.491	(1)		22853.40	+03	$a^3D_2 - z^1D_2^\circ$
J	*4373.563	(2)		22858.25	−15	$b^3F_4 - w^5D_4^\circ$
					−02	$b^3G_3 - x^3F_2^\circ$
U	4372.991	(1)		22861.24	−03	$c^3P_2 - x^3F_2^\circ$
B	4369.774	7	III	22878.07	−01	$a^1G_4 - z^1G_4^\circ$
J	4367.906	2	IIIA	22887.86	+01	$a^3F_2 - z^5G_2^\circ$
J	4367.581	5	IV	22889.56	−01	$b^3G_4 - z^3H_5^\circ$
U	4365.899	(1)		22898.38	00	$b^3G_4 - w^3D_3^\circ$
U	4360.810	(1)		22925.10	00	$b^3D_3 - u^3D_2^\circ$
B	4358.505	3	IV	22937.22	+02	$b^3G_5 - u^5D_4^\circ$
B	4352.737	9	III	22967.62	+01	$a^5P_1 - z^5S_2^\circ$
J	4351.549	3	IV	22973.88	−04	$b^3G_4 - x^3F_3^\circ$
J	4348.939	(1)		22987.67	−02	$b^3G_4 - z^3H_4^\circ$
U	4347.851	(1)		22993.43	−05	$z^5P_3^\circ - g^5D_2$
V	4347.239	(1)		22996.67	−02	$a^5D_4 - z^7F_4^\circ$
J	4346.558	(2)		23000.27	−01	$b^3H_4 - v^3G_4^\circ$
J	4343.699	(2)		23015.40	+02	$a^1G_4 - w^5G_4^\circ$
J	4343.257	(2)		23017.75	+10	$a^3D_3 - v^3D_2^\circ$
W	*4340.5	(1)		23032.4	0	$a^3G_3 - x^3D_2^\circ$
					+1	$z^5F_1^\circ - f^5D_2$
J	4338.260	(2)		23044.26	00	$a^5P_3 - x^5F_4^\circ$
B	4337.049	10	II	23050.69	−01	$a^3F_3 - z^5G_3^\circ$
U	4335.89	(1)		23056.86	+06	$z^3D_3^\circ - e^3G_3$
U	4330.959	(1)		23083.10	−01	$b^3H_5 - y^3H_6^\circ$
W	4327.92	(2)		23099.31	−11	$b^3H_4 - y^3H_4^\circ$
J	4327.100	3	V	23103.69	−03	$a^1D_2 - y^1D_2^\circ$
U	4326.760	(2)		23105.51	−04	$b^3G_5 - x^3F_4^\circ$

Ref	λ I A	Int	T C	Wave Number Observed	o−c	Desig
B	4325.765	35	II	23110.82	00	$a^3F_2 - z^3G_3^\circ$
					−13	$(a^5D_4 - z^7F_3^\circ)$
U	4324.966	(1)		23115.09	−02	$a^5P_2 - x^5F_3^\circ$
U	4320.52	(1)		23138.88	−19	$z^5F_2^\circ - f^5D_2$
SS	4320.376	(1)		23139.65	−08	$z^5F_3^\circ - f^5D_3$
SS	4317.067	(1)		23157.39	−12	$a^1D_2 - x^1D_2^\circ$
B	4315.087	10	III	23168.01	+01	$a^5P_2 - z^5S_2^\circ$
U	4309.380	4	IV	23198.69	−04	$b^3G_5 - z^3H_6^\circ$
J	4309.036	(2)		23200.54	−01	$a^1I_6 - y^3I_6^\circ$
B	4307.906	35	II	23206.63	−01	$a^3F_3 - z^3G_4^\circ$
SS	4306.601	(1)		23213.66	−13	$z^5F_1^\circ - f^5D_1$
B	4305.455	3	IV	23219.84	+01	$c^3P_2 - y^3S_1^\circ$
U	4305.20	(1)		23221.22	+03	$a^1D_2 - u^3G_3^\circ$
J	4304.552	(1)		23224.71	−06	$b^3G_5 - z^3H_5^\circ$
J	4302.191	(2)		23237.46	−03	$a^1G_4 - x^3G_4^\circ$
U	4300.825	(1)		23244.84	+01	$z^3F_2^\circ - f^3F_2$
U	4299.635	(1)		23251.27	00	$b^3G_3 - w^5G_4^\circ$
I	4299.242	18	III	23253.39	−01	$z^7D_4^\circ - e^7D_5$
					+03	$(b^3H_5 - y^3H_5^\circ)$
B	4298.040	(2)	IV	23259.90	−03	$a^1G_4 - x^3G_5^\circ$
V	4294.939	(1w)		23276.69	−14	$b^3H_5 - v^3G_4^\circ$?
B	4294.128	15	II	23281.09	+01	$a^3F_4 - z^5G_4^\circ$
U	4292.290	(1)		23291.06	+02	$a^5P_2 - x^5F_2^\circ$
I	*4291.466	4	IA	23295.53	+06	$a^3F_3 - z^5G_2^\circ$
					00	$a^5D_3 - z^7P_4^\circ$
J	4290.870	(1)		23298.76	−02	$b^3P_2 - w^5P_3^\circ$
J	4290.382	(2)		23301.41	+03	$b^3G_4 - w^5G_5^\circ$
U	4288.965	(1)		23309.11	−01	$b^3F_3 - w^5D_2^\circ$
J	4288.148	(2)		23313.55	−01	$a^3G_3 - y^3G_3^\circ$
U	4286.992	(1)		23319.84	−04	$z^3F_3^\circ - f^3F_3$
U	4286.437	(1)		23322.86	−03	$b^3G_5 - z^3H_4^\circ$
U	4285.829	(1)		23326.17	+02	$b^3D_2 - t^3D_2^\circ$
B	4285.445	3	IV	23328.26	+01	$b^3H_6 - y^3H_6^\circ$
U	4284.415	(1)		23333.87	−05	$b^3G_4 - z^1G_4^\circ$
B	4282.406	12	III	23344.81	00	$a^5P_3 - z^5S_2^\circ$
U	4280.53	(1)		23355.04	+07	$b^3H_6 - v^3G_5^\circ$
J	4279.864	(1)		23358.68	+03	$b^3P_0 - w^5P_1^\circ$
V	4279.480	(1)		23360.77	+02	$z^3D_3^\circ - f^3F_4$
J	4278.234	(1)		23367.58	−02	$z^5F_4^\circ - f^5D_3$
U	4277.68	(1)		23370.60	−02	$a^3H_5 - z^5H_5^\circ$
J	4276.684	(1)		23376.05	−07	$z^3F_4^\circ - f^3F_4$
U	4275.72	(1)		23381.32	−09	$b^3F_4 - w^5F_4^\circ$
W	4273.87	(1)		23391.44	+01	$c^3P_1 - v^5P_2^\circ$
B	4271.764	35	II	23402.97	−01	$a^3F_4 - z^3G_5^\circ$
J	4271.159	20	III	23406.28	−01	$z^7D_3^\circ - e^7D_4$
J	4268.744	2	IV	23419.53	+01	$a^3D_2 - w^3P_1^\circ$
B	4267.830	5	IV	23424.54	+01	$c^3P_0 - x^3P_1^\circ$
J	4266.968	3	IV	23429.27	−02	$a^3G_4 - y^3G_4^\circ$
J	*4265.260	(2)		23438.66	+06	$z^3D_2^\circ - f^3F_3$
					−11	$z^3D_1^\circ - e^3P_1$
U	4264.743	(1)		23441.50	+01	$z^3D_1^\circ - f^3F_2$
J	4264.209	(2)		23444.43	−04	$z^5F_4^\circ - e^7P_3$
H	4260.479	35	III	23464.96	00	$z^7D_5^\circ - e^7D_5$
V	4260.135	(1)		23466.85	+01	$c^3P_1 - v^5F_2^\circ$
U	4260.003	(2)		23467.58	−04	$z^5F_5^\circ - e^7F_6$
					−27	$(a^5P_3 - x^5F_2^\circ)$
J	4258.956	(1)		23473.34	−04	$b^3G_3 - x^3G_4^\circ$
J	4258.619	(1)		23475.20	−05	$b^3P_2 - w^5P_2^\circ$
J	4258.320	2	IA	23476.85	−02	$a^5D_2 - z^7P_3^\circ$
U	4256.79	(1)		23485.29	+10	$y^5F_3^\circ - i^5D_3$
V	4256.212	(3)		23488.48	−02	$z^5F_2^\circ - f^7D_1$
U	4255.496	(1)		23492.43	+03	$b^3G_3 - w^5G_2^\circ$
V	*4254.938	(1)		23495.51	+05	$b^3G_3 - x^3G_3^\circ$
					+11	$c^3P_2 - w^5G_2^\circ$
B	4250.790	25	II	23518.44	00	$a^3F_3 - z^3G_3^\circ$

TABLE B—(*Continued*)

Ref	λ I A	Int	T C	Wave Number Observed	o−c	Desig
J	4250.125	25	III	23522.12	00	$z^7D_2^\circ - e^7D_3$
J	4248.228	4	IV	23532.62	−03	$c^3P_1 - x^3P_2^\circ$
I	4247.432	12	III	23537.03	−05	$z^5F_4^\circ - e^5G_5$
SS	4246.572	(1)		23541.80	+12	$z^5F_1^\circ - e^7F_1$
J	4246.090	3	V	23544.47	00	$b^3D_3 - v^3P_2^\circ$
M	4245.358	tr?	IV?	23548.53	−09	$z^5F_5^\circ - f^5D_4$
I	4245.258	6	III	23549.09	00	$b^3P_0 - z^3S_1^\circ$
V	4243.786	(1w)		23557.25	+17	$z^3D_3^\circ - e^3P_2$
U	4243.370	(2)		23559.57	+02	$b^3D_2 - v^3P_2^\circ$
J	4242.730	(2)		23563.12	−02	$a^3D_2 - w^3P_2^\circ$
U	4242.592	(1)		23563.89	+03	$a^3G_4 - y^3G_3^\circ$
V	4241.112	(1)		23572.11	+03	$b^3P_2 - w^5P_1^\circ$
J	4240.372	(2)		23576.22	+01	$a^1D_2 - t^3D_1^\circ$?
J	*4239.847	2	III	23579.14	{+05	$a^3G_5 - y^3G_5^\circ$
					−06	$a^5F_3 - z^3F_4^\circ$
U	4239.735	3	IV	23579.76	+02	$b^3G_5 - w^5G_6^\circ$
I	4238.816	10n	IV	23584.88	−04	$z^5F_3^\circ - e^5G_4$
J	*4238.027	4	IV	23589.27	{−02	$z^5F_2^\circ - e^5S_2$
					00	$z^5F_2^\circ - e^7F_3$
M	4237.085	(2)	IIIA	23594.52	−05	$a^5F_3 - z^3D_3^\circ$
U	4236.76	(1)		23596.32	+04	$b^3D_1 - v^3P_2^\circ$
I	4235.942	25	III	23600.88	+01	$z^7D_4^\circ - e^7D_4$
I	4233.608	18	III	23613.89	00	$z^7D_1^\circ - e^7D_2$
U	4232.732	1	IA	23618.78	−02	$a^5D_1 - z^7P_2^\circ$
V	4231.525	(1)		23625.51	−07	$a^3D_3 - v^3G_3^\circ$
U	4230.584	(1)		23630.77	−06	$c^3P_2 - v^5P_3^\circ$
J	4229.760	(1)	III	23635.37	−02	$a^3F_4 - z^5G_3^\circ$
J	*4229.516	(1gn)		23636.73	{+15	$b^3G_5 - w^5G_5^\circ$
					−03	$a^3D_1 - w^3P_1^\circ$
SS	4228.722	(1)		23641.17	−05	$z^5F_4^\circ - f^7D_4$
J	4227.434	30	III	23648.37	{−01	$z^5F_5^\circ - e^5G_6$
					−07	$(z^5F_2^\circ - e^7F_1)$
J	4226.426	3	IV	23654.01	−01	$b^3P_1 - z^3S_1^\circ$
J	4225.956	3	IV	23656.65	+01	$a^1G_4 - w^3G_5^\circ$
J	4225.460	6n	IV	23659.42	−04	$z^5F_2^\circ - e^5G_3$
U	4224.517	3n	IV	23664.70	−03	$z^5F_1^\circ - e^7F_2$
J	4224.176	6n	IV	23666.61	−04	$z^5F_4^\circ - e^7F_5$
J	4222.219	12	III	23677.58	−01	$z^7D_3^\circ - e^7D_3$
J	4220.347	4	IV	23688.09	+03	$c^3P_1 - x^3P_0^\circ$
B	4219.364	12	IV	23693.60	−01	$a^1H_5 - y^3I_6^\circ$
J	4217.551	7n	IV	23703.79	−03	$z^5F_1^\circ - e^5G_2$
B	4216.186	8	I	23711.46	−01	$a^5D_4 - z^7P_4^\circ$
U	4215.970	(1)		23712.68	−01	$a^3G_5 - y^3G_4^\circ$
J	*4215.430	2	IV	23715.72	{+20	$a^3G_3 - x^5G_2^\circ$
					−05	$b^3G_4 - x^3G_5^\circ$
B	4213.650	5	IV	23725.73	+02	$b^3P_1 - y^3P_0^\circ$
J	4210.352	15	III	23744.32	−01	$z^7D_1^\circ - e^7D_1$
J	*4208.610	3n	V	23754.15	{−02	$z^5F_3^\circ - e^7F_3$
					−04	$z^5F_3^\circ - e^5S_2$
J	4207.130	4	IV	23762.50	−02	$b^3P_2 - z^3S_1^\circ$
J	4206.702	3	IA	23764.92	−02	$a^5D_3 - z^7P_3^\circ$
J	4205.546	(2)		23771.45	−04	$z^5F_2^\circ - e^7F_2$
B	4203.987	10	III	23780.27	−01	$b^3P_1 - y^3P_2^\circ$
V	4203.953	(1)		23780.46	−10	$a^1I_6 - z^1I_6^\circ$
V	4203.570	(1)		23782.63	+01	$a^5F_1 - z^3D_1^\circ$
U	4203.30	(1)		23784.15	−02	$b^3G_3 - v^5F_3^\circ$
V	*4202.755	(1)		23787.24	{+07	$c^3P_2 - v^5F_3^\circ$
					+01	$a^1G_4 - w^3G_4^\circ$
B	4202.031	30	I	23791.33	00	$a^3F_4 - z^3G_4^\circ$
W	4201.73	(1)		23793.04	−06	$a^1H_5 - w^3H_6^\circ$
J	4200.930	3n	V	23797.57	−05	$z^5F_3^\circ - e^7F_4$
W	4199.97	1	IA	23803.01	+09	$a^5D_2 - z^7P_2^\circ$
J	4199.098	20	III	23807.95	+01	$a^1G_4 - z^1H_5^\circ$
J	4198.645	4n	V	23810.52	−07	$z^5F_2^\circ - e^5G_2$
J	4198.310	20	III	23812.42	−01	$z^7D_5^\circ - e^7D_4$

Ref	λ I A	Int	T C	Wave Number Observed	o−c	Desig
V	4198.268	(1)		23812.66	−13	$z^5F_4^\circ - e^5G_4$
V	4196.533	(1)		23822.51	+06	$b^3G_5 - v^5F_5^\circ$
J	4196.218	4	IV	23824.29	−07	$z^5F_3^\circ - e^5G_3$
J	4195.615	(3)		23827.72	+03	$c^3P_2 - v^5P_2^\circ$
J	4195.337	5	IV	23829.30	−05	$z^5F_5^\circ - e^5G_5$
J	4191.685	(2)		23850.05	−05	$b^3P_0 - y^3P_1^\circ$
J	4191.436	15	III	23851.47	00	$z^7D_2^\circ - e^7D_1$
U	4189.564	(2)		23862.13	−04	$b^1G_4 - y^1F_3^\circ$
J	4187.802	20	III	23872.17	−01	$z^7D_4^\circ - e^7D_3$
U	4187.589	(1)		23873.38	−03	$z^5F_1^\circ - e^7G_2$
J	4187.044	20	III	23876.49	−01	$z^7D_3^\circ - e^7D_2$
B	4184.895	10	III	23888.75	−03	$b^3P_2 - y^3P_2^\circ$
U	4184.22	(1)		23892.61	+03	$a^3G_5 - x^5G_6^\circ$
V	4183.025	(1)		23899.43	−13	$z^5F_3^\circ - e^3D_3$
U	4182.770	(2b)		23900.89	−04	$z^5F_2^\circ - e^7G_3$
J	4182.384	4	IV	23903.09	−01	$c^3P_2 - v^5F_2^\circ$
J	4181.758	15	III	23906.67	00	$b^3P_2 - u^5D_3^\circ$
J	4177.597	4	IIA	23930.48	−02	$a^5F_4 - z^3F_4^\circ$
SS	4177.084	(1)		23933.42	−07	$z^5F_5^\circ - f^7D_4$
J	4176.571	7n	IV	23936.36	{−04	$z^5F_4^\circ - f^5F_5$
					−01	$(z^5F_3^\circ - e^7F_2)$
B	4175.640	10	III	23941.70	−02	$b^3P_1 - u^5D_2^\circ$
J	4174.917	5	IIA	23945.85	−01	$a^5F_4 - z^3D_3^\circ$
U	4174.419	(1)		23948.70	−10	$a^1H_5 - w^3H_4^\circ$
J	4173.926	2	IIA	23951.53	−03	$a^5F_2 - z^3D_1^\circ$
J	4173.322	2	IV	23955.00	−03	$b^3P_1 - y^3P_1^\circ$
J	4172.749	4	IIA	23958.29	−02	$a^5F_3 - z^3D_2^\circ$
V	4172.641	(1)		23958.91	−01	$z^5F_5^\circ - e^7F_5$
J	4172.126	5	IV	23961.86	+02	$a^3D_3 - w^3P_2^\circ$
V	4171.904	(2)		23963.14	−02	$a^3D_2 - z^1F_3^\circ$
J	4171.696	(2)		23964.34	+02	$b^1G_4 - x^1F_3^\circ$
B	4170.906	5	IV	23968.87	−03	$c^3P_2 - x^3P_2^\circ$
U	4169.766	(1)		23975.43	−05	$z^5F_3^\circ - e^5G_2$
U	4168.946	(1w)		23980.14	−03	$z^5F_2^\circ - e^7G_2$
V	4168.625	(1w)		23981.99	−04	$z^5F_4^\circ - e^7F_3$
V	4167.862	(2)		23986.38	00	$b^3H_4 - x^1G_4^\circ$
U	4164.80	(1)		24004.01	−11	$b^3G_4 - v^5F_3^\circ$
V	*4163.676	(1)		24010.49	{−07	$z^5F_2^\circ - e^7S_3$
					+08	$a^3G_5 - x^5G_5^\circ$
V	4161.488	(1)		24023.12	00	$b^3G_3 - w^3G_4^\circ$
V	4161.080	(1)		24025.47	−01	$z^5F_4^\circ - e^7F_4$
V	4160.561	(1)		24028.47	−06	$b^3G_5 - x^3G_4^\circ$
J	4158.798	5n	V	24038.66	−03	$z^5F_1^\circ - f^5F_2$
J	4157.788	8n	IV	24044.49	−06	$z^5F_2^\circ - f^5F_3$
B	4156.803	12	III	24050.19	−03	$b^3P_2 - u^5D_2^\circ$
V	4156.670	(1)		24050.96	−01	$b^3G_5 - x^3G_5^\circ$
V	4156.460	(1)		24052.18	−04	$z^5F_4^\circ - e^5G_3$
J	4154.812	9n	IV	24061.72	−04	$z^5F_4^\circ - e^7G_5$
J	4154.502	12	III	24063.51	−02	$b^3P_2 - y^3P_1^\circ$
J	4154.109	(1)		24065.79	−04	$z^5F_3^\circ - e^7G_3$
J	4153.906	10n	IV	24066.97	−03	$z^5F_3^\circ - f^5F_4$
J	4152.172	4	IIA	24077.02	−01	$a^5F_3 - z^3F_3^\circ$
V	4151.957	(1)		24078.26	−06	$a^1D_2 - t^3D_2^\circ$
J	4150.258	(4)		24088.12	−05	$z^5F_1^\circ - f^5F_1$
SS	4149.767	(1)		24090.97	−03	$a^5D_3 - z^7P_2^\circ$
J	4149.372	5n	V	24093.26	−05	$z^5F_5^\circ - e^7G_6$
B	4147.673	10	III	24103.13	−01	$a^3F_4 - z^3G_3^\circ$
U	4146.070	(2)		24112.45	−03	$b^3G_4 - w^3G_5^\circ$
U	4145.206	(1)		24117.48	+01	$a^3G_5 - x^5G_4^\circ$
B	4143.871	30	I	24125.24	−01	$a^3F_3 - y^3F_4^\circ$
J	4143.418	15	III	24127.88	00	$a^1G_4 - y^1G_4^\circ$
U	4142.625	(1N)		24132.50	−14	$y^5F_1^\circ - g^5G_2$
V	4141.862	(1)		24136.95	+01	$b^3G_3 - w^3G_3^\circ$
U	4141.352	(1)		24139.92	−02	$c^3P_2 - w^3G_3^\circ$

TABLE B—(*Continued*)

Ref	λ I A	Int	T C	Wave Number Observed	o−c	Desig
V	*4140.441	(1)		24145.23	−22 +16	$z^5F_2^\circ - f^5F_2$ $z^5F_3^\circ - e^7G_2$
J	4139.933	2	IIA	24148.20	−02	$a^5F_2 - z^3F_2^\circ$
U	4138.84	0		24154.57	+03	$a^3P_2 - x^5P_3^\circ$
J	4137.002	7	IV	24165.30	−03	$a^1P_1 - y^1D_2^\circ$
J	4136.512	(1)		24168.16	+06	$z^5F_4^\circ - e^7G_4$
U	4135.77	(1)		24172.50	−06	$y^5D_2^\circ - i^5D_2$
B	4134.681	12	IV	24178.87	−01	$b^3P_2 - w^3D_3^\circ$
V	*4134.433	(1)		24180.32	−05 −03	$z^5F_2^\circ - e^3D_2$ $c^3P_2 - x^3P_1^\circ$
U	4134.340	(1)		24180.86	−02	$a^5D_4 - z^7P_3^\circ$
J	4133.869	(2)		24183.62	−05	$z^5F_4^\circ - g^5D_4$
J	4132.903	8	III	24189.27	−01 +25	$b^3P_1 - w^3D_2^\circ$ $(a^3F_2 - y^5P_2^\circ)$
B	4132.060	25	II	24194.20	00	$a^3F_2 - y^3F_3^\circ$
U	*4130.035	(1)		24206.06	+02 +02	$a^3F_3 - y^5P_3^\circ$ $c^3P_0 - v^3D_1^\circ$
U	4129.22	(1)		24210.84	−14	$z^5F_2^\circ - g^5D_3$
J	*4127.807	3n	V	24219.13	−14 +01	$z^5D_1^\circ - f^5D_2$ $a^1P_1 - x^1D_2^\circ$
B	4127.612	7	IV	24220.27	−03	$b^3P_0 - w^3D_1^\circ$
U	4126.88	(1)		24224.57	−09	$b^3P_1 - u^5D_0^\circ$
J	4126.192	3n	IV	24228.61	−07	$z^5F_5^\circ - f^5F_5$
J	4125.884	(2)		24230.42	+03	$b^3P_1 - u^5D_1^\circ$
J	4125.622	(1)		24231.96	00	$y^5F_4^\circ - g^5G_5$
J	*4123.748	(1)		24242.97	+10 −10	$b^3F_2 - x^3D_2^\circ$ $b^3G_4 - w^3G_4^\circ$
J	4122.522	4	IV	24250.18	−03	$b^3P_1 - x^3F_2^\circ$
B	4121.806	5	IV	24254.39	−03	$b^3P_2 - x^3F_3^\circ$
J	4120.211	5	IV	24263.78	00	$b^3G_4 - z^1H_5^\circ$
V	4118.904	(1)		24271.48	−12	$z^5D_2^\circ - e^7P_3$
B	4118.549	15	IV	24273.57	−05	$a^1H_5 - z^1I_6^\circ$
V	*4117.872	(1)		24277.56	−12 −01	$z^5F_2^\circ - e^5P_3$ $y^5F_2^\circ - g^5G_2$
U	4117.71	(1)		24278.52	+01	$z^5P_2^\circ - f^3D_3$
U	4117.32	(1)		24280.82	−05	$c^3P_1 - 2_2^\circ$
U	4116.97	(1)		24282.88	−11	$z^5D_3^\circ - f^5D_4$
J	4114.957	(1w)		24294.76	−11	$z^5F_4^\circ - f^5F_4$
B	4114.449	5	IV	24297.76	−02	$b^3P_2 - w^3D_2^\circ$
J	4112.972	3n	V	24306.48	+02	$y^5F_5^\circ - g^5G_6$
V	4112.35	(1)		24310.16	−19	$z^5F_3^\circ - f^5F_2$
W	4111.1?	(1)		24317.6	−2	$z^5F_5^\circ - e^7F_4$
J	4109.808	9	IV?	24325.20	−04	$b^3P_1 - w^3D_1^\circ$
J	4109.070	(1)		24329.57	−09	$z^5D_0^\circ - f^5D_1$
B	4107.492	12	III	24338.91	+02	$b^3P_2 - u^5D_1^\circ$
V	4106.437	(1)		24345.17	−09	$z^5F_3^\circ - e^3D_2$
U	4106.265	(1)		24346.19	−03	$b^3F_3 - x^3D_3^\circ$
U	4104.97	(1)		24353.86	−17	$z^5F_5^\circ - e^7G_5$
K	*4104.132	3	V	24358.84	−12 +13	$z^5D_2^\circ - f^5D_2$ $b^3P_2 - x^3F_2^\circ$
U	4101.681	(1)		24373.39	+07	$a^3P_0 - w^5D_1^\circ$
J	4101.272	(1)		24375.82	−05	$z^5F_3^\circ - g^5D_3$
J	4100.745	3	IIA	24378.96	−03	$a^5F_5 - z^3F_4^\circ$
U	*4099.08	(1)		24388.86	−22 −13	$b^3H_4 - u^5F_5^\circ$? $a^3D_3 - x^1G_4^\circ$
J	4098.183	4n	IV	24394.20	−04	$z^5D_3^\circ - f^5D_3$
J	4097.099	(1)		24400.65	−11	$z^5D_1^\circ - f^5D_1$
U	4096.114	(1)		24406.52	+02	$b^3D_2 - x^1F_3^\circ$
I	4095.975	4	IV	24407.35	−02	$b^3F_3 - x^3D_2^\circ$
V	4092.512	(1)		24428.00	−32	$a^5F_4 - z^3F_3^\circ$
J	4091.561	(1)		24433.68	−06	$b^3P_2 - w^3D_1^\circ$
V	4090.984	(1w)		24437.12	00	$z^5F_4^\circ - f^5F_3$
U	4090.077	(1)		24442.54	−04	$z^5F_3^\circ - e^5P_3$
J	4089.225	(1)		24447.64	−04	$b^3G_5 - w^3G_5^\circ$
V	4088.567	(1)		24451.57	−02	$b^3D_2 - v^3P_1^\circ$
J	4087.099	(1)		24460.35	−03	$z^5F_5^\circ - e^7G_4$
W	4085.98	(1)		24467.05	+03	$y^5D_3^\circ - i^5D_2$
J	4085.312	4	IV	24471.05	−05	$z^5D_3^\circ - e^7P_3$
J	4085.011	4	IV	24472.85	−01	$b^3P_1 - 1_2^\circ$
J	4084.498	6	IV	24475.93	−01	$z^5F_5^\circ - g^5D_4$
J	4083.780	(1)	IV	24480.23	−11	$z^5F_2^\circ - e^3D_1$
V	4083.554	(1)		24481.59	−02	$a^3P_2 - x^5P_2^\circ$
U	4082.432	(2)		24488.31	−01	$b^3D_1 - v^3P_1^\circ$
J	4082.125	(1)		24490.15	−07	$z^5F_2^\circ - g^5D_2$
V	4080.886	(1w)		24497.59	−01	$z^5D_0^\circ - f^7D_1$
J	4080.226	2n	IV	24501.55	−08	$z^5D_1^\circ - f^5D_0$
J	4079.848	4	IV	24503.82	−02	$b^3P_0 - y^3S_1^\circ$
J	4078.365	4	IV	24512.73	−08	$b^3F_2 - x^3D_1^\circ$
SS	4076.884	(1)		24521.64	+03	$z^5D_2^\circ - e^7P_2$
J	4076.810	(1w)		24522.08	−06	$z^5D_2^\circ - f^7D_3$
J	4076.636	8n	IV	24523.13	−05	$z^5D_4^\circ - f^5D_4$
V	4076.498	(1)		24523.96	−04	$b^3F_2 - y^3G_3^\circ$
J	4076.232	(1)		24525.56	−04	$c^3P_1 - v^3D_1^\circ$
J	4074.794	5	IV	24534.22	−03	$a^1G_4 - w^3F_4^\circ$
K	4073.760	4n	IV	24540.44	00	$z^5D_2^\circ - f^5D_1$
V	4072.518	(2)		24547.93	−04	$z^5F_1^\circ - g^5D_1$
H	4071.740	40	II	24552.62	+01	$a^3F_2 - y^3F_2^\circ$
U	4071.52	(1)		24553.94	00	$b^3F_3 - y^3G_4^\circ$
J	4070.766	5n	III	24558.49	+02	$z^5D_3^\circ - f^5D_2$
U	4069.08	(1)		24568.67	−04	$z^5D_1^\circ - f^7D_1$
J	4067.984	8n	III	24575.29	−03	$z^5D_4^\circ - e^7P_4$
B	4067.275	4	III	24579.57	−02	$b^3F_4 - x^3D_3^\circ$
B	4066.979	6	III	24581.36	00	$b^3P_2 - 1_2^\circ$
U	4066.590	(1)		24583.71	−01	$b^3G_4 - y^1G_4^\circ$
U	4065.392	(2)		24590.95	−02	$z^5F_1^\circ - g^5D_0$
U	4064.45	(2)		24596.66	+02	$a^3F_3 - y^5P_2^\circ$
H	4063.597	45	II	24601.82	00	$a^3F_3 - y^3F_3^\circ$
J	4063.286	(3)		24603.70	−04	$z^5F_4^\circ - g^5D_3$
J	4062.446	10	III	24608.78	+01	$b^3P_1 - y^3S_1^\circ$
V	4059.726	3	V	24625.27	−06	$a^1D_2 - z^1P_1^\circ$
V	4058.766	3	IV	24631.10	−05	$a^3P_1 - w^5D_2^\circ$
K	**4058.227	4n	IV	24634.37	−06	$z^5D_4^\circ - f^5D_3$
U	4057.346	2	V	24639.72	−03	$a^3G_3 - x^3F_4^\circ$
U	4056.53	(1)		24644.67	+08	$z^7F_3^\circ - e^5F_3$
W	4055.98	(1)		24648.02	−02	$b^3D_2 - 11_3^\circ$
U	4055.039	3	V	24653.74	+02	$b^3F_4 - y^3G_5^\circ$
V	4054.883	3	V	24654.68	−05	$z^5F_2^\circ - g^5D_1$
V	4054.833	(1)		24654.99	−13	$z^5F_3^\circ - g^5D_2$
W	4054.18	(1)		24658.96	−02	$z^5D_2^\circ - f^7D_2$
W	4053.82	(1)		24661.15	+03	$c^3P_1 - w^3F_2^\circ$
SS	4052.724	(1)		24667.82	−04	$z^5D_3^\circ - f^7D_4$
V	4052.664	(1)		24668.18	−08	$a^1G_4 - w^3F_3^\circ$
V	4052.466	(1)		24669.39	−11	$z^5D_1^\circ - e^5S_2$
J	*4052.312	(1)		24670.33	−11 +07	$z^5F_4^\circ - e^5P_3$ $a^1I_6 - t^3G_5^\circ$
J	4051.923	(2)		24672.70	−05	$z^5F_3^\circ - e^5P_2$
U	4049.331	(1)		24688.49	−01	$b^3F_3 - y^3G_3^\circ$
V	*4047.315	(1)		24700.79	−04 +05	$a^3P_2 - x^5P_1^\circ$ $a^1I_6 - 12_5^\circ$
V	4046.629	(1)		24704.97	−07	$c^3P_1 - z^1D_2^\circ$
B	4045.815	60r	II	24709.94	00	$a^3F_4 - y^3F_4^\circ$
V	4045.139	(1)		24714.07	−06	$b^3G_3 - 2_2^\circ$
J	4044.614	6	IV	24717.28	+01	$b^3P_2 - y^3S_1^\circ$
V	*4043.901	5n	IV	24721.64	−04 −02	$a^3G_4 - u^5D_4^\circ$ $z^5D_3^\circ - f^7D_3$
S	4041.911	(−)		24733.81	−01	$b^3H_4 - t^5D_3^\circ$
V	*4041.288	(1)		24737.62	−04 −11	$b^3H_4 - v^3F_3^\circ$ $a^3D_2 - t^5D_3^\circ$
J	4040.650	4	V	24741.53	−04	$a^3D_2 - v^3F_3^\circ$
W	4039.94	(1)		24745.88	+02	$a^3G_4 - u^5D_3^\circ$

TABLE B—(*Continued*)

Ref	λ I A	Int	T C	Wave Number Observed	o−c	Desig
Q	*4038.622	(−)		24753.95	+11 −01	$b^3H_4 - u^5F_4^\circ$ $a^1P_1 - u^3D_2^\circ$
V	4037.725	(1)		24759.45	+15	$a^3P_2 - y^5G_3^\circ$?
SS	4033.190	(1)		24787.29	−02	$b^3F_4 - y^3G_4^\circ$
U	4032.630	4	III	24790.73	−01	$a^3F_4 - y^5P_3^\circ$
U	4032.469	(1)		24791.72	−10	$z^7F_1^\circ - e^5F_2$
U	4031.965	4	V	24794.82	−02	$a^3D_1 - v^3F_2^\circ$
V	4031.243	(2)		24799.26	−02	$c^3P_2 - v^3D_3^\circ$
J	4030.499	(6)	IV	24803.84	−07	$z^5D_4^\circ - e^5G_5$
V	4030.194	(3)		24805.72	−02	$a^5P_2 - x^5P_3^\circ$
V	*4029.640	3n	V	24809.13	−05 −03	$z^5D_2^\circ - e^5S_2$ $z^5D_2^\circ - e^7F_3$
J	4024.735	6n	V	24839.36	−07	$z^5D_3^\circ - e^5G_4$
J	4024.109	(1)		24843.22	−10	$a^3G_3 - x^3F_3^\circ$
U	*4022.744	(1)		24851.65	−05 00	$z^5D_1^\circ - e^7F_2$ $a^3D_3 - t^5D_4^\circ$
W	4022.45	(1)		24853.47	00	$a^3H_6 - w^5F_5^\circ$
I	4021.869	12	III	24857.06	−02	$a^3G_3 - z^3H_4^\circ$
V	*4021.622	(1')		24858.59	+10 −10	$z^5D_3^\circ - f^7D_2$ $a^3P_1 - w^5D_1^\circ$
V	4020.490	(1)		24865.59	−01	$b^3D_3 - t^3G_4^\circ$
U	4019.05	(1)		24874.50	−01	$b^3F_2 - x^5G_3^\circ$
J	4018.282	(4)		24879.25	−10	$z^5D_2^\circ - e^5G_3$
J	4017.156	6	III	24886.22	−01	$a^1G_4 - v^3G_5^\circ$
U	4017.093	(1)		24886.61	−07	$a^3G_3 - w^3D_2^\circ$
W	4016.54	(1)		24890.04	00	$a^3G_4 - x^3F_4^\circ$
U	4016.429	(2)		24890.73	−06	$z^5D_1^\circ - e^5G_2$
B	4014.534	10	III	24902.48	+02	$a^1H_5 - y^1H_5^\circ$
W	*4014.28	(1)		24904.05	+17 −09	$b^3G_3 - v^3D_2^\circ$ $b^3G_3 - w^3F_3^\circ$
J	4013.822	2	V	24906.89	+01	$c^3P_2 - v^3D_2^\circ$
V	4013.798	(1)		24907.04	−11	$c^3P_2 - w^3F_3^\circ$
J	4013.641	(2)		24908.02	−03	$z^5D_4^\circ - f^7D_4$
W	4012.16	(1)		24917.21	−03	$b^3H_6 - x^3H_6^\circ$
W	4011.71	(1)		24920.01	+04	$z^7D_3^\circ - e^5D_4$
U	4011.412	(1)		24921.86	−02	$b^3F_4 - y^3G_3^\circ$
W	*4010.77	(1)		24925.85	+13 −11	$z^7F_3^\circ - e^5F_2$ $b^3F_2 - x^5G_2^\circ$
W	4010.18	(1)		24929.51	+01	$b^3D_3 - 13_4^\circ$
I	4009.714	10	III.	24932.41	−01	$a^5P_1 - x^5P_2^\circ$
J	4007.277	6	IV	24947.57	−04	$a^3G_3 - x^3F_2^\circ$
V	4007.233	(1gn)		24947.85	+09	$a^3P_2 - z^5H_3^\circ$
V	4006.768	(1w)		24950.74	−19	$z^7F_0^\circ - e^5F_1$
J	4006.631	2	IV	24951.60	−03	$c^3P_0 - w^3P_1^\circ$
J	4006.314	3	IV	24953.57	+01	$b^3H_5 - v^3F_4^\circ$
B	4005.246	25	II	24960.23	00	$a^3F_3 - y^3F_2^\circ$
J	*4004.976	(1)		24961.91	+05 +06	$c^3P_2 - v^3D_1^\circ$ $z^5D_4^\circ - f^7D_3$
J	4004.832	(1)		24962.81	−01	$b^3H_6 - x^3H_5^\circ$
J	4003.764	2	V	24969.46	00	$a^1P_1 - u^3D_1^\circ$
V	*4002.665	(1)		24976.32	−16 +02	$z^7F_1^\circ - e^5F_1$ $a^3D_3 - v^3F_2^\circ$
J	4001.666	5	III	24982.55	00	$a^5P_3 - x^5P_3^\circ$
J	4000.466	2	V	24990.05	−04	$b^3G_4 - w^3F_4^\circ$
J	4000.266	(1)		24991.30	−08	$z^5D_2^\circ - e^7F_2$
W	4000.02	(1)		24992.83	−01	$b^3P_2 - w^5G_2^\circ$
I	**3998.054	10	III	25005.12	+04	$a^3G_5 - u^5D_4$
I	3997.394	15	III	25009.25	−01	$a^3G_4 - z^3H_5^\circ$
J	3996.968	2	V	25011.92	+06	$b^1G_4 - w^1G_4^\circ$
J	3995.996	4	IV	25018.00	−07	$a^3G_4 - w^3D_3^\circ$
J	3995.199	(1w)		25022.99	+12	$b^3H_5 - u^3G_5^\circ$
J	3994.117	2	V	25029.77	+01	$a^1G_4 - y^3H_5^\circ$
U	3992.395	(1)		25040.57	−01	$b^3H_4 - u^3G_4^\circ$
J	3990.379	2	V	25053.22	−01	$a^1G_4 - v^3G_4^\circ$
J	3989.859	(2d)	V	25056.48	−04	$a^1D_2 - y^1F_3^\circ$

Ref	λ I A	Int	T C	Wave Number Observed	o−c	Desig
J	3986.176	5	IV	25079.63	+01 +01	$a^3D_3 - v^3F_4^\circ$ ($z^5D_4^\circ - e^5G_4$)
J	3985.393	3	IV	25084.56	−04	$a^3D_2 - y^1D_2^\circ$
I	3983.960	10	III	25093.58	−03	$a^3G_4 - x^3F_3^\circ$
U	3983.35	(1)		25097.42	+04	$c^3P_2 - w^3F_2^\circ$
J	3981.775	7	III	25107.35	−03	$a^3G_4 - z^3H_4^\circ$
U	3981.104	(1)		25111.58	−06	$a^3P_1 - v^5D_2^\circ$
W	3980.65	(1)		25114.45	−10	$z^7D_4^\circ - e^5D_4$
U	3979.630	(1)		25120.88	+06	$z^5D_2^\circ - e^7G_3$
U	3978.464	(1)		25128.25	−02	$b^3P_2 - v^5P_3^\circ$
I	3977.743	12	III	25132,80	−01	$a^5P_2 - x^5P_2^\circ$
J	*3976.865	(1)		25138.35	+05 −04	$b^3G_3 - z^1D_2^\circ$ $a^3D_2 - x^1D_2^\circ$
J	3976.615	4	IV	25139.93	00	$a^1P_1 - t^3D_2^\circ$
U	3976.562	(1)		25140.26	−01	$a^3D_3 - v^3F_3^\circ$
U	3976.390	(1)		25141.35	+05	$c^3P_2 - z^1D_2^\circ$
U	3975.842	(1)		25144.82	+05	$z^3F_4^\circ - 2$
W	3975.21	(1)		25148.82	00	$z^7D_2^\circ - e^5D_3$
U	3974.764	(1)		25151.64	00	$a^5P_1 - x^5P_1^\circ$
J	3974.397	(1)		25153.96	−11	$z^5D_3^\circ - e^3D_3$
J	3973.655	3	V	25158.66	−01	$a^1D_2 - x^1F_3^\circ$
U	3972.918	(1)		25163.32	00	$a^1H_5 - t^3G_5^\circ$
U	3971.82	(1)		25170.28	+02	$a^3G_3 - 1_2^\circ$
I	3971.325	9	III	25173.42	−02	$a^3G_5 - x^3F_4^\circ$
J	3970.391	4	IV	25179.34	+04	$c^3P_1 - w^3P_0^\circ$
J	3969.628	(1)		25184.18	+03	$a^3D_3 - 4_4^\circ$
B	3969.261	30	II	25186.51	00	$a^3F_4 - y^3F_3^\circ$
J	3967.964	4n	IV	25194.74	−01	$z^5D_3^\circ - e^7G_4$
B	3967.423	8	IV	25198.17	+01	$b^3H_4 - u^3G_3^\circ$
J	3966.824	(1)		25201.98	−09	$a^3D_2 - u^3G_3^\circ$
J	*3966.630	10n	IV	25203.21	−03 −10	$z^5D_4^\circ - f^5F_5$ $a^3G_3 - z^1G_4^\circ$
V	*3966.532	(1n)		25203.83	−03 +07 −21	$a^3D_2 - u^5F_2^\circ$ $a^1D_2 - v^3P_2^\circ$ $z^5D_0^\circ - f^5F_1$
B	3966.066	10	III	25206.80	−01	$a^3F_2 - y^3D_3^\circ$
J	3965.511	(1)		25210.32	+01	$z^5D_3^\circ - g^5D_4$
U	3965.431	(1)		25210.83	−04	$a^3D_3 - 5^\circ$
J	3964.522	3	V	25216.61	−02	$b^3P_1 - v^5P_2^\circ$
J	3963.108	6n	V	25225.61	−05	$z^5D_1^\circ - f^5F_2$
V	3962.353	(2)		25230.42	−03	$z^5D_2^\circ - e^7S_3$
J	3961.147	(2)		25238.10	00	$b^3P_0 - v^5P_1^\circ$
J	3960.284	(1)		25243.60	+03	$b^3D_2 - t^3G_3^\circ$
W	3957.62	(1)		25260.59	+02	$z^5D_1^\circ - e^3D_2$
J	3957.027	4n	IV	25264.37	−07	$z^5D_2^\circ - f^5F_3$
B	3956.681	12	III	25266.58	−03	$a^3G_5 - z^3H_6^\circ$
J	3956.459	9	IV	25268.00	−01	$b^3H_6 - u^3G_5^\circ$
J	3955.956	2	V	25271.21	+02	$c^3P_1 - w^3P_1^\circ$
J	3955.352	(3)	IV	25275.07	−07	$z^5D_1^\circ - f^5F_1$
U	3954.715	(1)		25279.14	00	$b^3H_5 - 6_5^\circ$
U	3953.861	(1)		25284.60	−01	$b^3P_2 - v^5F_3^\circ$
J	3953.156	4	IV	25289.11	−01	$b^3G_3 - v^3G_4^\circ$
U	3952.702	(1)		25292.02	−02	$b^3P_1 - v^5F_2^\circ$
I	3952.606	8	IV	25292.63	−03	$a^3G_5 - z^3H_5^\circ$
I	3951.164	9	IV	25301.86	+02	$a^3D_1 - y^1D_2^\circ$
I	3949.954	10	III	25309.61	−01	$a^5P_3 - x^5P_2^\circ$
U	3949.156	(1)		25314.73	−11	$a^1P_1 - 8_1^\circ$
B	3948.779	10	IV	25317.14	+01	$b^3H_5 - u^3G_4^\circ$
J	3948.105	6n	IV	25321.47	−04	$z^5D_3^\circ - f^5F_4$
J	*3947.533	5	IV	25325.14	+01 −15	$b^3P_2 - v^5P_2^\circ$ $b^3G_5 - w^3F_4^\circ$
U	3947.391	(1)		25326.05	−07	$z^7D_5^\circ - e^5D_4$
J	3947.002	4n	IV	25328.54	−05	$z^5D_4^\circ - e^7G_5$
J	3945.119	4	IV	25340.63	+02	$a^3G_3 - w^5G_4^\circ$
J	3944.890	3	IV	25342.10	+03	$b^3G_4 - v^3G_5^\circ$

TABLE B—(*Continued*)

Ref	λ I A	Int	T C	Wave Number Observed	o−c	Desig
J	3944.748	(2)		25343.02	−01	$b^3P_1 - v^5P_1^\circ$
J	3943.339	2	IV	25352.07	+04	$a^5P_2 - x^5P_1^\circ$
B	3942.443	6	IV	25357.83	−02	$b^3P_1 - x^3P_2^\circ$
J	3941.283	(3)		25365.29	−05	$z^5D_2^\circ - f^5F_2$
B	**3940.882	5	II	25367.87	−02	$a^5F_3 - y^5D_4^\circ$
V	3940.044	(1)		25373.27	−06	$a^1P_1 - v^3P_2^\circ$
J	3937.329	3	IV	25390.77	00	$a^3G_5 - z^3H_4^\circ$
B	3935.815	8	III	25400.53	−01	$b^3P_2 - v^5F_2^\circ$
					+27	$(z^5D_2^\circ - e^3D_2)$
U	3935.306	(2)		25403.82	+06	$b^3P_1 - v^5F_1^\circ$
J	*3933.606	(2)	IV	25414.80	−01	$c^3P_1 - w^3P_2^\circ$
					−03	$z^5D_2^\circ - f^5F_1$
					+01	$a^3D_1 - u^5F_2^\circ$
J	*3932.629	4	IV	25421.11	+05	$a^5G_4 - w^5G_5^\circ$
					−26	$(z^7D_2^\circ - e^5D_2)$
J	3931.122	(3)		25430.86	−01	$z^5D_2^\circ - g^5D_3$
B	3930.299	25R	I	25436.18	−01	$a^5D_2 - z^5D_3^\circ$
J	3929.208	(1)		25443.24	+05	$a^3D_3 - u^3G_4^\circ$
J	3929.114	(1)		25443.85	+02	$a^3G_3 - w^5G_3^\circ$
J	3928.085	(1)		25450.52	+02	$z^5D_4^\circ - g^5D_4$
B	3927.922	30R	I	25451.57	−01	$a^5D_1 - z^5D_2^\circ$
					+04	$(b^3P_2 - v^5P_1^\circ)$
V	3926.001	(1)		25464.03	+08	$z^5D_3^\circ - f^5F_3$
J	3925.946	6	IV	25464.38	+02	$b^3P_0 - x^3P_1^\circ$
J	3925.646	4	IV	25466.33	−02	$b^3P_2 - x^3P_2^\circ$
V	3925.201	(1)		25469.22	+01	$z^5D_0^\circ - e^5P_1$
B	3922.914	25R	I	25484.06	−01	$a^5D_3 - z^5D_4^\circ$
U	3921.27	(1)		25494.75	−03	$b^3F_4 - z^3I_5^\circ$
J	3920.839	(1)		25497.55	−02	$z^5D_2^\circ - e^5P_3$
U	3920.645	(1)		25498.81	−07	$z^7D_4^\circ - e^5D_3$
B	3920.260	20r	I	25501.32	00	$a^5D_0 - z^5D_1^\circ$
J	3919.069	3	IV	25509.07	00	$b^3G_4 - v^3G_4^\circ$
J	3918.644	6	IV	25511.83	+02	$b^3G_3 - v^3G_3^\circ$
J	3918.418	4	IV	25513.30	+04	$b^3P_1 - x^3P_0^\circ$
J	3918.319	3		25513.95	−02	$a^3P_0 - x^3D_1^\circ$
B	3917.185	8	II	25521.33	−02	$a^5F_2 - y^5D_3^\circ$
I	3916.733	6	IV	25524.28	00	$b^3H_6 - 6_5^\circ$
W	3914.73	(1)		25537.34	+25	$a^3D_3 - x^1D_2^\circ$
V	3914.273	(1)		25540.32	+01	$z^5D_1^\circ - e^5P_1$
J	3913.635	4	III	25544.48	00	$a^3P_2 - w^5D_3^\circ$
U	3911.699	(1)		25557.13	+04	$a^3D_2 - t^3D_1^\circ$?
SS	3911.005	(1)		25561.66	−04	$z^5D_4^\circ - f^5F_4$
J	3910.845	(3)	IV	25562.71	−01	$a^3G_3 - x^3G_4^\circ$
J	3909.830	3	III	25569.34	+05	$b^3P_1 - x^3P_1^\circ$
J	3909.664	(1)	V	25570.43	00	$z^5D_1^\circ - g^5D_2$
B	3907.937	4	IV	25581.73	−01	$a^3G_3 - w^5G_2^\circ$
J	3907.464	(1)	III	25584.83	+03	$a^3G_3 - x^3G_3^\circ$
J	3906.748	2	V	25589.52	−04	$a^3D_2 - t^3D_3^\circ$
B	3906.482	8	I	25591.26	00	$a^5D_1 - z^5D_1^\circ$
J	3903.902	5	IV	25608.17	−03	$b^3G_4 - y^3H_4^\circ$
B	**3902.948	20	II	25614.43	00	$a^3F_3 - y^3D_3^\circ$
J	3900.519	2	V	25630.38	00	$z^5D_3^\circ - g^5D_3$
B	3899.709	30R	I	25635.70	00	$a^5D_2 - z^5D_2^\circ$
J	3899.037	2	IV	25640.12	−06	$a^3H_4 - y^3G_4^\circ$
K	3898.012	10	II	25646.86	−01	$a^5F_1 - y^5D_2^\circ$
J	3897.896	8	IV	25647.62	00	$a^3G_5 - w^5G_6^\circ$
J	3897.449	(2)	IV	25650.57	+02	$b^3G_5 - y^3H_6^\circ$
B	3895.658	25r	I	25662.36	00	$a^5D_1 - z^5D_0^\circ$
SS	3895.450	(1)		25663.73	−10	$z^5D_0^\circ - g^5D_1$
U	3894.49	(1)		25670.06	−10	$z^5D_4^\circ - e^7S_3$
J	3894.005	(2)	III	25673.25	+02	$a^3D_2 - u^3D_2^\circ$
J	3893.924	(2)	IV	25673.79	−03	$a^3H_5 - y^3G_5^\circ$
I	3893.391	7	IV	25677.30	+03	$b^3G_5 - v^3G_5^\circ$
V	3893.316	(1)		25677.80	+01	$b^3P_2 - x^3P_1^\circ$
U	3892.98	(1)		25680.01	+01	$z^5D_2^\circ - e^5P_1$

Ref	λ I A	Int	T C	Wave Number Observed	o−c	Desig
U	3892.894	(1)		25680.58	00	$a^3G_3 - v^5F_4^\circ$
J	3891.928	3	V	25686.95	+01	$a^1P_1 - z^1P_1^\circ$
J	3890.844	2	IV	25694.11	−01	$a^3G_4 - w^5G_3^\circ$
U	3890.39	(1)		25697.11	+02	$z^5D_3^\circ - e^5P_3$
SSb	3889.931	(1)		25700.14	−09	$z^5D_2^\circ - e^3D_1$
V	3888.825	3	IV	25707.45	00	$c^3P_2 - w^3P_1^\circ$
B	3888.517	20	II	25709.49	00	$a^3F_2 - y^3D_2^\circ$
SS	3888.424	(1)		25710.10	−01	$z^5D_2^\circ \quad g^5D_2$
B	3887.051	15	I	25719.18	−01	$a^5F_4 - y^5D_4^\circ$
B	3886.284	40R	I	25724.26	00	$a^5D_3 - z^5D_3^\circ$
I	3885.512	5	III	25729.37	−03	$a^3P_1 - x^3D_2^\circ$
V	3885.154	(1)		25731.74	−02	$b^3G_4 - v^3G_3^\circ$
W	3884.66	(1)		25735.01	+07	$z^5D_1^\circ - g^5D_1$
J	3884.359	3	IV	25737.00	00	$a^3G_5 - z^1G_4^\circ$
J	3883.282	(4)		25744.14	+03	$a^3D_3 - u^3D_3^\circ$
U	3878.740	(2)		25774.29	−04	$a^3D_1 - t^3D_1^\circ$?
U	3878.676	(8)		25774.71	−04	$a^3H_4 - y^3G_3^\circ$
B	3878.575	100r	II	25775.39	00	$a^5D_2 - z^5D_1^\circ$
B	3878.021	60	II	25779.07	00	$a^5F_3 - y^5D_3^\circ$
W	3876.67	(1)		25788.05	+03	$a^3P_2 - w^5F_3^\circ$
J	3876.043	4	III	25792.22	−02	$a^5F_1 - z^3P_2^\circ$
V	3874.053	(1)		25805.47	+05	$a^3P_2 - w^5D_2^\circ$
B	3873.763	8	IV	25807.40	−02	$a^3H_5 - y^3G_4^\circ$
V	3872.923	1	IV	25813.00	−01	$a^3G_4 - x^3G_4^\circ$
B	3872.504	60	II	25815.79	−01	$a^5F_2 - y^5D_2^\circ$
J	3871.750	4	IV	25820.82	+02	$b^3G_5 - y^3H_5^\circ$
U	3869.609	(1)	IV	25835.11	+02	$a^3G_4 - x^3G_3^\circ$
K	3869.562	(2)	IV	25835.42	−04	$a^3G_4 - x^3G_5^\circ$
V	3868.243	(1)		25844.23	−04	$b^3G_5 - v^3G_4^\circ$
V	3867.925	1	IV	25846.36	+03	$b^3F_3 - u^5D_4^\circ$
B	3867.219	7	IV	25851.07	00	$c^3P_2 - w^3P_2^\circ$
B	3865.526	30	II	25862.40	00	$a^5F_1 - y^5D_1^\circ$
V	3863.745	2	IV	25874.32	+02	$a^3G_5 - w^5G_4^\circ$
U	3861.60	(1)		25888.69	−04	$a^3D_2 - u^3D_1^\circ$
J	*3861.341	2	IV	25890.42	+09	$a^3G_5 - v^5F_5^\circ$
					−05	$a^3D_1 - u^3D_2^\circ$
B	3859.913	300R	I	25900.01	+01	$a^5D_4 - z^5D_4^\circ$
I	3859.214	10	III	25904.69	+03	$a^3H_6 - y^3G_5^\circ$
B	3856.373	50r	IA	25923.78	00	$a^5D_3 - z^5D_2^\circ$
J	3855.846	(1w)		25927.32	+06	$z^5D_3^\circ - e^5P_2$
V	3855.329	(1w)		25930.80	−07	$a^3G_4 - v^5F_4^\circ$
U	3854.375	(1)		25937.22	−06	$z^5D_4^\circ - e^5P_3$
V	3853.462	(1)		25943.36	−04	$b^3G_5 - y^3H_4^\circ$
I	3852.574	6	IV	25949.34	+01	$a^5P_3 - w^5D_4^\circ$
B	3850.820	12	II	25961.16	−01	$a^5F_2 - z^3P_2^\circ$
B	3849.969	40	II	25966.90	00	$a^5F_1 - y^5D_0^\circ$
V	3846.949	(1)		25987.28	−03	$a^3H_5 - x^5G_6^\circ$
B	3846.803	8	IV	25988.27	+01	$a^3D_3 - t^3D_3^\circ$
J	3846.412	2	IV	25990.91	+07	$a^1H_5 - w^1G_4^\circ$
V	3846.001	(1w)		25993.69	00	$z^5F_4^\circ - f^5P_3$
J	3845.692	(1)		25995.78	+04	$a^1D_2 - t^3G_3^\circ$
K	3845.170	(5)		25999.30	−04	$a^3P_1 - x^3D_1^\circ$
B	3843.259	8	IV	26012.23	+03	$a^1G_4 - z^1F_3^\circ$
U	3842.975	(1)		26014.16	+10	$b^3F_3 - u^5D_2^\circ$
B	3841.051	80r	II	26027.19	00	$a^3F_2 - y^3D_1^\circ$
B	3840.439	80r	II	26031.33	00	$a^5F_2 - y^5D_1^\circ$
U	3839.630	(2w)		26036.82	+01	$z^3D_1^\circ - i^5D_2$
B	3839.259	7	IV	26039.33	00	$a^1G_4 - x^1G_4^\circ$
J	3837.132	1	IV	26053.77	+01	$b^3F_2 - x^3F_3^\circ$
I	3836.332	4	IV	26059.20	00	$a^3D_2 - t^3D_2^\circ$
B	3834.225	100r	II	26073.52	00	$a^5F_3 - y^5D_2^\circ$
J	3833.311	5	IV	26079.74	+03	$b^3F_4 - u^5D_4^\circ$
J	3830.850	1	IV	26096.49	+08	$a^3G_5 - x^3G_4^\circ$
J	3830.757	1	IV	26097.12	00	$b^3F_2 - w^3D_2^\circ$
U	3829.771	(2)		26103.84	−05	$b^3F_4 - u^5D_3^\circ$

TABLE B—(*Continued*)

Ref	λ I A	Int	T C	Wave Number Observed	Wave Number o−c	Desig
V	*3829.458	1	IV	26105.98	+01	$a^3D_1 - u^3D_1^\circ$
					−09	$b^3P_1 - 2_2^\circ$
J	3829.125	(1)		26108.25	−01	$b^1G_4 - s^3G_5^\circ$
V	3828.510	(1n)		26112.44	−02	$a^3G_3 - w^3G_4^\circ$
B	3827.825	75r	II	26117.11	00	$a^3F_3 - y^3D_2^\circ$
J	3820.572	1	IV	26118.84	−01	$a^3G_5 - x^3G_5^\circ$
J	3826.836	1	IV	26123.86	+06	$a^3G_4 - v^5F_3^\circ$
B	3825.884	200R	II	26130.36	−01	$a^5F_4 - y^5D_3^\circ$
J	3825.404	(1)		26133.64	00	$a^3P_2 - y^5S_2^\circ$
B	3824.444	50r	IA	26140.20	+01	$a^5D_4 - z^5D_3^\circ$
V	3824.306	2	IV?	26141.14	00	$b^3H_4 - w^3H_4^\circ$
J	3824.074	1	IV	26142.73	+01	$b^3F_3 - w^3D_3^\circ$
J	3821.834	3	IV	26158.05	00	$b^3F_2 - x^3F_2^\circ$
I	3821.181	10	IV	26162.52	+02	$b^3H_5 - y^3I_6^\circ$
I	3820.428	250R	II	26167.68	00	$a^5F_5 - y^5D_4^\circ$
U	3817.650	3n	IV	26186.72	00	$z^5F_5^\circ - g^5F_5$
J	3816.340	4	IV	26195.71	+03	$a^5P_2 - w^5D_3^\circ$
B	3815.842	100r	II	26199.13	+01	$a^3F_4 - y^3D_3^\circ$
J	3814.526	5	IIIA	26208.17	+01	$a^5F_1 - z^3P_1^\circ$
V	3813.891	2	V	26212.53	−01	$a^1I_6 - x^1H_5^\circ$
J	3813.638	2	IV	26214.27	00	$a^3G_5 - v^5F_4^\circ$
J	3813.059	5?	IV	26218.25	−01	$b^3F_3 - x^3F_3^\circ$
					+10	$(a^3H_6 - x^5G_6^\circ)$
G	3812.964	40	II	26218.92	+03	$a^5F_3 - z^3P_2^\circ$
J	3811.892	2	IV	26226.27	−01	$a^3G_3 - w^3G_3^\circ$
U	*3811.05	(1)		26232.07	+05	$b^3F_3 - z^3H_4^\circ$
					−09	$a^3G_4 - w^3G_5^\circ$
J	3810.759	2	IV	26234.07	−04	$a^3D_2 - 8_1^\circ$
V	3809.043	(1)		26245.89	+02	$b^3P_0 - v^3D_1^\circ$
J	3808.731	4	IV	26248.04	−02	$b^3F_4 - x^3F_4^\circ$
J	3808.286	(1)		26251.11	+02	$c^3P_2 - z^1F_3^\circ$
G	3807.534	7	III	26256.31	+08	$a^5P_1 - w^5D_2^\circ$
G	3806.697	10	III	26262.08	+09	$b^3H_5 - w^3H_5^\circ$
					+46	$(b^3F_3 - w^3D_2^\circ)$
J	3806.203	2	IV	26265.47	+10	$a^1P_1 - v^3P_1^\circ$
B	3805.345	12	IV	26271.40	−02	$b^3H_4 - y^3I_5^\circ$
J	3804.013	(2)		26280.59	+02	$z^5F_5^\circ - h^5D_4$
J	3802.283	(1)		26292.55	−05	$a^3D_2 - v^3P_2^\circ$
J	3801.975	(2w)		26294.68	+03	$z^5F_5^\circ - f^5G_6$
J	3801.804	1	IV	26295.86	+04	$b^3P_1 - v^3D_2^\circ$
J	3801.681	3	IV	26296.71	−01	$b^3P_2 - v^3D_3^\circ$
B	3799.549	50	II	26311.47	00	$a^5F_3 - y^5F_4^\circ$
B	3798.513	40	II	26318.65	+01	$a^5F_4 - y^5F_5^\circ$
J	3797.948	(1)		26322.56	+01	$b^3F_3 - x^3F_2^\circ$
B	3797.517	12	III	26325.55	+03	$b^3H_6 - w^3H_6^\circ$
U	3796.90	(1)		26329.83	−12	$a^3D_2 - s^3D_3^\circ$
U	3796.00	(1)		26336.07	+09	$a^3H_6 - x^5G_5^\circ$
B	3795.004	60	II	26342.98	00	$a^5F_2 - y^5F_3^\circ$
J	3794.340	8	III	26347.59	−05	$a^3H_4 - z^3I_5^\circ$
J	3793.872	1	IV	26350.84	+04	$b^3P_1 - v^3D_1^\circ$
J	3793.478	(1)		26353.58	+02	$z^7P_3^\circ - f^5D_3$
U	3793.360	1	IV	26354.40	+01	$z^7P_2^\circ - e^7P_2$
U	3792.833	1	IV	26358.06	−01	$a^5P_1 - w^5F_2^\circ$
J	3792.156	2	IV	26362.77	+02	$a^3G_4 - w^3G_4^\circ$
U	3791.73	(1)		26365.73	+07	$z^5F_2^\circ - f^5P_1$
J	3791.504	(1)		26367.30	+02	$b^3F_4 - z^3H_5^\circ$
J	*3790.756	1	IVA	26372.50	+01	$a^5P_3 - w^5D_3^\circ$
					−08	$a^3P_0 - w^5P_1^\circ$
J	3790.656	(1)		26373.20	−02	$z^7P_2^\circ - f^5D_1$
B	3790.095	12	II	26377.10	00	$a^5F_2 - z^3P_1^\circ$
					+100	$(b^3F_4 - w^3D_3^\circ)$
J	3789.570	(1)		26380.75	+05	$b^3F_2 - 1_2^\circ$
J	3789.178	3	IV	26383.48	+02	$a^3G_4 - z^1H_5^\circ$
B	3787.883	50	II	26392.50	−01	$a^5F_1 - y^5F_2^\circ$
J	3787.164	(1)		26397.51	00	$b^3D_2 - w^1D_2^\circ$

Ref	λ I A	Int	T C	Wave Number Observed	Wave Number o−c	Desig
J	3786.678	8	III	26400.90	−02	$a^5F_1 - z^3P_0^\circ$
J	**3786.176	4	IV	26404.40	+08	$b^3P_2 - v^3D_2^\circ$
G	3785.950	6	IV	26405.98	+04	$a^3H_5 - z^3I_6^\circ$
J	3785.706	(1)		26407.68	+04	$b^3H_6 - y^3I_6^\circ$
J	3782.608	(1)		26429.31	+04	$c^3P_1 - v^3F_2^\circ$
J	3782.450	1	IV	26430.41	−02	$z^7P_3^\circ - e^7P_3$
J	3781.938	(1)		26433.99	+05	$b^3D_2 - w^1F_3^\circ$
J	3781.188	2	IV	26439.23	+01	$a^5P_2 - w^5F_3^\circ$
J	3779.486	2	IV	26451.14	+13	$a^5P_1 - w^5F_1^\circ$
V	3779.444	(2n)	IV	26451.43	+08	$a^3D_1 - 8_1^\circ$
V	3779.424	(1)		26451.57	−06	$b^3F_4 - x^3F_3^\circ$
U	3779.213	(1)		26453.05	−06	$a^3G_3 - y^1G_4^\circ$
J	3778.697	(1)		26456.66	+04	$a^5P_2 - w^5D_2^\circ$
J	3778.509	4	IV	26457.98	+08	$a^3D_3 - t^3D_2^\circ$
U	3778.320	(1)		26459.30	00	$b^3P_2 - v^3D_1^\circ$
J	3777.448	2	IV	26465.41	+01	$b^3F_4 - z^3H_4^\circ$
J	3777.061	(1)		26468.12	+08	$b^3G_4 - z^1F_3^\circ$
G	3776.454	6	IV	26472.38	+04	$a^5P_3 - w^5F_4^\circ$
J	3775.860	(1)		26476.54	−03	$a^3G_4 - w^3G_3^\circ$
G	3774.823	5	IV	26483.81	+04	$a^5P_1 - w^5D_1^\circ$
J	3773.699	1	IV	26491.70	−06	$z^7P_2^\circ - f^7D_2$
V	3773.364	(1)		26494.05	−03	$a^1G_4 - x^3H_5^\circ$
V	3770.405	3	IV	26514.84	−04	$a^3H_5 - z^3I_5^\circ$
V	3770.305		IV	26515.55	−01	$a^3G_5 - w^3G_5^\circ$
V	3769.995	4	IV	26517.73	−06	$z^7P_3^\circ - f^5D_2$
W	3768.23	(1)		26530.15	−09	$b^3P_1 - z^1D_2^\circ$
J	3768.030	3	IV	26531.56	+01	$a^5P_1 - w^5D_0^\circ$
B	3767.194	80r	II	26537.44	−01	$a^5F_1 - y^5F_1^\circ$
V	3766.665	1	IV	26541.17	00	$z^7P_2^\circ - f^7D_1$
V	3766.092	(1)		26545.21	+01	$b^3F_3 - 1_2^\circ$
W	3765.70	(1)		26547.97	00	$b^3H_5 - y^3I_5^\circ$
B	3765.542	20	IV	26549.09	00	$b^3H_6 - y^3I_7^\circ$
B	3763.790	100r	II	26561.44	00	$a^5F_2 - y^5F_2^\circ$
V	3762.205	(1)		26572.63	−02	$z^5F_4^\circ - e^3G_5$
J	3761.416	1	IV	26578.21	−04	$b^3F_3 - z^1G_4^\circ$
G	3760.534	6	III	26584.44	−01	$a^5P_1 - y^5S_2^\circ$
B	3760.052	8	III	26587.85	+01	$a^3H_6 - z^3I_7^\circ$
V	3759.155	(1)		26594.20	+02	$a^1I_6 - s^3G_5^\circ$
B	3758.235	150R	II	26600.70	00	$a^5F_3 - y^5F_3^\circ$
J	3757.459	1	IV	26606.20	−01	$a^3D_2 - z^1P_1^\circ$
J	3756.939	4	IV	26609.88	−03	$a^1H_5 - v^3H_5^\circ$
J	3756.069	1	IVA	26616.04	+01	$a^5P_3 - w^5F_3^\circ$
J	3754.506	1	IV	26627.12	−06	$z^7P_3^\circ - f^7D_4$
G	**3753.610	8	III	26633.48	+05	$a^5P_3 - w^5D_2^\circ$
J	3753.154	(1gn)		26636.71	−07	$a^3H_6 - z^3I_6^\circ$
J	*3752.420	(1w)		26641.92	−02	$z^7P_2^\circ - e^7F_3$
					−04	$z^7P_2^\circ - e^5S_2$
J	3751.820	(1)		26646.19	+04	$a^3G_5 - w^3G_4^\circ$
J	3751.059	(1)		26651.59	+05	$a^3D_2 - s^3D_2^\circ$
V	3750.677	(1)		26654.31	+04	$b^3F_2 - w^5G_3^\circ$
B	3749.487	200R	II	26662.76	−01	$a^5F_4 - y^5F_4^\circ$
J	3748.969	5	IV	26666.45	00	$z^7P_4^\circ - f^7D_5$
V	§3748.492	7	IV?	26669.84	+01	$a^1H_5 - v^3H_6^\circ$
B	3748.264	60R	IA	26671.46	−01	$a^5D_1 - z^5F_2^\circ$
J	3746.931	6	IV	26680.95	−02	$z^7P_3^\circ - f^7D_3$
J	3745.486	1	IV	26684.12	−04	$a^5P_2 - w^5D_1^\circ$
J	3745.901	40r	IA	26688.29	00	$a^5D_0 - z^5F_1^\circ$
J	3745.561	100R	I	26690.71	+01	$a^5D_2 - z^5F_3^\circ$
J	3744.105	4	IV	26701.09	−02	$z^7P_2^\circ - e^7F_1$
SS	3743.781	(0)		26703.40	00	$a^3G_4 - y^1G_4^\circ$
J	3743.468	6	IV?	26705.63	+03	$a^1H_5 - x^1H_5^\circ$
G	3743.364	20	IIA	26706.38	00	$a^5F_2 - y^5F_1^\circ$
V	3742.937	(0)		26709.42	+06	$z^5F_1^\circ - f^5G_2$
J	3742.621	4	IV	26711.68	−04	$z^7P_4^\circ - f^5D_4$
W	3742.07	(1)		26715.61	+06	$b^3F_3 - w^5G_4^\circ$

TABLE B—(*Continued*)

Ref	λ I A	Int	T C	Wave Number Observed	o−c	Desig
J	3740.247	3	IV	26728.63	−02	$a^3D_3 - s^3D_3^\circ$
V	*3740.061	(1)		26729.96	+04 00	$z^5F_3^\circ - g^7D_4$ $a^1G_4 - v^3F_4^\circ$
V	3739.317	1	IV	26735.28	+01	$a^5P_3 - w^5F_2^\circ$
J	3739.120	1	IV	26736.69	−03	$a^5P_1 - v^5D_2^\circ$
B	3738.308	10	IV	26742.49	−02	$b^3H_5 - z^1I_6^\circ$
B	3737.133	150R	I	26750.90	00	$a^5D_3 - z^5F_4^\circ$
J	3735.325	6	IV	26763.85	00	$z^7P_4^\circ - e^7P_4$
B	3734.867	300r	II	26767.13	−01	$a^5F_5 - y^5F_5^\circ$
B	3733.319	40r	IA	26778.23	00	$a^5D_1 - z^5F_1^\circ$
B	3732.399	10	III	26784.83	−01	$a^5P_2 - y^5S_2^\circ$
J	3731.374	2	IV	26792.19	+01	$b^3F_2 - w^5G_2^\circ$
J	3730.945	2	IV	26795.27	+03	$b^3F_2 - x^3G_3^\circ$
G	3730.386	3	IV	26799.28	+01	$a^1G_4 - u^3G_5^\circ$
J	3728.668	1	IV	26811.63	00	$b^3F_4 - z^1G_4^\circ$
J	3727.809	3	IV	26817.81	00	$z^7P_3^\circ - f^7D_2$
B	3727.621	50r	II	26819.16	00	$a^5F_3 - y^5F_2^\circ$
J	3727.096	4	IV	26822.94	−03	$z^7P_4^\circ - f^5D_3$
J	3726.927	6	IV	26824.16	00 −26	$z^7P_2^\circ - e^7F_2$ $(a^5P_2 - v^5D_3^\circ)$
J	3725.498	(1)		26834.45	−04	$a^1G_4 - 4_4^\circ$
B	3724.380	8	III	26842.50	−02	$a^3P_2 - x^3D_3^\circ$
B	3722.564	50r	IA	26855.59	−01	$a^5D_2 - z^5F_2^\circ$
J	3722.028	(1)		26859.46	−02	$a^3G_3 - w^3F_4^\circ$
V	3721.606	(1)		26862.51	−02	$b^3G_3 - v^3F_2^\circ$
U	3721.510	2	IV	26863.20	−06	$z^7P_2^\circ - e^5G_2$
V	3721.396	1	IV	26864.02	−01	$a^3P_0 - y^3P_1^\circ$
V	*3721.278	2	IV	26864.87	−06 −05	$z^5F_5^\circ - e^3G_5$ $a^5P_3 - v^5D_4^\circ$
V	3721.189	(1)		26865.50	−01	$c^3P_2 - v^3F_2^\circ$
B	3719.935	250R	I	26874.57	+01	$a^5D_4 - z^5F_5^\circ$
J	3718.407	3	IV	26885.62	00	$a^3G_3 - v^3D_3^\circ$
G	*3716.442	12	IV	26899.83	−01 +10	$z^7P_4^\circ - e^7P_3$ $z^5F_4^\circ - e^3G_4$
G	3715.911	4	IV	26903.68	+01	$a^3P_2 - x^3D_2^\circ$
J	3711.411	2	IV	26936.29	+02	$c^3P_1 - y^1D_2^\circ$
J	3711.225	3	IV	26937.65	−01	$b^3F_3 - x^3G_4^\circ$
U	3709.665	(1)		26948.97	+04	$b^3F_4 - w^5G_4^\circ$
J	3709.535	(1)		26949.92	00	$b^3G_4 - x^3H_5^\circ$
G	3709.246	75r	II	26952.02	+02	$a^5F_4 - y^5F_3^\circ$
V	*3708.602	(1)		26956.70	−05 +02	$a^3H_4 - u^5D_3^\circ$ $b^3F_3 - w^5G_2^\circ$
G	3707.918	8	III	26961.67	+02	$a^5P_3 - y^5S_2^\circ$
U	3707.824	20	I	26962.35	−01	$a^5D_2 - z^5F_1^\circ$
I	*3707.048	8	IV	26968.00	+01 −02	$z^7P_3^\circ - e^7F_3$ $z^7P_3^\circ - e^5S_2$
B	3705.567	100r	I	26978.77	00	$a^5D_3 - z^5F_3^\circ$
B	3704.463	10	IV	26986.81	+01	$a^3G_5 - y^1G_4^\circ$
V	3704.336	(1)		26987.74	+09	$b^3H_6 - z^1I_6^\circ$
U	3704.021	(1)		26990.04	−02	$c^3P_1 - x^1D_2^\circ$
J	3703.824	3	IV	26991.47	+01	$b^3P_0 - w^3P_1^\circ$
J	3703.697	3	IV	26992.40	−04	$z^7P_4^\circ - e^5G_5$
J	***3703.556	5	IV?	26993.42	−07 +20	$a^3G_3 - w^3F_3^\circ$ $a^3G_3 - v^3D_2^\circ$
J	*3702.500	1	IVA	27001.12	−19 −11	$a^3F_2 - x^5D_3^\circ$ $a^5P_3 - v^5D_3^\circ$
J	3702.033	3	IV	27004.53	+03	$b^3P_1 - w^3P_0^\circ$
G	3701.086	20	IV	27011.44	00	$z^7P_3^\circ - e^7F_4$
J	3699.147	1	IV	27025.60	−06	$c^3P_2 - t^5D_3^\circ$
J	3698.611	2	IV	27029.51	+01	$c^3P_2 - v^3F_3^\circ$
U	3697.536	(1w)		27037.37	−03	$a^3D_2 - y^1F_3^\circ$
G	3697.426	6n	IV	27038.18	00	$z^7P_3^\circ - e^5G_3$
U	3696.03	(1)		27048.39	00	$a^3P_1 - z^3S_1^\circ$
V	*3695.507	(1)		27052.21	+07 +09	$b^3F_4 - w^5G_3^\circ$ $z^5F_2^\circ - g^7D_2$
B	*3695.054	8	IV	27055.53	+01 −01	$b^3F_3 - v^5F_4^\circ$ $a^1G_4 - 6_5^\circ$
G	3694.005	20	IV	27063.21	−02	$z^7P_2 - e^7S_3$
J	3693.008	1	IV	27070.52	+14	$b^3G_3 - 4_4^\circ$
J	3690.730	4	IV	27087.23	−01	$a^1H_5 - s^3G_5^\circ$
V	*3690.450	(1)		27089.29	+09 −04	$c^3P_0 - t^3D_1^\circ$? $z^5D_1^\circ - f^5P_2$
S	3690.095	(−)		27091.89	−22	$b^3F_3 - v^5P_3^\circ$
V	3689.897	(1w)		27093.34	−19	$a^1G_4 - u^3G_4^\circ$
G	*3689.457	12	IV	27096.58	−01 +19	$z^7P_4^\circ - f^7D_4$ $b^3P_1 - w^3P_1^\circ$
V	3688.877	(1)		27100.84	−09	$a^3H_4 - x^3F_4^\circ$
V	3688.476	(1w)		27103.78	−06	$a^3D_3 - 9_4^\circ$
J	3687.656	4	III	27109.81	+04	$a^3G_4 - w^3F_4^\circ$
B	3687.458	40r	I	27111.27	+01	$a^5F_5 - y^5F_4^\circ$
J	3687.100	2	IV	27113.90	−02	$a^5P_3 - v^5D_2^\circ$
J	3686.260	2	IV	27120.07	−01	$a^3P_1 - y^3P_0^\circ$
G	3685.998	15n	IV	27122.01	−01	$z^7P_4^\circ - e^7F_5$
G	3684.108	15	IV	27135.91	00	$a^3G_4 - v^3D_3^\circ$
V	*3683.616	(1)		27139.54	+15 −01	$a^3P_0 - u^5D_1^\circ$ $a^3D_2 - x^1F_3^\circ$
G	3683.054	10	IA	27143.68	+02	$a^5D_3 - z^5F_2^\circ$
J	3682.226	20	IV	27149.78	+10	$a^1D_2 - w^1D_2^\circ$
U	3681.88	(1)		27152.34	+03	$b^1G_4 - v^1G_4^\circ$
U	3681.651	(1)		27154.03	−03	$z^7P_3^\circ - e^7G_4$
U	3680.675	2n	IV	27161.23	−05	$z^5D_4^\circ - g^5F_5$
B	3679.915	40r	IA	27166.83	−01	$a^5D_4 - z^5F_4^\circ$
W	*3679.53	(1)		27169.68	+05 −02	$z^7P_3^\circ - g^5D_4$ $c^3P_1 - t^5D_0^\circ$
W	3679.33	(1)		27171.15	+11	$b^3F_4 - x^3G_4^\circ$
W	3678.98	(1)		27173.74	+12	$a^3P_2 - x^3D_1^\circ$
J	3678.863	3	IV	27174.60	−04	$a^3P_1 - y^3P_2^\circ$
B	3677.630	12	IV	27183.71	−01	$a^3G_3 - w^3F_2^\circ$
V	3677.477	(2)		27184.85	+04	$a^3P_2 - y^3G_3^\circ$
J	3677.309	2	IV	27186.09	−02	$a^1D_2 - w^1F_3^\circ$
V	3676.879	(1)		27189.27	−04	$z^7P_3^\circ - e^5G_2$
B	3676.314	6	IV	27193.44	−04	$b^3F_4 - x^3G_5^\circ$
J	3674.766	2	IV	27204.90	+01	$b^3P_2 - w^3P_1^\circ$
U	3672.722	1	IV	27220.04	−11	$a^3H_4 - z^3H_5^\circ$
W	3671.51	(1)		27229.02	+01	$z^5D_2^\circ - f^5P_2$
V	3670.810	1	IV	27234.22	−02	$a^3P_0 - w^3D_1^\circ$
J	3670.071	3	IV	27239.70	+16	$b^3G_5 - x^3H_6^\circ$
U	3670.028	3	IV	27240.02	+01	$b^3P_1 - w^3P_2^\circ$
B	3669.523	10	IV	27243.77	−01	$a^3G_4 - w^3F_3^\circ$
J	3669.151	3	IV	27246.53	+08	$b^3G_4 - v^3F_3^\circ$
U	3668.893	(1)		27248.45	00	$b^3F_3 - v^3F_3^\circ$
W	3668.6	(1)		27250.6	−3	$b^3F_2 - v^5P_1^\circ$?
U	3668.214	(1)		27253.49	−03	$z^5D_3^\circ - g^5F_4$
V	*3667.999	1	IV	27255.09	−04 −02	$z^5D_4^\circ - h^5D_4$ $b^3G_4 - u^3G_5^\circ$
G	3667.252	3n	IV	27260.64	+11	$z^5D_4^\circ - f^5P_3$
U	3666.944	(1)		27262.93	+14	$a^3F_2 - x^5D_2^\circ$
W	*3666.24	1	IV	27268.17	00 +01	$a^3H_5 - x^3F_4^\circ$ $z^7P_4^\circ - e^5G_4$
U	3664.694	(1)		27279.67	+02	$z^7P_3^\circ - e^7G_3$
G	3664.537	2	IV	27280.83	00	$z^7P_3^\circ - f^5F_4$
W	3663.95	(1)		27285.21	+09	$b^3G_5 - x^3H_5^\circ$
V	*3663.458	1	IV	27288.87	−05 −03	$b^3F_3 - v^5P_2^\circ$ $b^3F_4 - v^5F_4^\circ$
W	3663.25	(1)		27290.42	+09	$b^3G_4 - 4_4^\circ$
W	3661.36	(1)		27304.50	00	$a^3H_4 - x^3F_3^\circ$
W	3660.33	(1)		27312.19	−05	$z^7F_3^\circ - f^5D_4$
G	3659.516	8	IV	27318.27	+01	$a^3H_4 - z^3H_4^\circ$
U	3658.55	(1)		27325.48	−01	$b^3F_4 - v^5P_3^\circ$
W	3657.89	1	IV	27330.41	+06	$z^7P_2^\circ - e^5P_3$

TABLE B—(*Continued*)

Ref	λ I A	Int	T C	Wave Number Observed	Wave Number o−c	Desig
U	3657.139	1	IV	27336.02	−07	$a^3P_1 - u^5D_2^\circ$
J	3655.465	4	IV	27348.54	+03	$b^3P_2 - w^3P_2^\circ$
W	3654.66	(1)		27354.56	+08	$a^5P_1 - x^3D_2^\circ$
V	3653.763	1	IV	27361.28	−07	$a^3H_5 - z^3H_6^\circ$
B	3651.469	20	IV	27378.47	+01	$a^3G_3 - v^3G_4^\circ$
W	*3651.10	(1)		27381.23	00 +14	$z^7F_4^\circ - f^7D_5$ $a^3D_2 - 11_3^\circ$
J	3650.280	5	IV	27387.38	−01	$a^3H_5 - z^3H_5^\circ$
J	3650.031	4	IV	27389.25	−03	$z^7P_3^\circ - e^7S_3$
B	3649.508	12	IV	27393.18	+01	$a^3G_5 - w^3F_4^\circ$
J	3649.304	5	IA	27394.71	+01	$a^5D_4 - z^5F_3^\circ$
B	3647.844	100R	I	27405.67	00	$a^5F_4 - z^5G_5^\circ$
J	3647.427	3	IV	27408.81	−12	$a^3F_3 - x^5D_3^\circ$
J	3645.822	6	IV	27420.87	+03	$c^3P_0 - u^3D_1^\circ$
V	*3645.494	1	IV	27423.34	−15 +02 +07	$z^7F_3^\circ - f^5D_3$ $b^3G_3 - x^1D_2^\circ$ $z^7P_3^\circ - f^5F_3$
V	*3645.090	2	IV	27426.38	−12 +06	$z^7F_4^\circ - f^5D_4$ $c^3P_2 - x^1D_2^\circ$
U	3644.798	(1)		27428.58	+05	$z^5D_3^\circ - f^5P_2$
SS	*3643.812	(1)		27435.99	+02 −11	$a^3F_2 - x^5D_1^\circ$ $a^3D_3 - y^1F_3^\circ$
V	3643.716	1	IV	27436.72	00	$b^3F_2 - w^3G_3^\circ$
J	3643.627	2	IV	27437.39	−01	$z^7P_4^\circ - e^7F_3$
G	3640.388	15	IV	27461.80	+05	$a^3G_4 - v^3G_5^\circ$
G	3638.296	12	IV	27477.59	−01	$a^3G_3 - y^3H_4^\circ$
J	3637.862	3n	IV	27480.87	+02	$z^7P_4^\circ - e^7F_4$
W	3637.73	(1)		27481.87	+04	$b^3F_4 - v^5F_3^\circ$
V	3637.251	1	IV	27485.49	−02	$a^3H_5 - z^3H_4^\circ$
J	3636.995	2	IV	27487.42	+02	$b^3F_3 - w^3G_4^\circ$
V	3636.650	1	IV	27490.03	+03	$c^3P_2 - u^3G_3^\circ$
U	3636.234	(1)		27493.17	−07	$a^1D_2 - s^3G_3^\circ$
V	*3636.186	2	IV	27493.54	−18 −17	$a^5P_2 - x^3D_3^\circ$ $z^5D_4^\circ - g^5F_4$
W	3635.19	2	IVA	27501.07	00	$c^3P_2 - t^5D_1^\circ$
G	3634.326	6n	IV	27507.61	+02	$z^7P_4^\circ - e^5G_3$
U	3633.833	1	IV	27511.34	−04	$b^3G_4 - 6_5^\circ$
SS	3633.077	(1)		27517.06	−07	$z^7P_4^\circ - e^7G_5$
J	3632.979	3	IV	27517.80	+03	$a^3P_0 - y^3S_1^\circ$
J	3632.558	3	IV	27521.00	00	$b^3G_5 - v^3F_4^\circ$
J	3632.042	10	IV	27524.90	+10	$c^3P_1 - u^3D_2^\circ$
B	3631.464	125R	I	27529.28	+01	$a^5F_3 - z^5G_4^\circ$
J	3631.103	7	IV	27532.02	−02	$z^7F_5^\circ - f^7D_5$
J	3630.353	4	IV	27537.71	−04	$z^7F_4^\circ - f^5D_3$
V	3628.094	1	IV	27554.85	−02	$a^5P_2 - x^3D_2^\circ$
U	3627.06	(1)		27562.71	00	$a^1H_5 - u^3H_5^\circ$
G	3625.140	6	IV	27577.31	+01	$z^7F_5^\circ - f^5D_4$
U	3624.31	(1)		27583.62	−03	$a^3P_1 - w^3D_2^\circ$
U	3623.772	2	IV	27587.72	00	$z^7F_3^\circ - f^5D_2$
G	*3623.440	1	IV	27590.24	+05 −07	$b^3F_4 - w^3G_5^\circ$ $b^3G_5 - u^3G_5^\circ$
G	3623.187	8	IV	27592.17	−02	$a^3H_6 - z^3H_6^\circ$
G	3622.001	12	IV	27601.21	+06	$a^3G_3 - v^3G_3^\circ$
V	3621.718	(2)		27603.36	+01	$a^1H_5 - u^3H_4^\circ$
B	3621.463	15	IV	27605.30	+02	$a^3G_4 - y^3H_5^\circ$
U	3620.228	(1)		27614.72	+10	$z^7F_4^\circ - e^7P_3$
U	3619.772	(1)		27618.20	−03	$a^3H_6 - z^3H_5^\circ$
B	3618.769	125R	I	27625.86	+01	$a^5F_2 - z^5G_3^\circ$
J	*3618.392	2	IV	27628.74	−01 −09	$a^3G_4 - v^3G_4^\circ$ $z^5D_3^\circ - f^5G_4$
B	3617.788	12	IV	27633.35	+01	$c^3P_2 - u^3D_3^\circ$
W	3617.09	(1)		27638.68	+08	$a^1G_4 - t^3D_3^\circ$
U	3616.326	(1)		27644.52	−06	$a^3P_1 - x^3F_2^\circ$
SS	3616.157	(1)		27645.81	−07	$z^5D_4^\circ - h^5D_3$
U	3615.665	(1)		27649.57	00	$a^3F_4 - x^5D_4^\circ$

Ref	λ I A	Int	T C	Wave Number Observed	Wave Number o−c	Desig
W	3615.19	(1)		27653.21	+12	$z^5D_1^\circ - h^5D_1$
W	3613.15	(1)		27668.82	+01	$z^7F_2^\circ - e^7P_2$
J	*3612.940	1	IVA	27670.43	+02 −10	$a^3F_3 - x^5D_2^\circ$ $a^5P_3 - x^3D_3^\circ$
G	3612.068	8	IV	27677.11	+05	$z^7F_5^\circ - e^5G_6$
J	3610.703	2	IV	27687.57	−07	$z^7F_2^\circ - f^5D_1$
G	3610.159	20	III	27691.75	00	$z^7F_6^\circ - e^7F_6^\circ$
B	3608.861	100r	I	27701.70	00	$a^5F_1 - z^5G_2$
J	*3608.146	3	IV	27707.19	−04 +24	$z^7F_4^\circ - e^5G_5$ $b^3G_4 - u^3G_3^\circ$
G	3606.679	20	III	27718.46	+03	$a^3G_5 - y^3H_6^\circ$
G	**3605.450	15	IV	27727.91	+02 +42	$a^3G_4 - y^3H_4^\circ$ $(z^7F_6^\circ - f^7D_6)$
U	3604.383	(1)		27736.12	−05	$z^7F_1^\circ - f^5D_0$
J	3603.828	1	IV	27740.39	−01	$c^3P_1 - u^3D_1^\circ$
U	3603.572	(1)		27742.36	+01	$a^3H_6 - w^5G_6^\circ$
G	3603.205	10	IV	27745.18	+03	$a^3G_5 - v^3G_5^\circ$
G	*3602.534	3	IV	27750.36	−01 +12	$z^7F_3^\circ - e^7P_2$ $z^7P_4^\circ - f^5F_4$
U	3602.46	2	IV	27750.92	+02	$z^7F_3^\circ - f^7D_3$
U	3602.08	1	IVA	27753.85	+01	$z^7F_1^\circ - f^7D_2$
G	3599.624	3	IV	27772.79	+01	$a^1H_5 - u^3F_4^\circ$
U	3598.98	(1)		27777.75	+04	$z^7F_0^\circ - f^7D_1$
W	3598.93?	(1)		27778.14	+01	$z^5D_1^\circ - g^5F_2$
U	3598.721	1	IVA	27779.75	−04	$a^3D_3 - 11_3^\circ$
W	3597.05	3n	IV	27792.66	−11	$z^5D_2^\circ - h^5D_1$
U	3596.198	1	IV	27799.24	+05	$a^3H_5 - w^5G_5^\circ$
U	3595.857	(1)		27801.88	+09	$a^3H_4 - w^5G_4^\circ$
U	3595.66	(1)		27803.40	+15	$z^7F_1^\circ - f^7D_1$
U	3595.308	2	IV	27806.12	−06	$z^7F_2^\circ - f^7D_2$
G	3594.632	8	IV	27811.35	−02	$z^7F_4^\circ - f^7D_4$
U	3593.329	(1)		27821.44	−03	$z^5D_2^\circ - f^5G_3$
U	3592.881	(1)		27824.91	+09	$a^5P_2 - x^3D_1^\circ$
W	3592.68	(1)		27826.46	−07	$z^5D_3^\circ - h^5D_2$
U	3592.486	(1)		27827.96	−09	$b^3F_3 - y^1G_4^\circ$
U	3591.485	(1)		27835.72	+06	$z^5D_0^\circ - g^5F_1$
U	3591.345	(1)		27836.80	00	$z^7F_4^\circ - e^7F_5$
W	3590.99	(1)		27839.56	+07	$z^5D_4^\circ - e^5G_5$
W	3590.66	(1)		27842.12	+02	$b^1G_4 - t^3F_3^\circ$
U	3590.086	(1)		27846.57	−01	$b^3G_5 - 6_5^\circ$
G	3589.456	3	IV	27851.46	+02	$a^3G_4 - v^3G_3^\circ$
B	3589.107	8	III	27854.16	−01	$a^5F_5 - z^5G_5^\circ$
J	3588.918	2	IV	27855.63	+04	$z^7F_2^\circ - f^7D_1$
G	3588.615	3	IV	27857.98	−05	$z^7F_5^\circ - e^5G_5$
J	3587.424	2	IV	27867.23	00	$a^3P_1 - 1_2^\circ$
U	3587.240	2	IV	27868.66	−02	$z^7F_3^\circ - e^5G_4$
G	3586.985	30	II	27870.64	+01	$a^5F_2 - z^5G_2^\circ$
SS	3586.751	(2)		27872.46	−05	$z^7F_6^\circ - e^5G_6$
B	3586.114	10	IV	27877.41	+06 −08	$b^3H_6 - t^3G_5^\circ$ $(c^3P_2 - t^3D_3^\circ)$
J	3585.708	20	II	27880.57	00	$a^5F_4 - z^5G_4^\circ$
B	3585.320	30	II	27883.58	00	$a^5F_3 - z^5G_3^\circ$
V	3585.193	(2)		27884.57	00	$b^3G_5 - u^3G_4^\circ$
J	*3584.960	4	IV	27886.38	+01 −19	$b^3H_5 - t^3G_4^\circ$ $z^7P_3^\circ - e^5P_2$
J	3584.790	1	IV	27887.71	−03	$z^7F_3^\circ - f^7D_2$
B	3584.663	8	IV	27888.70	+02	$a^3G_5 - y^3H_5^\circ$
J	3583.337	2	IV	27899.01	+18	$z^5D_0^\circ - f^3D_1$?
W	3582.69	(2)		27904.05	+01	$z^7F_1^\circ - e^5S_2$
W	3582.56	(1)		27905.06	+05	$a^3H_4 - w^5G_3^\circ$
J	3582.201	5	IV	27907.86	+03	$b^3H_6 - 12_5^\circ$
U	3581.816	(1)		27910.86	−01	$c^3P_1 - t^3D_2^\circ$
J	3581.645	1	IV	27912.19	+04	$a^3G_5 - v^3G_4^\circ$
B	3581.195	250R	I	27915.70	00	$a^5F_5 - z^5G_6^\circ$
U	3578.380	(1)		27937.66	+01	$z^7F_0^\circ - e^7F_1$

TABLE B—(*Continued*)

Ref	λ I A	Int	T C	Wave Number Observed	o−c	Desig	Ref	λ I A	Int	T C	Wave Number Observed	o−c	Desig
B	3576.760	2	IV	27950.31	+04	$b^3H_5 - 13_4^\circ$	G	3530.385	2	IV	28317.46	+01	$z^7F_6^\circ - e^7G_6$
G	*3575.976	2	IV	27956.45	+09	$z^7F_2^\circ - e^7F_3$	G	3529.818	6	III	28322.01	00	$z^7F_1^\circ - e^7G_1$
					+07	$z^7F_2^\circ - e^5S_2$	U	3529.531	(1)		28324.31	−06	$a^1G_4 - y^3I_5^\circ$
J	3575.374	4	III	27961.15	−01	$c^3P_2 - u^3D_2^\circ$	G	3527.792	4	IV	28338.27	+02	$z^7F_4^\circ - e^7G_4$
J	3575.249	2	IV	27962.13	−04	$z^7F_5^\circ - f^7D_4$	J	3526.673	5	IV	28347.26	00	$z^7F_2^\circ - e^7G_2$
U	3575.118	(1)		27963.15	−04	$z^7F_1^\circ - e^7F_1$	J	3526.465	4	IV	28348.93	+01	$a^3P_2 - y^3P_2^\circ$
G	3573.896	4	IV	27972.71	00	$b^3H_4 - t^3G_3^\circ$	J	3526.377	4	IV	28349.64	+06	$z^7F_3^\circ - e^7G_3$
U	3573.836	3	IV	27973.18	−01	$a^3H_6 - w^5G_6^\circ$	W	3526.23	(3)		28350.82	+06	$z^7F_3^\circ - f^5F_4$
U	3573.400	2	IV	27976.60	−02	$a^3D_2 - t^3G_3^\circ$	J	3526.167	15	II	28351.33	+01	$a^5F_3 - z^3G_3^\circ$
U	3572.60	(1)		27982.86	−08	$z^7F_4^\circ - e^5G_4$	J	3526.039	20	I	28352.36	+02	$a^5D_2 - z^5P_3^\circ$
G	3571.995	6	IV	27987.60	00	$z^7F_5^\circ - e^7F_5$	V	3526.016	1	IV	28352.54	−10	$b^3F_3 - 3_3^\circ$
V	3571.228	2	IVA	27993.61	−01	$a^3F_4 - x^5D_3^\circ$	U	3525.856	(1)		28353.83	+01	$z^7F_4^\circ - g^5D_4$
V	3570.243	20	III	28001.33	00	$z^7F_6^\circ - e^7G_7$	G	3524.236	4	IV	28366.87	+06	$a^3P_2 - u^5D_3^\circ$
G	3570.100	100R	I	28002.46	00	$a^5F_4 - z^3G_5^\circ$	J	3524.075	3	IV	28368.16	00	$b^3F_3 - v^3D_2^\circ$
W	3569.99	(1)		28003.32	+18	$a^3P_1 - y^3S_1^\circ$	W	3523.30	(1)		28374.40	+05	$z^7F_2^\circ - e^7G_1$
J	3568.977	4	IV	28011.26	−03	$a^3G_5 - y^3H_4^\circ$	U	3522.896	(1)		28377.65	00	$z^7F_2^\circ - e^7S_3$
V	3568.828	(2)		28012.43	00	$a^3D_3 - t^3G_4^\circ$	G	3522.268	(3)		28382.71	−01	$z^7F_5^\circ - e^7G_5$
U	3568.423	(1)		28015.61	+08	$z^7F_2^\circ - e^7F_1$	J	3521.833	2	IVA	28386.22	+02	$a^5P_1 - w^5P_2^\circ$
W	3567.36	(1)		28023.96	+06	$a^3H_4 - x^3G_4^\circ$	B	3521.264	25	II	28390.81	−01	$a^5F_4 - z^3G_4^\circ$
U	3567.038	2	IV	28026.49	−06	$z^7F_2^\circ - e^5G_3$	U	3520.855	(1)		28394.10	−06	$b^3F_2 - w^3F_2^\circ$
W	3566.59	(1)		28030.01	−02	$a^3H_6 - w^5G_5^\circ$	W	3518.86	(2)		28410.20	+08	$a^5P_2 - w^5P_3^\circ$
J	*3565.583	3	IV	28037.93	+01	$z^7F_3^\circ - e^7F_3$	W	3518.68	(1)		28411.65	+01	$z^7F_2^\circ - f^5F_3$
					−01	$z^7F_3^\circ - e^5S_2$	W	3516.55	(1)		28428.86	+04	$z^7F_3^\circ - e^7G_2$
B	3565.381	60r	II	28039.52	00	$a^5F_3 - z^3G_4^\circ$	G	3516.403	5	IV	28430.05	+07	$b^3G_3 - w^3H_4^\circ$
W	3564.11	(1)		28049.51	+02	$a^3F_2 - x^5F_2^\circ$	U	3514.626	(1)		28444.42	00	$a^3H_6 - x^3G_5^\circ$
V	3560.705	5	IV	28076.34	+01	$a^3D_3 - 13_4^\circ$	B	3513.820	30	II	28450.95	−01	$a^5F_5 - z^3G_5^\circ$
J	3559.506	2	IV	28085.79	+01	$c^3P_1 - 8_1^\circ$	U	3513.065	(1)		28457.06	−05	$a^3F_3 - x^5F_2^\circ$
B	3558.518	30	II	28093.59	−01	$a^5F_2 - z^3G_3^\circ$	U	3512.97	(1)		28457.83	−05	$c^3P_1 - z^1P_1^\circ$
G	3556.877	7	IV	28106.55	00	$z^7F_4^\circ - f^5F_5$	U	3512.239	(1)		28463.76	−08	$z^7F_4^\circ - e^7G_3$
W	3556.68	(1)		28108.11	00	$z^7F_3^\circ - e^5G_3$	W	3512.08	(1)		28465.04	+02	$z^7F_4^\circ - f^5F_4$
G	3554.922	40	III	28122.01	+02	$z^7F_5^\circ - e^7G_6$	U	3511.748	(1)		28467.74	−06	$b^3F_4 - w^3F_4^\circ$
W	3554.50	2	IV	28125.35	+01	$z^7F_1^\circ - e^5G_2$	U	3510.446	(2)		28478.29	00	$a^3P_0 - x^3P_1^\circ$
J	3554.122	4	IIIA	28128.34	−01	$a^5F_3 - z^5G_2^\circ$	U	3509.870	(1)		28482.97	−06	$a^5P_1 - w^5P_1^\circ$
G	3553.741	6	IV	28131.36	+07	$a^1H_5 - v^1G_4^\circ$	W	3509.12	(1)		28489.05	−01	$z^7F_5^\circ - e^7G_4$
G	3552.828	3	IV	28138.58	00	$z^7F_2^\circ - e^7F_2$	W	3508.52	(1)		28493.93	−01	$b^3F_4 - v^3D_3^\circ$
U	3552.42	(1)		28141.81	+05	$a^3H_4 - v^5F_4^\circ$	J	3508.494	5	IV	28494.14	−09	$b^3G_4 - w^3H_5^\circ$
V	3552.112	1	IV	28144.25	−02	$c^3P_1 - v^3P_2^\circ$	W	§3507.39	(1)		28503.11	−10	$c^3P_1 - s^3D_2^\circ$
J	3549.868	4	III	28162.04	−01	$a^3F_2 - x^5F_1^\circ$	G	3506.498	6	IV	28510.36	−01	$a^3P_2 - u^5D_2^\circ$
U	3548.037	(2)		28176.58	−08	$c^3P_2 - u^3D_1^\circ$	W	3506.23	(1)		28512.54	00	$z^7F_2^\circ - f^5F_2$
J	*3547.203	(2)		28183.20	+14	$z^7F_6^\circ - e^7F_5$	V	3505.065	2	IV	28522.01	−03	$c^3P_2 - 8_1^\circ$
					+02	$b^3H_4 - w^1G_4^\circ$	U	3504.859	2	IV	28523.69	+01	$a^3P_2 - y^3P_1^\circ$
U	3546.21	(1)		28191.09	−05	$a^3H_5 - x^3G_4^\circ$	U	3504.455	(1)		28526.97	+03	$b^3P_2 - v^3F_3^\circ$
U	3545.832	(1)		28194.10	+01	$a^1G_4 - w^3H_4^\circ$	G	3500.564	2	IV	28558.69	+03	$b^3F_3 - w^3F_2^\circ$
G	3545.639	5	IV	28195.63	00	$z^7F_4^\circ - e^7F_4$	A	3497.843	40	I	28580.90	00	$a^5D_1 - z^5P_2^\circ$
U	3544.631	(2)		28203.65	−01	$b^3F_2 - v^3D_2^\circ$	V	3497.137	(1)		28586.67	+08	$a^5P_2 - w^5P_2^\circ$
J	3543.669	(4)		28211.31	+02	$a^1P_1 - w^1D_2^\circ$	J	3497.110	10	III	28586.89	−04	$a^5P_3 - w^5P_3^\circ$
U	3543.392	(1)		28213.51	−08	$a^3H_5 - x^3G_5^\circ$	U	3496.19	(1)		28594.41	+06	$a^3H_4 - z^1H_5^\circ$
V	3542.243	1	IV	28222.66	−01	$a^3P_2 - z^3S_1^\circ$	G	3495.285	8	IV	28601.81	00	$b^3F_4 - w^3F_3^\circ$
G	3542.076	15	IV	28224.00	+01	$z^7F_3^\circ - e^7G_4$	U	3494.170	(1)		28610.94	−06	$a^3P_1 - v^5P_2^\circ$
G	3541.083	15	IV	28231.91	00	$z^7F_4^\circ - e^7G_5$	U	3493.698	(2)		28614.81	−04	$a^3G_4 - x^1G_4^\circ$
J	3540.709	3	IIIA	28234.89	+02	$a^5F_4 - z^5G_3^\circ$	U	3493.290	(1)		28618.15	−03	$a^3F_4 - x^5F_4^\circ$
G	3540.121	3	IV	28239.58	+02	$z^7F_3^\circ - g^5D_4$	A	3490.575	100r	I	28640.41	00	$a^5D_3 - z^5P_3^\circ$
U	3538.79	(1)		28250.20	−07	$a^1H_5 - x^3I_6^\circ$	J	3489.670	4	IV	28647.84	+02	$b^3G_5 - w^3H_6^\circ$
W	3538.55	(1)		28252.12	+09	$a^3P_0 - v^5P_1^\circ$	U	3486.556	(1)		28673.42	−05	$a^5P_1 - z^3S_1^\circ$
W	3538.31	(1)		28254.04	00	$a^1D_2 - u^3F_2^\circ$	A	3485.342	7	IV	28683.41	−01	$a^5P_2 - w^5P_1^\circ$
J	3537.896	4	IV	28257.34	−02	$z^7F_5^\circ - f^5F_5$	U	3484.972	(1)		28686.45	+04	$a^3P_1 - v^5F_2^\circ$
J	3537.729	3	IV	28258.67	+03	$b^3F_2 - v^3D_1^\circ$	U	3484.858	(1)		28687.39	−07	$a^3H_4 - w^3G_3^\circ$
J	§3537.491	1	IV	28260.58	+02	$b^3F_3 - v^3D_3^\circ$	G	3483.006	3	IIIA	28702.64	+02	$a^5F_4 - z^3G_3^\circ$
G	3536.556	15	IV	28268.05	+03	$z^7F_2^\circ - e^7G_3$	V	3481.558	(1)		28714.58	+02	$a^3P_2 - x^3F_3^\circ$
U	*3534.914	(1)		28281.18	+07	$a^3F_4 - x^5F_5^\circ$	V	*3479.683	(1)		28730.06	+12	$b^3G_5 - y^3I_6^\circ$
					00	$a^3F_3 - x^5F_3^\circ$						−05	$a^1H_5 - t^3F_4^\circ$
U	3534.53	(1)		28284.25	−03	$a^1H_5 - x^3I_5^\circ$	V	3478.788	(1)		28737.45	+05	$a^3P_1 - v^5P_1^\circ$
G	3533.201	10	IV	28294.89	−03	$z^7F_1^\circ - e^7G_2$	U	3478.374	(1gn)		28740.87	−01	$a^3H_5 - w^3G_4^\circ$
J	3533.008	5	IV	28296.43	−04	$z^7F_0^\circ - e^7G_1$	U	3477.856	(2)		28745.15	−01	$a^5P_1 - y^3P_0^\circ$
U	3531.446	(1)		28308.95	−05	$a^3H_5 - v^5F_4^\circ$	V	3477.007	(1)		28752.17	−04	$a^3P_1 - x^3P_2^\circ$

TABLE B—(*Continued*)

Ref	λ I A	Int	T C	Wave Number Observed	o−c	Desig
J	3476.853	(2)		28753.44	+04	$b^3F_3 - v^3G_4^\circ$
A	3476.704	40	I	28754.68	00	$a^5D_0 - z^5P_1^\circ$
V	*3476.336	(2w)		28757.71	−22	$a^3P_2 - w^3D_2^\circ$
					+08	$z^5P_3^\circ - i^5D_3$
V	*3475.867	(1)		28761.60	+10	$a^3H_5 - z^1H_5^\circ$
					+13	$b^3P_1 - y^1D_2^\circ$
J	3475.651	6	IV	28763.38	−02	$a^5P_3 - w^5P_2^\circ$
G	3475.450	70r	I	28765.04	+01	$a^5D_2 - z^5P_2^\circ$
V	3473.497	(1)		28781.22	+02	$a^5F_2 - y^5P_3^\circ$
U	3471.350	6	IV	28799.02	−01	$a^3P_2 - u^5D_1^\circ$
V	3471.27	5	IV	28799.68	−05	$a^5P_1 - y^3P_2^\circ$
V	3469.834	2	IV	28811.60	+01	$b^3F_2 - v^3G_3^\circ$
V	3469.390	(1)		28815.29	+03	$b^3P_1 - x^1D_2^\circ$
V	3469.012	(2)		28818.43	+01	$b^3H_4 - v^3H_4^\circ$
V	3468.849	4	IV	28819.78	00	$b^3F_4 - v^3G_5^\circ$
V	3466.501	3	IIIA	28839.30	−01	$a^5F_5 - z^3G_4^\circ$
V	3466.279	(1)		28841.15	+02	$a^3H_6 - w^3G_5^\circ$
A	3465.863	60r	I	28844.61	−01	$a^5D_1 - z^5P_1^\circ$
V	3464.914	(1)		28852.51	−03	$b^3F_3 - y^3H_4^\circ$
V	3463.305	2	IV	28865.91	+04	$a^3F_4 - x^5F_3^\circ$
V	3462.808	(1)		28870.06	+09	$b^3P_2 - y^1D_2^\circ$
J	3462.353	2	IV	28873.85	−02	$a^5P_2 - z^3S_1^\circ$
G	3459.911	4	IV	28894.23	+09	$c^3P_2 - z^1P_1^\circ$
					+35	$(a^3P_2 - w^3D_1^\circ)$
V	3459.429	(2)		28898.26	+01	$a^3G_5 - x^1G_4^\circ$
J	3458.304	4	IV	28907.66	+03	$a^3P_1 - x^3P_0^\circ$
V	3457.512	(1)		28914.28	−01	$a^3H_4 - y^1G_4^\circ$
V	*3457.090	(3w)		28917.81	−01	$z^5P_3^\circ - i^5D_2$
					+01	$b^3P_2 - 7_2^\circ$
J	3453.022	(2)		28951.87	00	$a^3G_3 - v^3F_2^\circ$
G	3452.273	10	III	28958.15	+02	$a^5F_3 - y^3F_4^\circ$
G	3451.915	10	IV	28961.16	−01	$a^5P_1 - u^5D_2^\circ$
J	3451.628	2	IV	28963.57	−09	$a^3P_1 - x^3P_1^\circ$
					+26	$(b^3F_4 - y^3H_5^\circ)$
G	3450.328	10	IV	28974.48	00	$a^5P_1 - y^3P_1^\circ$
V	3448.869	(1)		28986.74	−04	$b^3F_4 - v^3G_4^\circ$
V	3448.786	(1)		28987.44	00	$b^3P_2 - u^3G_3^\circ$
U	3448.472	(1)		28990.07	00	$b^3G_3 - 9_4^\circ$
G	3447.278	8	IV	29000.12	00	$a^5P_2 - y^3P_2^\circ$
U	3446.947	(1)		29002.90	+03	$a^5F_1 - y^5P_2^\circ$
V	3446.791	(1)		29004.21	−02	$b^3F_2 - w^3P_1^\circ$
A	3445.151	20	III	29018.02	+01	$a^5P_2 - u^5D_3^\circ$
A	3443.878	50r	I	29028.74	−01	$a^5D_2 - z^5P_1^\circ$
V	*3442.979	(1)		29036.32	+01	$c^3P_1 - v^3P_1^\circ$
					−13	$a^1D_2 - t^3F_3^\circ$
J	3442.672	3	IIIA	29038.91	−02	$a^5F_3 - y^5P_3^\circ$
V	3442.364	5	IV	29041.51	+01	$a^3P_2 - 1_2^\circ$
J	3440.989	75R	I	29053.12	+02	$a^5D_3 - z^5P_2^\circ$
J	3440.610	150R	I	29056.32	−02	$a^5D_4 - z^5P_3^\circ$
U	3439.039	(1)		29069.59	−01	$a^5G_4 - x^3H_5^\circ$
U	3437.952	(2)		29078.78	−02	$b^3H_5 - v^3H_5^\circ$
V	3437.631	(1)		29081.49	−04	$a^3H_5 - y^1G_4^\circ$
G	3437.046	3	IV	29086.44	00	$a^1G_4 - y^1F_3^\circ$
V	3436.045	(1)		29094.92	−05	$b^3H_5 - v^3H_4^\circ$
V	3434.029	(1w)		29112.00	00	$a^3G_3 - t^5D_3^\circ$
V	3432.023	(1)		29129.01	−02	$b^3P_0 - t^3D_1^\circ$?
J	*3431.815	3	IV	29130.78	00	$b^3P_2 - u^3D_3^\circ$
					+22	$a^3D_2 - w^1D_2^\circ$
V	3428.746	(2)		29156.85	+02	$z^5P_3^\circ - 4_2$
G	3428.192	8	III	29161.56	00	$a^5P_2 - u^5D_2^\circ$
A	3427.121	20	III	29170.66	+02	$a^5P_3 - u^5D_4^\circ$
J	3427.002	2	IIIA	29171.69	−11	$a^5F_2 - y^5P_2^\circ$
J	3426.637	5	IV	29174.80	−08	$a^5P_2 - y^3P_1^\circ$
J	*3426.383	5d	IIIA	29176.96	+03	$a^5P_3 - y^3P_2^\circ$
					−02	$a^5F_2 - y^3F_3^\circ$
U	3426.337	(2)		29177.35	−06	$a^3P_2 - y^3S_1^\circ$
G	3425.009	4	IV	29188.67	+08	$a^1G_4 - x^1F_3^\circ$
G	3424.284	10	III	29194.84	+02	$a^5P_3 - u^5D_3^\circ$
G	3422.656	7	IV	29208.73	00	$a^5P_1 - w^3D_2^\circ$
J	3422.499	3	IV	29210.07	+05	$b^3G_4 - 9_4^\circ$
V	3419.706	(1)		29233.93	−03	$b^3P_1 - t^3D_1^\circ$?
U	3419.154	(1)		29238.65	00	$z^5D_3^\circ - f^3F_2$
G	3418.507	10	III	29244.18	+07	$a^5P_1 - u^5D_0^\circ$
U	3418.176	(2w)		29247.01	−14	$z^5D_1^\circ - e^3P_0$?
G	3417.842	12	III	29249.87	+03	$a^5P_1 - u^5D_1^\circ$
J	3417.273	(1gn)		29254.74	−11	$a^5F_1 - y^5P_1^\circ$
U	3416.679	(1)		29259.83	+03	$a^3P_0 - v^3D_1^\circ$
G	3415.530	4	IV	29269.67	+01	$a^5P_1 - x^3F_2^\circ$
A	3413.135	15	III	29290.21	−01	$a^5P_2 - w^3D_3^\circ$
G	3411.353	3	IV	29305.51	+03	$a^3G_4 - v^3F_4^\circ$
V	3411.134	(1)		29307.39	−03	$a^3G_5 - x^3H_6^\circ$
V	3410.905	(1)		29309.36	−07	$a^5F_4 - y^3F_4^\circ$
G	3410.171	3	IV	29315.66	+01	$a^1P_1 - u^3F_2^\circ$
U	3410.031	(1)		29316.87	+02	$a^1G_4 - 10_3^\circ$
U	3409.218	(2)		29323.86	−08	$b^3H_6 - v^3H_5^\circ$
A	3407.461	20d	III	29338.98	+08	$a^5P_3 - x^3F_4^\circ$
J	3406.803	6	IV	29344.64	−04	$a^5P_1 - w^3D_1^\circ$
J	3406.442	3	IV	29347.75	−05	$a^3D_1 - w^1D_2^\circ$
W	3405.83	(2)		29353.02	+02	$a^3G_5 - x^3H_5^\circ$
U	3404.923	(1)		29360.85	−06	$a^3G_5 - t^5D_4^\circ$
V	3404.755	(1)		29362.29	00	$a^3G_4 - t^5D_3^\circ$
G	3404.357	6	IV	29365.73	−03	$a^5P_2 - x^3F_3^\circ$
V	*3404.301	3	IIIA	29366.21	+08	$a^3G_4 - v^3F_3^\circ$
					−25	$a^5F_1 - y^3F_2^\circ$
U	3403.299	(2)		29374.86	+07	$a^3G_4 - u^3G_5^\circ$
G	3402.256	5	IV	29383.86	00	$b^3H_6 - v^3H_6^\circ$
A	3401.521	6	III	29390.21	−01	$a^5F_4 - y^5P_3^\circ$
A	3399.336	15	III	29409.10	−02	$a^5P_2 - w^3D_2^\circ$
V	3399.230	(1)		29410.02	+01	$a^3G_4 - 4_4^\circ$
U	3398.220	(1)		29418.76	00	$a^3G_3 - u^3G_4^\circ$
V	3397.642	2	IIIA	29423.76	−02	$a^5F_2 - y^5P_1^\circ$
V	3397.560	(1)		29424.47	−01	$b^3G_3 - x^1F_3^\circ$
V	3397.221	(1)		29427.41	−07	$c^3P_2 - x^1F_3^\circ$
A	3396.978	4	IIIA	29429.59	−01	$a^5F_3 - y^5P_2^\circ$
V	3396.386	(1)		29434.64	−06	$a^5F_3 - y^3F_3^\circ$
G	3394.583	5	IV	29450.28	+05	$a^5P_2 - u^5D_1^\circ$
V	3394.085	(1)		29454.60	−07	$a^3H_4 - w^3F_3^\circ$
V	3393.915	(1)		29456.07	+03	$a^3P_2 - x^3G_3^\circ$
V	*3393.609	(1w)		29458.73	+13	$b^3P_2 - u^3D_2^\circ$
					−14	$a^3G_3 - y^1D_2^\circ$
V	3393.382	(1)		29460.70	+03	$b^3P_0 - u^3D_1^\circ$
G	3392.652	15	III	29467.04	+01	$a^5P_3 - w^3D_3^\circ$
G	3392.304	8	IV	29470.06	00	$a^5P_2 - x^3F_2^\circ$
U	3392.014	2	IV	29472.58	+01	$c^3P_2 - v^3P_1^\circ$
V	3389.748	2	IV	29492.28	−03	$a^5P_1 - 1_2^\circ$
V	3388.966	(1w)		29499.09	+04	$c^3P_1 - t^5P_1^\circ$
W	3388.8	(1)		29500.5	+1	$a^3P_1 - 2_2^\circ$
V	3387.410	2	IV	29512.64	−02	$a^3G_3 - x^1D_2^\circ$
C	3383.981	8	IV	29542.54	−03	$a^5P_3 - x^3F_3^\circ$
G	*3383.692	5	IV	29545.07	−01	$a^5P_2 - w^3D_1^\circ$
					−15	$b^3G_5 - 9_4^\circ$
V	3383.387	(1)		29547.73	−14	$b^3F_2 - z^1F_3^\circ$?
G	3382.403	3	IV	29556.32	−01	$a^5P_3 - z^3H_4^\circ$
V	*3381.340	(2)		29565.61	+01	$b^3P_1 - u^3D_1^\circ$
					−08	$a^3D_3 - w^1F_3^\circ$
C	3380.111	8	IV	29576.36	+02	$a^3G_3 - u^3G_3^\circ$
V	3380.004	(1)		29577.30	−18	$z^5F_5^\circ - 2$
G	3379.017	6	IV	29585.94	+01	$a^5P_3 - w^3D_2^\circ$
G	3378.676	6	IV	29588.92	+04	$a^3G_5 - v^3F_4^\circ$
V	3374.221	(1)		29627.99	−23	$a^5P_1 - y^3S_1^\circ$

TABLE B—(*Continued*)

Ref	λ I A	Int	T C	Wave Number Observed	o−c	Desig
V	3373.874	(1)		29631.04	−02	$a^3G_4 - 6_5^\circ$
U	3372.352	(1)		29644.41	−02	$b^3G_4 - x^1F_3^\circ$
G	3372.070	3	IV	29646.89	+03	$a^5P_3 - x^3F_2^\circ$
A	3370.786	10	IV	29658.18	−01	$a^3G_5 - u^3G_5^\circ$
G	**3369.549	8	III	29669.07	+02 +05	$a^3G_4 - u^3G_4^\circ$ $(c^3P_2 - 11_3^\circ)$
V	3368.983	(1)		29674.05	−05	$b^3P_2 - u^3D_1^\circ$
U	3367.159	(1)		29690.13	−06	$a^3P_1 - v^3D_2^\circ$
U	3366.867	5	IV	29692.70	00	$a^5P_2 - 1_2^\circ$
U	3366.789	5	IV	29693.39	−02	$a^3G_5 - 4_4^\circ$
V	3364.639	(1)		29712.37	00	$b^3F_3 - z^1F_3^\circ$
V	3363.815	(1)		29719.64	−04	$a^3G_3 - u^3D_3^\circ$
V	3361.959	(1)		29736.05	−02	$b^3P_1 - t^3D_2^\circ$
U	3360.922	(1)		29745.22	+05	$a^3P_1 - v^3D_1^\circ$
V	3359.814	(2)		29755.03	−02	$b^3H_4 - u^3H_5^\circ$
U	3359.491	2	IIIA	29757.89	−03	$a^5F_5 - y^3F_4^\circ$
H	3356.407	3	IV	29785.03	−24	$a^3P_2 - v^5P_2^\circ$
U	3356.323	1	IVA	29785.98	−02	$a^5F_4 - y^3F_3^\circ$
V	3355.517	(1)		29793.14	+02	$a^5F_3 - y^3F_2^\circ$
C	3355.228	6	IV	29795.70	+01	$b^3H_4 - u^3H_4^\circ$
U	3354.064	3	IV	29806.04	−01	$b^3P_0 - 8_1^\circ$
U	3353.267	(1)		29813.13	−03	$a^3H_5 - y^3H_6^\circ$
V	3352.929	(1)		29816.13	−04	$a^3H_4 - y^3H_5^\circ$
V	3351.750	3	IV	29826.62	−01	$a^3G_4 - u^3G_3^\circ$
V	3351.529	2	IV	29828.58	−03	$a^5P_2 - y^3S_1^\circ$
V	*3350.284	(3)		29839.67	+03 −21	$a^3H_4 - v^3G_4^\circ$ $a^3H_5 - v^3G_5^\circ$
V	3349.739	(1)		29844.52	−05	$b^3P_2 - t^3D_2^\circ$
A	3347.927	6	IV	29860.68	−01	$a^3P_2 - v^5F_2^\circ$
V	3347.507	(1)		29864.42	−03	$b^3G_4 - t^3G_5^\circ$
U	3346.936	1	IV	29869.52	+01	$a^5P_3 - 1_2^\circ$
V	3345.679	(1)		29880.74	+05	$a^3P_1 - w^3F_2^\circ$
V	3343.678	(1)		29898.62	−04	$b^3G_3 - t^3G_4^\circ$
U	3343.240	(1)		29902.54	−02	$a^5P_3 - z^1G_4^\circ$
V	3342.298	4	V	29910.96	−02	$b^3P_1 - 8_1^\circ$
U	3342.216	5	IV	29911.70	+03	$a^3P_2 - v^5P_1^\circ$
G	3341.906	5	IIIA	29914.48	+02	$a^3G_5 - 6_5^\circ$
A	3340.566	6	IV	29926.47	−02	$a^3P_2 - x^3P_2^\circ$
U	3339.582	(1w)		29935.29	−02	$c^3P_2 - t^5P_1^\circ$
U	*3339.195	2	IV	29938.76	−02 −03	$a^3H_4 - y^3H_4^\circ$ $b^3G_5 - y^1H_5^\circ$
U	3338.638	(3w)		29943.75	+01	$z^7P_3^\circ - g^7D_4$
C	3337.666	6	IV	29952.48	+03	$a^3G_5 - u^3G_4^\circ$
U	3336.254	(3)		29965.15	+03	$b^3H_4 - u^3F_4^\circ$
V	3335.776	4	IV	29969.44	−03	$b^3P_1 - v^3P_2^\circ$
V	3335.513	(1)		29971.81	00	$a^3F_3 - x^5P_3^\circ$
V	3335.403	(1)		29972.80	−08	$b^3F_4 - x^1G_4^\circ$
V	3334.278	(1)		29982.91	−01	$b^3H_5 - u^3H_6^\circ$
V	3334.223	(3)		29983.40	−01	$a^3H_5 - y^3H_5^\circ$
V	3331.778	(2)		30005.40	+01	$a^3P_0 - w^3P_1^\circ$
U	3331.612	3	IV	30006.90	+02	$a^3H_5 - v^3G_4^\circ$
V	3329.532	(2)		30025.64	−02	$a^1G_4 - t^3G_3^\circ$
C	3328.867	5	IV	30031.64	+04	$b^3H_5 - u^3H_5^\circ$
V	3327.961	(1)		30039.82	−04	$a^5P_3 - w^5G_4^\circ$
V	3327.498	(2)		30044.00	00	$a^3H_6 - y^3H_6^\circ$
V	3325.468	4	IV	30062.34	+01	$a^3H_4 - v^3G_3^\circ$
V	3324.541	4	IV	30070.72	00	$a^3H_6 - v^3G_5^\circ$
V	3324.372	(2)		30072.25	+01	$b^3H_5 - u^3H_4^\circ$
C	3323.737	7	IV	30077.99	+02	$b^3P_2 - v^3P_2^\circ$
G	3322.474	5n	IV	30089.42	−01	$z^7P_4^\circ - g^7D_5$
U	3320.779	(2n,gn)		30104.78	−01	$z^7P_2^\circ - g^7D_2$
V	3320.650	(2)		30105.95	−07	$a^3H_5 - y^3H_4^\circ$
V	3319.258	2	IV	30118.58	−03	$b^3G_4 - t^3G_4^\circ$
G	3317.121	3	IV	30137.98	+05	$a^3P_2 - x^3P_1^\circ$
C	3314.742	7	IV	30159.61	+01	$a^3D_2 - u^3F_3^\circ$

Ref	λ I A	Int	T C	Wave Number Observed	o−c	Desig
U	3314.441	(2)		30162.35	+04	$b^3F_2 - v^3F_2^\circ$
V	§3314.070	(1)		30165.73	−03	$a^1P_1 - t^3F_2^\circ$
V	3313.723	(1)		30168.89	−05	$a^3F_2 - y^5G_3^\circ$
U	3312.224	(1)		30182.54	+03	$b^3G_4 - 13_4^\circ$
V	3311.451	(1)		30189.58	−01	$a^5F_2 - y^3D_3^\circ$
V	3310.496	(3)		30198.29	−01	$a^3D_3 - u^3H_4^\circ$
V	3310.347	4	IV	30199.65	00	$b^3G_5 - t^3G_5^\circ$
G	3307.234	5	IV	30228.07	+01	$b^3H_6 - u^3H_6^\circ$
U	3307.008	(1)		30230.14	+01	$b^3G_5 - 12_5^\circ$
S	3306.703	(−)		30232.93	+07	$z^7P_3^\circ - g^7D_3$
U	3306.490	(1)		30234.88	−04	$a^3D_2 - u^3F_2^\circ$
C	3306.356	20	III	30236.10	+02 −03	$a^5P_1 - v^5P_2^\circ$ $(a^1G_4 - w^1G_4^\circ)$
C	3305.971	20	III	30239.62	+01	$a^5P_2 - v^5P_3^\circ$
V	3303.574	(2)		30261.56	+01	$b^3G_3 - t^3G_3^\circ$
U	3301.917	(1)		30276.75	+01	$b^3H_6 - u^3H_5^\circ$
V	3301.227	(2)		30283.08	00	$b^3P_1 - z^1P_1^\circ$
U	3299.509	(1)		30298.85	−03	$a^3F_3 - x^5P_2^\circ$
U	3299.077	(1w)		30302.81	−08	$z^5F_3^\circ - i^5D_4$
A	3298.133	6	IV	30311.49	00	$a^5P_1 - v^5F_2^\circ$
V	§3296.806	(1)		30323.69	+06	$b^3H_4 - v^1G_4^\circ$
U	3296.467	(1)		30326.80	−01	$b^3F_3 - v^3F_2^\circ$
U	3293.142	(1)		30357.43	+02	$a^3F_2 - z^5H_3^\circ$
G	3292.590	8	IV	30362.51	+03	$a^5P_1 - v^5P_1^\circ$
G	3292.022	8	IV	30367.76	+03	$a^3D_3 - u^3F_4^\circ$
G	3290.988	5	IV	30377.29	−01	$a^5P_1 - x^3P_2^\circ$
U	3290.714	(2)		30379.82	−01	$a^5P_3 - v^5F_4^\circ$
V	3289.442	(2)		30391.57	−01	$b^3P_2 - z^1P_1^\circ$
U	3288.967	2	IV	30395.96	+01	$a^5P_2 - v^5F_3^\circ$
U	3288.651	(2)		30398.88	+01	$a^3P_1 - w^3P_0^\circ$
V	3287.117	(1w)		30413.07	−08	$z^7P_4^\circ - g^7D_4$
A	3286.755	20	III	30416.42	00	$a^5P_3 - v^5P_3^\circ$
U	3286.444	(2w)		30419.30	+03	$z^5F_3^\circ - i^5D_3$
U	3286.022	2	IV	30423.20	−01	$a^5P_1 - v^5F_1^\circ$
U	3285.20	(1)		30430.81	−03	$z^7P_3^\circ - g^7D_2$
A	3284.588	5	IV	30436.48	+01	$a^5P_2 - v^5P_2^\circ$
V	3283.430	(1)		30447.21	−10	$a^5F_3 - y^3D_3^\circ$
G	3282.891	(2)		30452.21	+05	$a^3D_1 - u^3F_2^\circ$
U	3282.720	(1)		30453.80	−01	$b^3G_5 - t^3G_4^\circ$
V	3280.763	(1)		30471.96	−06	$b^3G_3 - w^1G_4^\circ$
C	3280.261	8	IV	30476.62	00	$b^3H_4 - x^3I_5^\circ$
U	3279.739	(21)		30481.48	−02	$b^3G_4 - t^3G_3^\circ$
V	*3278.741	4	IV	30490.76	00 −02	$a^3P_1 - w^3P_1^\circ$ $b^3F_3 - v^3F_3^\circ$
U	3276.468	4	IV	30511.91	+02	$a^5P_2 - v^5F_2^\circ$
V	3275.848	(1)		30517.68	−03	$b^3G_5 - 13_4^\circ$
V	3275.685	(1)		30510.20	−12	$a^3G_3 - w^3H_4^\circ$
U	3274.453	(2w)		30530.68	−07	$z^5F_4^\circ - i^5D_4$
U	3272.71	(1)		30546.95	+14	$z^5F_1^\circ - 4_2$
U	3271.683	(2)		30556.53	+03	$a^3F_4 - x^5P_3^\circ$
U	3271.487	(2)		30558.36	+06	$a^3D_3 - u^3F_3^\circ$
A	3271.002	15	III	30562.89	+02	$a^5P_2 - v^5P_1^\circ$
V	3269.964	(1)		30572.59	−17	$a^5P_3 - v^5F_3^\circ$
U	3269.235	(1w)		30579.41	−05	$z^5F_3^\circ - i^5D_2$
G	3268.234	5	IV	30588.78	+04	$a^5P_1 - x^3P_1^\circ$
G	3265.616	15	III	30613.30	+02	$a^5P_3 - v^5P_2^\circ$
G	3265.046	8	IA	30618.64	00	$a^5D_2 - z^3D_3^\circ$
U	3264.710	(2)		30621.79	−14	$z^7D_2^\circ - f^5D_3$
U	3264.512	5	IV	30623.65	+05	$a^5P_2 - v^5F_1^\circ$
V	3263.378	(2)		30634.29	−09	$a^3P_1 - w^3P_2^\circ$
V	3262.009	(2)		30647.15	+02	$z^5F_4^\circ - i^5D_3$
V	3261.332	(2)		30653.51	−06	$z^5F_2^\circ - 4_2$
U	3260.261	4	IV	30663.58	+07	$b^3F_4 - v^3F_4^\circ$
G	3259.991	6	IV	30666.12	−02	$z^7D_3^\circ - f^5D_4$
A	3257.594	8	IV	30688.68	−01	$a^5P_3 - v^5F_2^\circ$

TABLE B—(*Continued*)

Ref	λ I A	Int	T C	Wave Number Observed	o−c	Desig
V	*3257.244	2	IV	30691.98	+01	$b^3G_4 - w^1G_4^\circ$
					−29	$a^5F_2 - y^3D_2^\circ$
U	3254.734	(2)		30715.65	−05	$a^3G_5 - w^3H_6^\circ$
C	3254.363	10	IV	30719.15	−01	$b^3H_5 - x^3I_6^\circ$
V	3254.261	(1)		30720.11	−21	$b^3F_4 - t^5D_3^\circ$
U	3253.949	(2)		30723.06	−04	$b^3F_2 - x^1D_2^\circ$
U	3253.834	(1)		30724.15	−01	$b^3F_4 - v^3F_3^\circ$
V	3253.610	4	IV	30726.26	+02	$a^3D_3 - v^1G_4^\circ$
U	3252.926	4	IV	30732.72	−10	$b^3F_4 - u^3G_5^\circ$
G	3251.236	8	IV	30748.69	−03	$a^5P_2 - w^3G_3^\circ$
U	3250.623	4	IV	30754.49	−01	$a^5P_3 - x^3P_2^\circ$
U	*3250.394	(2)		30756.66	+08	$b^3P_0 - v^3P_1^\circ$
					−20	$a^3P_2 - v^3D_3^\circ$
U	3249.191	3	IV	30768.05	+01	$b^3F_4 - 4_4^\circ$
V	3249.037	(1)		30769.50	−11	$a^3G_4 - w^3H_4^\circ$
G	3248.206	10	IV	30777.38	−02	$z^7D_3^\circ - f^5D_3$
U	3247.278	3	IV	30786.17	+01	$z^7D_2^\circ - f^5D_2$
U	3246.962	6	IV	30789.17	+04	$a^5P_2 - x^3P_1^\circ$
U	3246.482	3	IV	30793.72	+02	$b^3F_3 - u^3G_4^\circ$
G	3246.005	8	I	30798.24	00	$a^5D_1 - z^3D_2^\circ$
V	3245.984	(2)		30798.44	−17	$a^5F_4 - y^3D_3^\circ$
A	3244.190	15	IV	30815.47	+01	$z^7D_4^\circ - f^7D_5$
V	*3243.406	3	IV	30822.92	−11	$z^5F_5^\circ - i^5D_4$
					+15	$b^3P_2 - y^1F_3^\circ$
U	3243.109	(1)		30825.75	+01	$a^3H_4 - x^1G_4^\circ$
V	3242.268	(1)		30833.74	−07	$b^3F_3 - y^1D_2^\circ$
U	3241.52	(1)		30840.86	−18	$a^5F_1 - y^3D_1^\circ$
U	3240.013	(1)		30855.20	00	$a^1G_4 - v^3H_5^\circ$
A	3239.436	15	IV	30860.70	−02	$z^7D_4^\circ - f^5D_4$
					+20	$(z^7D_1^\circ - f^5D_1)$
V	*3239.029	(1)		30864.57	+11	$a^3P_2 - v^3D_2^\circ$
					−17	$a^3P_2 - w^3F_3^\circ$
S	3238.535	(−)		30869.28	+09	$z^7P_2^\circ - e^3P_1$
V	3237.234	(1)		30881.69	+05	$b^3F_3 - 7_2^\circ$
A	3236.223	8	IA	30891.33	−01	$a^5D_3 - z^3F_4^\circ$
U	3235.592	(1)		30897.36	+05	$a^3G_5 - w^3H_5^\circ$
G	3234.614	7	IA	30906.70	−01	$a^5D_3 - z^3D_3^\circ$
G	3233.967	12	IV	30912.88	+02	$z^7D_4^\circ - e^7P_4$
V	3233.304	(1)		30919.22	−22	$a^3P_2 - v^3D_1^\circ$?
G	3233.053	8	IV	30921.62	+01	$b^3H_6 - x^3I_7^\circ$
U	3231.576	(1)		30935.75	+09	$a^3F_4 - y^5G_5^\circ$
G	3230.963	10	IV	30941.62	−01	$z^7D_3^\circ - f^5D_2$
V	3230.210	6	IV	30948.84	+03	$z^7D_2^\circ - e^7P_2$
U	3229.994	(3)		30950.91	+02	$a^1G_4 - x^1H_5^\circ$
W	3229.78?	(1)		30952.96	−11	$b^3F_3 - u^5F_2^\circ$
G	3229.123	4	IIA	30959.26	−02	$a^5D_0 - z^3D_1^\circ$
G	3228.900	3	IV	30961.39	+02	$z^7D_1^\circ - f^5D_0$
U	3228.254	5	IV	30967.59	−05	$z^7D_2^\circ - f^5D_1$
V	3228.003	(2)		30970.00	−01	$b^3P_2 - v^3P_1^\circ$
G	3227.798	15	IV	30971.96	−02	$z^7D_4^\circ - f^5D_3$
U	3227.063	3	IV	30979.02	−02	$z^7D_1^\circ - f^7D_2$
U	3226.720	2	IIIA	30982.31	−06	$a^5D_2 - z^3D_2^\circ$
A	3225.789	25	III	30991.25	−04	$z^7D_5^\circ - e^7F_6$
U	*3225.607	(1)		30993.00	+02	$a^3H_5 - x^1G_4^\circ$
					+10	$b^3D_3 - r^3G_3^\circ$
U	3223.844	(1)		31009.95	−02	$a^5F_2 - y^3D_1^\circ$
V	3223.273	(1)		31015.44	+04	$a^3F_4 - z^5H_5^\circ$
A	3222.069	20	III	31027.03	00	$z^7D_5^\circ - f^7D_5$
					−14	$(b^3G_5 - w^1G_4^\circ)$
U	3221.931	2	IV	31028.36	−09	$z^7D_1^\circ - f^7D_1$
G	3219.806	10	III	31048.84	00	$z^7D_4^\circ - e^7P_3$
					−39	$(a^5D_1 - z^3D_1^\circ)$
G	3219.581	12	IV	31051.01	00	$z^7D_3^\circ - f^7D_4$
A	3217.380	10	IV	31072.24	−05	$z^7D_5^\circ - f^5D_4$
A	3215.940	12	IV	31086.16	−02	$z^7D_2^\circ - f^7D_2$
V	3215.637	(3)		31089.09	+03	$z^7F_6^\circ - e^3G_5$
V	3214.624	(1)		31098.88	00	$a^3P_2 - z^1D_2^\circ$
G	3214.396	8	IA	31101.09	00	$a^5D_2 - z^3F_3^\circ$
V	*3214.044	20	III	31104.50	−16	$z^5F_4^\circ - g^5G_5$
					−31	$z^7D_3^\circ - f^7D_3$
					+22	$(z^7D_3^\circ - e^7P_2)$
U	3213.754	(1)		31107.30	+04	$b^3G_3 - v^3H_4^\circ$
G	3211.989	10	IV	31124.40	−03	$z^7D_5^\circ - e^7P_4$
V	*3211.872	4	IV	31125.53	+01	$a^5P_1 - 2_2^\circ$
					+05	$z^5F_3^\circ - g^5G_4$
U	3211.683	8	IV	31127.36	+04	$z^5F_5^\circ - g^5G_6$
U	3211.487	4	IV	31129.25	+01	$z^7D_1^\circ - e^5S_2$
G	3210.830	10	IV	31135.63	+04	$z^7D_2^\circ - f^7D_1$
G	3210.230	8	IV	31141.45	00	$z^7D_4^\circ - e^5G_5$
G	*3209.297	6	IV	31150.50	+01	$z^5F_2^\circ - g^5G_3$
					+03	$z^7F_6^\circ - g^7D_5$
V	3209.115	(1)		31152.27	−09	$a^5P_3 - y^1G_4^\circ$
G	3208.470	4	IV	31158.53	+08	$z^5F_1^\circ - g^5G_2$
V	3207.649	(1w)		31166.51	+05	$b^3P_2 - 11_3^\circ$
U	3207.089	2	IV	31171.95	−10	$z^7D_5^\circ - e^5G_6$
A	3205.400	15	IV	31188.37	−02	$z^7D_1^\circ - e^7F_1$
V	3202.562	2	IV	31216.01	−02	$a^1G_4 - w^1F_3^\circ$
S	3201.891	(−)		31222.55	−04	$z^7D_3^\circ - e^5G_4$
U	3200.784	2	IIIA	31233.35	00	$a^5D_2 - z^3D_1^\circ$
A	*3200.475	15	IV	31236.37	+01	$z^7D_2^\circ - e^7F_3$
					−01	$z^7D_2^\circ - e^5S_2$
G	3199.530	15	II	31245.59	00	$z^7D_4^\circ - f^7D_4$
					−29	$(a^5D_1 - z^3F_2^\circ)$
U	3198.266	(1)		31257.94	00	$b^3F_2 - u^3D_2^\circ$
U	3197.521	(1)		31265.22	+01	$z^5F_2^\circ - g^5G_2$
U	3196.977	(2)		31270.54	+10	$a^5D_3 - z^3D_2^\circ$
A	**3196.930	20	II	31271.00	−02	$z^7D_4^\circ - e^7F_5$
V	3196.147	2	IV	31278.66	−08	$z^7F_5^\circ - g^7D_4$
V	3194.422	3	IV	31295.55	+02	$z^7D_2^\circ - e^7F_1$
U	3193.303	8	IV	31306.52	−03	$z^7D_2^\circ - e^5G_3$
U	3193.228	10	IA	31307.25	−02	$a^5D_4 - z^3F_4^\circ$
G	3192.799	8	IV	31311.46	+02	$z^7D_1^\circ - e^7F_2$
					+42	$(b^3G_4 - v^3H_5^\circ)$
U	*3192.417	(1)		31315.20	−07	$a^5P_1 - v^3D_2^\circ$
					+19	$z^5F_3^\circ - g^5G_3$
A	3191.659	7	IA	31322.64	00	$a^5D_4 - z^3D_3^\circ$
S	3191.180	(−)		31327.34	+13	$b^3G_4 - v^3H_4^\circ$
U	3191.116	(1)		31327.97	−03	$b^3F_4 - u^3D_3^\circ$
U	3190.816	(2)		31330.91	+04	$a^1G_4 - s^3G_4^\circ$
U	3190.651	(2)		31332.54	+01	$a^1G_4 - s^3G_5^\circ$
W	3190.02	(1)		31338.74	−03	$b^3F_3 - t^3D_2^\circ$
G	3188.819	7	IV	31350.54	+01	$z^7D_1^\circ - e^5G_2$
G	3188.567	4	IV	31353.02	00	$z^7D_5^\circ - e^5G_6$
A	3184.896	7	IA	31389.15	−01	$a^5D_3 - z^3F_3^\circ$
U	*3184.622	3	IV	31391.85	+02	$z^7D_3^\circ - e^7F_3$
					00	$z^7D_3^\circ - e^5S_2$
G	3182.970	3	IV	31408.14	+08	$a^5P_2 - v^3D_3^\circ$
U	*3182.060	3	IV	31417.13	−04	$z^7D_4^\circ - e^5G_4$
					+08	$z^7F_4^\circ - g^7D_3$
U	*3181.922	(2)		31418.49	00	$c^3P_2 - w^1D_2^\circ$
					−09	$z^7D_2^\circ - e^7F_2$
U	3181.847	(3)		31419.22	+01	$z^7F_2^\circ - g^7D_2$
G	3181.522	4	IV	31422.44	00	$b^3F_3 - u^3D_2^\circ$
G	3180.756	5	IIA	31430.01	00	$a^5D_2 - z^3F_2^\circ$
G	3180.223	20	IV	31435.27	−01	$z^7D_3^\circ - e^7F_4$
U	§3179.479	(1)		31442.63	+02	$a^3F_2 - w^5D_1^\circ$
U	3178.967	3	IV	31447.70	−03	$a^3H_6 - x^3H_5^\circ$
V	3178.545	2	IV	31451.87	−05	$b^3G_3 - w^1F_3^\circ$
A	3178.015	10	IV	31457.11	−05	$z^7D_5^\circ - f^7D_4$
					−56	$(z^7D_2^\circ - e^5G_2)$

TABLE B—(*Continued*)

Ref	λ	Int	T C	Wave Number		Desig
	I A			Observed	o−c	
Q	3177.54	(2)	Fe II	31461.82	−20	($z^7D_3^\circ - e^5G_3$)
V	3176.366	2	IV	31473.44	00	$b^3F_2 - u^3D_1^\circ$
W	3175.97	(1)		31477.37	+02	$z^7F_0^\circ - g^7D_1$
A	3175.447	12	IV	31482.55	−04	$z^7D_5^\circ - e^7F_5$
V	3173.663	(3r)		31500.25	+11	$a^5P_2 - 3_3^\circ$
U	3173.608	(1)		31500.79	+02	$z^7F_3^\circ - g^7D_2$
W	3173.40	(1)		31502.86	−03	$z^7F_1^\circ - g^7D_1$
V	*3172.067	2	IV	31516.10	+17	$a^5P_2 - w^3F_3^\circ$
					−27	$a^3H_4 - v^3F_4^\circ$
U	3171.663	2	IV	31520.11	−01	$z^7D_1^\circ - e^7G_2$
V	*3171.353	5	IV	31523.19	−09	$a^3F_4 - w^5D_4^\circ$
					+03	$a^1G_4 - s^3G_3^\circ$
U	3168.858	2	IV	31548.01	−01	$z^7D_2^\circ - e^7G_3$
V	3167.907	(1)		31557.48	+08	$z^5D_3^\circ - i^5D_4$
G	3166.435	6	IV	31572.15	00	$b^3F_4 - t^3D_3^\circ$
G	3165.860	4	IV	31577.89	−01	$z^7D_3^\circ - e^7G_4$
V	3165.005	3	IV	31586.41	00	$z^7D_4^\circ - e^7F_3$
U	3164.308	(1)		31593.37	−09	$z^7D_3^\circ - g^5D_4$
U	*3162.335	2n	IV	31613.08	−06	$z^7D_3^\circ - e^5G_2$
					−02	$a^3G_5 - 9_4^\circ$
G	3161.949	8	IV	31616.94	−04	$z^7D_5^\circ - e^7G_6$
V	3161.370	4	IV	31622.73	+04	$a^3F_3 - w^5D_2^\circ$
A	3160.658	10	IV	31629.86	00	$z^7D_4^\circ - e^7F_4$
U	3160.344	(2)		31633.00	+01	$a^3H_6 - x^3H_6^\circ$
V	3160.200	(2n)		31634.44	−01	$z^5D_2^\circ - i^5D_2$
W	3158.99	(2)		31646.56	+32	$b^3G_5 - v^3H_5^\circ$?
U	3157.992	(2)		31656.56	−04	$z^7D_4^\circ - e^5G_3$
K	3157.88	6	IV	31657.64	−01	$z^7D_2^\circ - e^7S_3$
A	3157.040	8	IV	31666.10	−04	$z^7D_4^\circ - e^7G_5$
U	3156.464	(1)		31671.88	+01	$b^3G_4 - w^1F_3^\circ$
G	3156.275	5n	IV	31673.77	−01	$z^5D_3^\circ - i^5D_3$
V	3155.293	2	IV	31683.63	+02	$a^3H_5 - v^3F_4^\circ$
SS	3155.131	(1)		31685.26	−14	$z^7D_1^\circ - f^5F_2$
U	3154.505	2	IV	31691.55	−09	$z^7D_2^\circ - f^5F_3$
SS	3154.421	(1)		31692.39	−08	$a^5P_3 - v^3D_2^\circ$
U	3153.322	(1)		31703.44	−05	$z^7D_3^\circ - e^7G_3$
G	3153.200	5	IV	31704.66	−01	$z^7D_3^\circ - f^5F_4$
S	*3153.064	(−)		31706.03	−03	$b^3G_5 - v^3H_6^\circ$
					−13	$a^5P_2 - w^3F_2^\circ$
V	3151.867	(1)		31718.08	00	$a^5D_3 - z^3F_2^\circ$
G	3151.353	10	IV	31723.25	−02	$a^3G_4 - y^1H_5^\circ$
U	3150.304	(2n)		31733.81	+03	$z^5D_1^\circ - 4_2$
U	3148.420	(2)		31752.80	−12	$a^3H_5 - u^3G_5^\circ$
U	3147.793	(1)		31759.12	+07	$b^3G_3 - s^3G_3^\circ$
U	3146.475	(1)		31772.42	−06	$z^7D_4^\circ - e^7G_4$
V	3145.057	(2)		31786.74	+03	$b^3G_4 - s^3G_4^\circ$
V	3144.488	6n	IV	31792.50	−04	$z^7D_2^\circ - f^5F_2$
C	3143.990	8	IV	31797.54	−05	$z^5D_4^\circ - i^5D_4$
V	3143.242	2	IIIA	31805.10	00	$a^5D_4 - z^3F_3^\circ$
V	3142.888	5	IV	31808.69	+04	$a^3P_2 - w^3P_2^\circ$
U	3142.453	6	IV	31813.08	−04	$z^7D_3^\circ - e^7S_3$
U	3140.391	5n	V	31833.98	+01	$z^5D_3^\circ - i^5D_2$
U	3139.661	(1)		31841.38	−05	$z^7D_5^\circ - e^7F_4$
U	3135.863	(1)		31879.94	00	$a^3H_4 - u^3G_4^\circ$
A	**3134.111	10	III	31897.76	00	$a^5F_3 - x^5D_4^\circ$
V	3132.514	4n	V	31914.02	+05	$z^5D_4^\circ - i^5D_3$
V	3129.334	5	IV	31946.45	+01	$a^3F_4 - w^5D_3^\circ$
SS	3129.178	(1)		31948.04	+03	$z^7D_3^\circ - f^5F_2$
U	3128.901	1	IV	31950.87	−04	$a^3F_3 - y^5S_2^\circ$
C	*3125.653	15	III	31984.07	−01	$a^5F_2 - x^5D_3^\circ$
					+02	$z^7D_5^\circ - e^7G_4$
U	3124.099	(1)		31999.98	−08	$z^7D_1^\circ - e^5P_1$
U	3123.353	(1)		32007.62	−08	$z^7D_4^\circ - e^7S_3$
R	3122.665	(−)		32014.68	+07	$a^3G_4 - 12_5^\circ$
W	3121.76	(1)		32023.96	+01	$a^5P_1 - w^3P_0^\circ$
G	3120.435	6	IV	32037.56	+04	$a^3H_4 - u^3G_3^\circ$
G	3119.495	6	IV	32047.21	+03	$a^3H_5 - u^3G_4^\circ$
U	3117.640	1	IIIA	32066.27	−01	$a^5F_2 - y^7P_2^\circ$
A	§3116.633	12	III	32076.64	00	$a^5F_1 - x^5D_2^\circ$
U	3116.250	(1)		32080.58	+34	$z^7D_3^\circ - e^5P_3$?
V	3112.079	3	IV	32123.57	00	$b^3G_5 - s^3G_5^\circ$
U	3111.686	(2)		32127.63	−01	$b^3F_4 - w^3H_4^\circ$
U	3109.05	(1)		32154.87	−07	$z^7D_2^\circ - e^5P_2$
U	§3106.542	(1)		32180.83	−03	$a^3H_4 - u^3D_3^\circ$
U	3101.004	(2)		32238.30	+01	$a^3G_4 - t^3G_4^\circ$
V	3100.838	(2)		32240.02	−01	$a^3H_6 - 6_6^\circ$
G	3100.666	20	II	32241.81	00	$a^5F_3 - x^5D_3^\circ$
G	3100.304	20	II	32245.57	00	$a^5F_2 - x^5D_2^\circ$
U	3099.971	15	II	32249.04	−01	$a^5F_4 - x^5D_4^\circ$
U	3099.897	20	II	32249.81	−01	$a^5F_1 - x^5D_1^\circ$
U	3098.192	6	IV	32267.55	+02	$a^3G_5 - t^3G_5^\circ$
V	3095.270	(2)		32298.01	00	$a^3G_5 - 12_5^\circ$
U	3094.870	(1)		32302.19	00	$a^3G_4 - 13_4^\circ$
U	3093.883	(2ld)		32312.49	−05	$b^3F_4 - s^3D_3^\circ$
V	3093.806	3	IVA	32313.30	−02	$a^3F_2 - x^3D_2^\circ$
U	**3092.778	2	III?	32324.04	+03	$a^5F_3 - y^7P_2^\circ$
A	3091.578	20	II	32336.58	−01	$a^5F_1 - x^5D_0^\circ$
V	3090.209	(1)		32350.91	+02	$a^3G_3 - t^3G_3^\circ$
A	3083.742	20	II	32418.75	00	$a^5F_2 - x^5D_1^\circ$
U	3083.152	(1)		32424.95	−06	$a^3H_4 - t^3D_3^\circ$
V	3078.436	3	IV	32474.62	+02	$a^3P_0 - u^3D_1^\circ$
G	3078.014	4	IVA	32479.08	+03	$a^5F_3 - y^7P_3^\circ$
A	3075.721	25r	II	32503.29	00	$a^5F_3 - x^5D_2^\circ$
V	3074.157	(2)		32519.82	−03	$b^3G_3 - u^3F_2^\circ$
V	3073.982	(1)		32521.68	−01	$a^3G_5 - t^3G_4^\circ$
S	3073.244	(−)		32529.48	−09	$a^1G_4 - x^3I_5^\circ$
S	3068.927	(−)		32575.24	+06	$a^3F_4 - v^5D_3^\circ$
G	3068.175	8	IV	32583.22	−04	$a^3F_2 - x^3D_1^\circ$
V	3067.952	(1)		32585.59	00	$a^3G_5 - 13_4^\circ$
A	3067.244	30r	II	32593.12	+02	$a^5F_4 - x^5D_3^\circ$
U	3067.120	8	IV	32594.43	−02	$a^3F_2 - y^3G_3^\circ$
U	3066.483	3	IV	32601.20	+02	$a^3G_4 - t^3G_3^\circ$
U	3063.933	(2)		32628.34	+01	$a^3P_1 - t^3D_1^\circ$?
S	3063.149	(1)		32636.69	+03	$a^5P_3 - w^3P_2^\circ$
S	3062.872	(1)		32639.64	−04	$b^3G_5 - u^3H_4^\circ$
G	3060.984	4	IV	32659.73	−05	$a^3F_3 - x^3D_3^\circ$
V	3060.545	(1)		32664.45	−03	$b^3G_4 - u^3F_3^\circ$
A	3059.086	100R	I	32680.03	00	$a^5D_3 - y^5D_4^c$
A	3057.446	40R	II	32697.56	+01	$a^5F_5 - x^5D_4^\circ$
C	3055.263	12	III	32720.92	−02	$a^3F_3 - x^3D_2^\circ$
S	3054.949	(−)		32724.28	+02	$b^3F_2 - x^1F_3$
W	*3053.44	(2)		32740.46	+16	$a^5F_1 - z^5S_2^\circ$
					−11	$z^7P_4^\circ - 2$
G	3053.065	5	IV	32744.48	+01	$a^3P_1 - u^3D_2^\circ$
A	3047.605	100R	I	32803.14	00	$a^5D_2 - y^5D_3^\circ$
S	3047.201	(−)		32807.49	+06	$b^3P_1 - w^1D_2^\circ$
U	3047.050	(1)		32809.11	00	$b^3G_5 - u^3F_4^\circ$
U	3046.930	(1)		32810.41	−02	$a^3H_5 - w^3H_6^\circ$
S	3046.819	(−)		32811.60	−05	$a^3G_4 - w^1G_4^\circ$
V	3045.594	(1)		32824.80	00	$a^3H_4 - w^3H_5^\circ$
G	3045.077	5	III	32830.37	+03	$a^5F_4 - y^7P_3^\circ$
G	3042.666	15	III	32856.39	+06	$a^5F_2 - x^5F_3^\circ$
G	3042.020	15	III	32863.36	+02	$a^5F_1 - x^5F_2^\circ$
V	3041.745	15	III	32866.33	−07	$a^5F_3 - x^5F_4^\circ$
V	3041.639	10	IV	32867.48	−02	$a^3F_3 - y^3G_4^\circ$
C	3040.428	15	III	32880.57	−02	$a^5F_4 - x^5F_5^\circ$
V	3039.322	(2)		32892.53	−02	$a^3H_5 - y^3I_6^\circ$
V	3037.782	2	IVA	32909.21	−02	$a^5F_2 - z^5S_2^\circ$
A	3037.388	80R	I	32913.47	+01	$a^5D_1 - y^5D_2^\circ$
W	3034.51	(2n)		32944.69	−27	$a^3F_2 - x^5G_3^\circ$?

TABLE B—(*Continued*)

Ref	λ I A	Int	T C	Wave Number Observed	Wave Number o−c	Desig
U	3033.101	(1)		32959.99	+02	$a^3P_1 - u^3D_1^\circ$
G	3031.638	15	III	32975.90	00	$a^5F_1 - x^5F_1^\circ$
G	3031.213	12	IV	32980.52	+02	$a^3H_4 - w^3H_4^\circ$
S	3030.757	(−)		32985.48	+07	$b^3G_4 - x^3I_5^\circ$
C	3030.149	15	IV	32992.10	+06	$a^3H_5 - w^3H_5^\circ$
V	3029.237	3	IV	33002.08	+01	$a^3F_3 - y^3G_3^\circ$
G	3026.462	15	III	33032.29	+03	$a^5F_2 - x^5F_2^\circ$
K	3025.843	50R	I	33039.05	00	$a^5D_0 - y^5D_1^\circ$
U	3025.638	15	IV	33041.29	+02	$a^3H_6 - w^3H_6^\circ$
V	3025.283	3	III	33045.16	+05	$a^5F_4 - y^7P_4^\circ$
C	3024.033	15r	IA	33058.82	−01	$a^5D_1 - z^3P_2^\circ$
S	3023.583	(−)		33063.74	−07	$a^5P_3 - x^1G_4^\circ$
G	3021.074	150R	I	33091.20	−01	$a^5D_3 - y^5D_3^\circ$
U	3020.640	200R	I	33095.96	00	$a^5D_4 - y^5D_4^\circ$
U	3020.487	100R	II	33097.63	+04	$a^5D_2 - y^5D_2^\circ$
U	3019.290	(1)		33110.75	−03	$a^3H_4 - y^3I_5^\circ$
G	3018.983	15r	III	33114.12	+06	$a^5F_3 - x^5F_3^\circ$
U	3018.134	(1)		33123.44	+05	$a^3H_6 - y^3I_6^\circ$
G	3017.628	15r	IA	33129.00	00	$a^5D_1 - y^5D_1^\circ$
G	3016.186	12	III	33144.83	+01	$a^5F_2 - x^5F_1^\circ$
C	3015.913	4	IV	33147.83	+09	$a^3H_5 - w^3H_4^\circ$
U	3014.175	3	IV	33166.92	−03	$a^5F_3 - z^5S_2^\circ$
S	3014.120	(−)		33167.54	−08	$b^3G_5 - v^1G_4^\circ$
G	3011.482	7	IV	33196.60	00	$a^3G_3 - v^3H_4^\circ$
C	3009.570	25r	II	33217.69	00	$a^5F_4 - x^5F_4^\circ$
V	3009.098	3	IV	33222.90	+02	$a^3H_6 - w^3H_5^\circ$
G	3008.139	60R	I	33233.50	+01	$a^5D_1 - y^5D_0^\circ$
U	3007.281	12r	I	33242.97	+01	$a^5D_2 - z^3P_2^\circ$
U	3007.145	8	III	33244.47	−01	$a^3F_4 - x^3D_3^\circ$
G	3005.302	3	IV	33264.85	+01	$a^3H_6 - y^3I_7^\circ$
U	3004.620	(1)		33272.41	+12	$a^3F_3 - x^5G_4^\circ$
V	3004.119	(2)		33277.96	−06	$a^3H_5 - y^3I_5^\circ$
C	3003.031	10	III	33290.01	+02	$a^5F_3 - x^5F_2^\circ$
SS	3001.663	(1)		33305.18	−08	$c^3P_2 - t^3F_3^\circ$
G	3000.950	100R	I	33313.10	−02	$a^5D_2 - y^5D_1^\circ$
G	3000.452	8	III	33318.63	+03	$a^3F_4 - y^3G_5^\circ$
A	2999.512	30R	II	33329.07	−02	$a^5F_5 - x^5F_5^\circ$
G	2996.386	5	IV	33363.84	00	$a^3P_1 - v^3P_2^\circ$
U	2995.838	(1)		33369.91	−05	$b^3G_3 - t^3F_2^\circ$
U	2994.507	}100R	I	{33384.77	−05	$a^5D_0 - z^3P_1^\circ$
G	2994.427			33385.66	00	$a^5D_3 - y^5D_2^\circ$
C	2990.392	6	IV	33430.70	−02	$a^3G_4 - v^3H_5^\circ$
W	2989.4	(1)		33441.8	−1	$a^3F_2 - w^5P_1^\circ$?
S	2988.942	(−)		33446.93	+04	$a^3G_4 - v^3H_4^\circ$
G	2988.468	2	IV	33452.23	+03	$a^3F_4 - y^3G_4^\circ$
A	2987.292	10	III	33465.40	+05	$a^5F_4 - x^5F_3^\circ$
U	§2986.653	(1)		33472.56	00	$a^3H_5 - z^1I_6^\circ$
G	2986.456	3	III	33474.77	+01	$a^5D_1 - z^3P_1^\circ$
V	§2984.785	10	IV	33493.51	−10	$a^5F_5 - y^7P_4^\circ$
G	2983.574	125R	I	33507.10	−04	$a^5D_4 - y^5D_3^\circ$
U	§2982.234	(1)		33522.16	+15	$b^3G_4 - t^3F_3^\circ$
G	2981.852	6	IV	33526.46	−01	$a^5P_3 - t^5D_4^\circ$
A	2981.446	20r	I	33531.02	−01	$a^5D_3 - z^3P_2^\circ$
G	2980.532	5	IV	33541.29	+03	$a^3G_3 - w^1F_3^\circ$
U	2976.922	(1)		33581.97	−21	$z^7F_6^\circ - 1$
W	2976.5	(1)		33586.7	−1	$a^3F_4 - y^3G_3^\circ$
G	2976.126	5	IV	33590.95	+03	$a^3P_2 - u^3D_3^\circ$
W	2974.78	(1)		33606.15	−01	$z^7F_5^\circ - 2$
U	2973.237	60R	I	33623.59	−02	$a^5D_3 - y^5F_4^\circ$
U	2973.134	60R	I	33624.75	−02	$a^5D_2 - y^5F_3^\circ$
G	2972.277	3	IV	33634.45	+01	$a^5P_2 - t^5D_3^\circ$
G	*2970.106	40R	I	33659.00	{+11 {−10	$a^5D_2 - z^3P_1^\circ$ $a^5D_1 - y^5F_2^\circ$
G	2969.474	10	I	33666.19	00	$a^5F_5 - x^5F_4^\circ$
U	2969.362	5	II	33667.46	−05	$a^5D_1 - z^3P_0^\circ$

Ref	λ I A	Int	T C	Wave Number Observed	Wave Number o−c	Desig
U	2968.481	(2)		33677.46	+01	$a^3P_1 - z^1P_1^\circ$
G	2966.901	125R	II	33695.39	−03	$a^5D_4 - y^5F_5^\circ$
U	2966.26	(2)		33702.68	+02	$a^5P_1 - t^5D_2^\circ$
U	2965.811	2	IV	33707.77	−06	$a^3H_5 - 9_4^\circ$
A	2965.255	20	II	33714.09	{−01 {−03	$a^5D_0 - y^5F_1^\circ$ $(a^3G_5 - v^3H_5^\circ)$
U	2963.71	(1n)		33731.67	−08	$z^7F_3^\circ - 3$
W	2962.11	(2)		33749.89	−03	$a^3F_4 - x^5G_5^\circ$
U	2961.70	(1)		33754.56	+12	$a^5P_3 - v^3F_4^\circ$
U	2960.666	(2)		33766.35	−09	$b^3G_5 - t^3F_4^\circ$
U	2960.299	1	IV	33770.53	+02	$a^3P_0 - v^3P_1^\circ$
C	2959.992	10	IV	33774.03	−01	$a^3G_5 - v^3H_6^\circ$
G	2959.682	5		33777.57	−06	$z^7F_6^\circ - 1$
U	2957.491	(2)		33802.60	00	$a^3P_2 - t^3D_1^\circ$?
A	2957.365	30R	II	33804.04	00	$a^5D_1 - y^5F_1^\circ$
U	2956.86	(2n)		33809.81	00	$a^3G_5 - x^1H_5^\circ$
U	2956.71	(1)		33811.52	+27	$a^5P_3 - t^5D_3^\circ$
G	2954.651	5	IV	33835.08	+01	$a^3P_2 - t^3D_3^\circ$
A	2953.940	50R	II	33843.23	00	$a^5D_2 - y^5F_2^\circ$
U	2953.486	5	IV	33848.43	+04	$a^3G_3 - s^3G_3^\circ$
G	2950.240	20n	IV	33885.67	−02	$a^5P_3 - 5^\circ$
U	2948.733	(2)		33902.99	−06	$a^5P_2 - t^5D_2^\circ$
U	2948.433	4	IV	33906.44	+05	$a^3G_4 - s^3G_4^\circ$
G	2947.877	60R	I	33912.83	−01	$a^5D_3 - y^5F_3^\circ$
U	2947.363	(2)		33918.74	00	$a^3P_2 - u^3D_2^\circ$
U	2941.77	(1)		33983.24	−18	$z^7D_4^\circ - h^5D_3$
A	2941.343	15r	I	33988.16	−01	$a^5D_2 - y^5F_1^\circ$
G	2940.586	(3)		33996.92	+10	$z^7F_5^\circ - 3$
G	2939.072	(1)		34014.43	+08	$a^5P_1 - t^5D_0^\circ$
G	2937.806	10n	IV	34029.08	−06	$a^5P_2 - 7_2^\circ$
G	2936.904	60R	I	34039.53	−01	$a^5D_4 - y^5F_4^\circ$
W	2936.1	(1)		34048.9	+2	$a^3F_2 - w^3D_3^\circ$
U	2934.370	(1)		34068.93	00	$a^5P_3 - u^5F_3^\circ$
W	2931.8	(1)		34098.8	+1	$a^5G_4 - s^3G_3^\circ$
U	2931.420	(2)		34103.21	−05	$a^3H_4 - 10_3^\circ$
W	2930.6	(1)		34112.8	0	$z^7D_1^\circ - h^5D_1$?
V	2929.618	2	IV	34124.18	−02	$a^3F_2 - x^3F_3^\circ$
V	2929.118	6	IV	34130.01	−03	$b^3H_4 - t^3H_4^\circ$
A	2929.008	25r	I	34131.29	−01	$a^5D_3 - y^5F_2$
U	2928.753	(3)		34134.26	+02	$a^3P_2 - u^3D_1^\circ$
U	2928.105	(2)		34141.82	+05	$a^5P_3 - u^5P_3^\circ$
V	2925.899	4	IV	34167.56	−01	$a^3F_2 - w^3D_2^\circ$
G	2925.359	4	V	34173.86	−01	$a^3G_3 - u^3H_4^\circ$
W	2924.6	(1n)		34182.7	−2	$a^5P_1 - u^5P_1^\circ$
G	2923.851	7	IV	34191.49	+04	$a^3G_5 - s^3G_5^\circ$
G	2923.288	7	IV	34198.08	−01	$b^3H_5 - t^3H_6^\circ$
V	2922.62	(1n)		34205.89	−06	$a^5P_3 - 7_2^\circ$
U	2922.383	(1)		34208.66	−02	$a^3F_2 - u^5D_1^\circ$
C	2920.691	5	IV	34228.48	−02	$a^3F_2 - x^3F_2^\circ$
U	2920.29	(1)		34233.18	−07	$a^3P_0 - t^5P_1^\circ$
G	2919.838	(2)		34238.47	+03	$z^7D_4^\circ - g^7D_5$
V	2918.354	3	IV	34255.89	+01	$a^3P_1 - v^3P_1^\circ$
G	2918.023	10	IV	34259.77	+02	$b^3H_6 - t^3H_6^\circ$
G	2914.305	3	IV	34303.48	−04	$a^3F_2 - w^3D_1^\circ$
V	2912.257	3	IV	34327.60	−03	$a^3F_3 - u^5D_2^\circ$
A	2912.158	20r	I	34328.77	+01	$a^5D_4 - y^5F_3^\circ$
U	2910.930	(3)		34343.25	−05	$a^3G_3 - u^3F_4^\circ$
U	2909.313	(1)		34362.34	+08	$a^3H_5 - t^3G_5^\circ$
V	2908.864	(2)		34367.64	+06	$z^7D_3^\circ - g^7D_4$
G	*2907.518	5	V	34383.56	{+3 {+04	$a^5P_2 - u^5P_1^\circ$ $a^3G_4 - u^3H_5^\circ$
U	2905.57	(1)		34406.60	+01	$b^3H_5 - t^3H_4^\circ$
G	2901.910	5	IV	34450.00	−01	$z^7D_5^\circ - g^7D_6$
G	2901.381	5	IV	34456.27	−02	$a^3F_3 - w^3D_3^\circ$
C	2899.416	8	IV	34479.63	+01	$a^3P_2 - 8_1^\circ$

TABLE B—(*Continued*)

Ref	λ I A	Int	T C	Wave Number		Desig
				Observed	o − c	
W	2897.6	(1)		34501.2	0	$z^7D_2^\circ - g^7D_3$
C	2895.035	8	III	34531.80	−02	$a^3F_3 - x^3F_3^\circ$
C	2894.505	10	III	34538.12	+01	$a^3P_2 - v^3P_2^\circ$
V	2893.882	2	IV	34545.56	−03	$a^3F_3 - z^3H_4^\circ$
V	2893.763	1	IVA	34546.98	+02	$a^5F_2 - x^5P_3^\circ$
G	2892.479	(1)		34562.32	+16	$z^7D_4^\circ - g^7D_4$
W	2891.73	(2)		34571.26	−02	$b^3H_6 - q^3G_5^\circ$
U	2891.410	(1)		34575.09	−10	$a^3F_3 - w^3D_2^\circ$
U	2890.868	(2)		34581.57	−12	$a^3D_3 - q^3G_3^\circ$
V	2889.991	(2)		34592.07	00	$z^7D_1^\circ - g^7D_2$
W	2889.89	(3)		34593.28	+18	$a^3H_6 - t^3G_5^\circ$
U	2887.961	(1)		34616.38	−04	$a^3H_5 - t^3G_4^\circ$
G	2887.806	5	V	34618.24	00	$a^3G_5 - u^3H_6^\circ$
W	2887.36	(1)		34623.59	+01	$a^3H_6 - 12_5^\circ$
G	2886.316	3	IV	34636.11	−01	$a^3F_3 - x^3F_2^\circ$
V	2883.748	4	V	34666.95	+03	$a^3G_5 - u^3H_5^\circ$
G	2880.575	2	IV	34705.14	+04	$a^5F_1 - x^5P_2^\circ$
U	2879.461	(1)		34718.56	−06	$a^3P_1 - t^5P_1^\circ$
C	2877.300	8	III	34744.64	+05	$a^3F_4 - u^5D_4^\circ$
G	§2875.302	5	IV	34768.78	+01	$a^3F_4 - u^5D_3^\circ$
W	2874.89	(3)		34773.76	+03	$z^7D_5^\circ - g^7D_4$
C	2874.172	10	I	34782.44	−01	$a^5D_4 - z^5G_5^\circ$
U	2873.655	(2)		34788.70	−05	$b^3F_4 - v^3H_5^\circ$
W	2872.5	(1)		34802.7	0	$b^3P_2 - t^3F_3^\circ$
G	2872.333	7	III	34804.71	+02	$a^5F_3 - x^5P_3^\circ$
U	2871.73	(1)		34812.02	−05	$a^3H_4 - t^3G_3^\circ$
U	2871.31	(1)		34817.11	−17	$z^7F_4^\circ - i^5D_3$
U	2869.833	(2)		34835.03	−20	$z^7D_2^\circ - g^7D_1$
A	2869.308	10	I	34841.41	00	$a^5D_3 - z^5G_4^\circ$
G	§*2868.454	3	IV	34851.78	{+06	$a^3P_2 - z^1P_1^\circ$
					{+07	$z^7F_5^\circ - i^5D_4$
G	2868.213	(1)		34854.71	+03	$z^7D_3^\circ - g^7D_2$
U	2867.880	(1)		34858.76	−01	$a^3F_3 - 1_2^\circ$
G	2867.560	3	IV	34862.64	+01	$a^3F_2 - w^5G_2^\circ$
G	2867.311	3	IV	34865.68	−01	$a^3F_2 - x^3G_3^\circ$
G	2866.624	7	II	34874.02	−01	$a^5F_2 - x^5P_2^\circ$
C	2863.864	8	I	34907.63	−02	$a^5D_2 - z^5G_3^\circ$
G	2863.429	8	III	34912.94	−01	$a^3F_4 - x^3F_4^\circ$
G	2862.496	4	IV	34924.32	−01	$a^5F_1 - x^5P_1^\circ$
G	2858.896	5	II	34968.29	00	$a^5D_1 - z^5G_2^\circ$
W	§2857.20	(1)		34989.05	+13	$a^5P_1 - v^3P_2^\circ$
U	2853.774	(3)		35031.05	+01	$b^3F_3 - s^3G_4^\circ$
V	2853.685	(2)		35032.14	−03	$a^3F_4 - z^3H_5^\circ$
G	2852.952	(1)		35041.14	+16	$a^3F_4 - w^3D_3^\circ$
A	2851.798	15r	II	35055.32	00	$a^5F_1 - y^5G_2^\circ$
W	2851.52	(2)		35058.74	−09	$b^3F_2 - s^3G_3^\circ$
G	2848.713	5	III	35093.28	+30	$a^5F_2 - x^5P_1^\circ$
G	2846.830	3	IV	35116.50	−02	$a^3F_4 - x^3F_3^\circ$
U	2845.714	(2)		35130.26	−02	$a^3F_4 - z^3H_4^\circ$
C	2845.595	8	III	35131.74	−02	$a^5F_3 - x^5P_2^\circ$
U	2845.544	(1)		35132.37	+03	$a^3F_3 - w^5G_3^\circ$
G	2843.977	20r	II	35151.72	00	$a^5F_2 - y^5G_3^\circ$
U	2843.923	(3)		35152.39	−03	$a^5D_2 - z^5G_2^\circ$
G	2843.631	10	III	35155.99	+01	$a^5F_4 - x^5P_3^\circ$
U	2840.932	(3)		35189.39	+08	$a^5P_2 - v^3P_2^\circ$
G	2840.422	6	II	35195.72	00	$a^5D_3 - z^5G_3^\circ$
A	2838.120	10	III	35224.26	+01	$a^5F_2 - y^5G_2^\circ$
G	2836.315	(1)		35246.67	−07	$z^7F_6^\circ - h^7D_5$
G	2835.948	(1)		35251.24	+01	$a^3F_3 - x^3G_4^\circ$
G	2835.457	6	I	35257.33	−01	$a^5D_4 - z^5G_4^\circ$
G	2834.755	(2)		35266.07	−01	$b^3F_4 - s^3G_5^\circ$
U	*2834.414	(1)		35270.31	{−02	$a^3F_2 - v^5F_2^\circ$
					{+06	$a^3F_3 - w^5G_2^\circ$
U	2834.177	(1)		35273.26	−05	$a^5F_3 - x^3G_3^\circ$
U	2833.401	(2)		35282.92	+01	$a^3P_2 - y^1F_3^\circ$

Ref	λ I A	Int	T C	Wave Number		Desig
				Observed	o − c	
A	2832.436	25r	II	35294.94	+01	$a^5F_3 - y^5G_4^\circ$
G	2828.808	7	III	35340.21	+03	$a^5F_2 - z^5H_3^\circ$
G	2827.892	5	III	35351.65	−01	$a^5D_3 - z^3G_4^\circ$
W	2827.67	(2n)		35354.43	−05	$a^3G_5 - x^3I_6^\circ$
U	2826.50	(3)		35369.06	−03	$a^3F_3 - v^5F_4^\circ$
U	§2825.995	(2)		35375.38	−01	$a^5D_2 - z^3G_3^\circ$
V	2825.687	6	II	35379.23	−01	$a^5D_4 - z^3G_5^\circ$
G	2825.557	20	II	35380.87	00	$a^5F_3 - z^5H_4^\circ$
U	2824.70	(2)		35391.60	00	$a^3G_3 - t^3F_3^\circ$
A	2823.276	20	II	35409.45	+01	$a^5F_3 - y^5G_3^\circ$
U	2821.63	(1)		35430.10	−05	$a^3P_2 - v^3P_1^\circ$
G	2820.801	2	IV	35440.51	+02	$a^5D_3 - z^5G_2^\circ$
W	2819.5	(2)		35456.9	0	$b^3F_4 - s^3G_3^\circ$
G	§2819.286	(1)		35459.56	+26	$a^3G_3 - t^3F_2^\circ$
G	2817.505	6	III	35481.98	+01	$a^5F_3 - y^5G_2^\circ$
G	2815.506	3	IV	35507.17	00	$a^3F_2 - w^3G_3^\circ$
G	2815.017	(1)		35513.34	−05	$a^3P_2 - 10_3^\circ$
A	2813.288	30R	II	35535.15	+01	$a^5F_4 - y^5G_5^\circ$
U	2812.31	(1)		35547.51	−07	$a^3F_2 - x^3P_1^\circ$
G	2812.042	(1)		35550.90	−02	$a^3G_4 - t^3F_4^\circ$
U	2811.160	(1n)		35562.06	+04	$a^3F_3 - v^5F_3^\circ$
G	2808.328	2	III	35597.91	00	$a^5F_3 - z^5H_3^\circ$
U	2807.96	(1)		35602.58	+04	$a^3F_3 - v^5P_2^\circ$
U	2807.245	2	III	35611.65	00	$a^5D_4 - z^5G_3^\circ$
C	2806.984	20	II	35614.96	+08	$a^5F_4 - z^5H_5^\circ$
W	2806.5	(1n)		35621.1	0	$z^7F_6^\circ - g^5G_5$
G	2806.072	(1)		35626.54	−06	$a^3P_2 - 11_3^\circ$
G	2805.808	(1)		35629.88	+04	$a^3F_4 - v^5F_5^\circ$
V	2804.865	(2)		35641.86	−03	$a^3G_4 - t^3F_3^\circ$
A	2804.521	20	II	35646.23	+01	$a^5F_4 - y^5G_4^\circ$
G	2803.613	(2)		35657.78	00	$a^3H_4 - v^3H_4^\circ$
G	2803.169	(1)		35663.43	−03	$a^5D_3 - z^3G_3^\circ$
C	2797.775	15	III	35732.18	+01	$a^5F_4 - z^5H_4^\circ$
G	2796.871	(1)		35743.72	−03	$a^3F_3 - x^3P_2^\circ$
G	2795.540	(2)		35760.75	+01	$a^5F_4 - y^5G_3^\circ$
G	2795.006	3	III	35767.58	−01	$a^5D_4 - z^3G_4^\circ$
G	2794.700	(1)		35771.49	+02	$a^5F_3 - w^5D_4^\circ$
U	2794.157	(1)		35778.44	−22	$a^5P_3 - 9_4^\circ$
G	2792.397	1	III	35800.99	+02	$a^3F_3 - w^3G_4^\circ$
G	2791.786	(2)		35808.82	+07	$a^3H_5 - v^3H_5^\circ$
G	2789.803	(3)		35834.28	−04	$a^3G_5 - t^3F_4^\circ$
G	2789.477	(2)		35838.47	00	$a^5P_3 - t^5P_3^\circ$
G	2788.106	30	II	35856.09	−02	$a^5F_5 - y^5G_6^\circ$
U	2787.935	5	II	35858.29	−07	$a^3F_4 - x^3G_5^\circ$
U	2787.12	(1)		35868.77	00	$a^3H_5 - v^3H_6^\circ$
U	2786.18	(1)		35880.88	−08	$a^5P_1 - v^3P_1^\circ$
U	2784.346	(2)		35904.51	−03	$a^3H_5 - x^1H_5^\circ$
U	2784.017	(2)		35908.75	−06	$b^3F_3 - u^3F_3^\circ$
U	2782.055	(1)		35934.07	−04	$a^5P_2 - y^1F_3^\circ$
C	2781.835	4	III	35936.92	+02	$a^5F_2 - w^5D_3^\circ$
U	2780.700	1	III	35951.58	−04	$b^3F_4 - u^3F_4^\circ$
A	2778.221	20	III	35983.66	+02	$a^5F_5 - y^5G_5^\circ$
G	§2774.730	3	III	36028.93	+01	$a^5F_1 - w^5D_2^\circ$
U	2774.15	(1)		36036.46	+20	$a^5P_2 - x^1F_3^\circ$
U	2773.907	(1)		36039.62	−07	$a^3H_6 - v^3H_5^\circ$
W	2772.86	(2)		36053.22	+01	$b^3G_4 - r^3G_4^\circ$
G	2772.113	1	III	36062.94	+06	$a^5D_2 - y^5P_3^\circ$
V	2772.083	20	II	36063.33	−05	$a^5F_5 - z^5H_5^\circ$
G	2770.695	(1)		36081.39	+04	$a^5P_2 - v^3P_1^\circ$
G	2769.670	1	III	36094.74	+02	$a^5F_5 - y^5G_4^\circ$
G	2769.297	(6)		36099.60	−01	$a^3H_6 - v^3H_6^\circ$
G	2768.432	(2)		36110.88	−04	$a^5P_3 - y^1F_3^\circ$
A	§2767.523	20	III	36122.75	−01	$a^5F_4 - w^5D_4^\circ$
G	2766.909	2	III	36130.76	00	$a^5F_1 - w^5F_2^\circ$
U	2766.03	(1)		36142.24	+10	$b^3F_4 - u^3F_3^\circ$

TABLE B—(*Continued*)

Ref	λ I A	Int	T C	Wave Number Observed	o−c	Desig
U	2765.70	(1)		36146.56	−15	$a^3F_4 - v^5F_3^\circ$
G	2764.323	3	III	36164.56	+04	$a^5P_2 - 10_3^\circ$
C	2763.108	4	III	36180.46	+02	$a^5F_2 - w^5F_3^\circ$
					−20	$(a^5F_5 - z^5H_4^\circ)$
G	2762.770	(3)		36184.89	00	$a^5P_1 - t^5P_2^\circ$
G	2762.027	15	III	36194.62	−01	$a^5F_3 - w^5D_3^\circ$
G	§2761.780	18	III	36197.86	+02	$a^5F_2 - w^5D_2^\circ$
W	2761.5	(1)		36201.5	−3	$a^3P_1 - w^1D_2^\circ$
G	2759.814	4	III	36223.65	−05	$a^5F_1 - w^5F_1^\circ$
G	2757.315	10	III	36256.47	+01	$a^5F_1 - w^5D_1^\circ$
G	2756.329	20	I	36269.45	−01	$a^5D_1 - y^5P_2^\circ$
U	2756.264	(3)		36270.30	+03	$a^5D_3 - y^3F_4^\circ$
U	2755.184	(3)		36284.51	−01	$a^3H_5 - s^3G_4^\circ$
G	2754.427	2	III	36294.48	00	$a^5F_3 - w^5F_4^\circ$
G	2754.030	3	III	36299.71	+03	$a^5F_2 - w^5F_2^\circ$
G	2753.687	3	III	36304.23	−01	$a^5F_1 - w^5D_0^\circ$
G	2750.872	5		36341.38	+05	$a^5P_3 - 10_3^\circ$
U	2750.708	(1)		36343.55	−15	$a^5P_1 - t^5P_1^\circ$
G	2750.140	25r	II	36351.06	00	$a^5D_3 - y^5P_3^\circ$
G	2747.553	(3)		36385.29	+01	$a^5P_2 - t^5P_2^\circ$
C	§2746.982	20	III	36392.84	00	$a^5F_5 - z^5H_6^\circ$
					+22	$(a^5F_2 - w^5F_1^\circ)$
G	2744.526	8	III	36425.42	+04	$a^5F_2 - w^5D_1^\circ$
G	2744.068	10	II	36431.50	00	$a^5D_0 - y^5P_1^\circ$
G	2743.564	3	III	36438.18	+01	$a^5F_3 - w^5F_3^\circ$
G	2742.406	30r	II	36453.57	−02	$a^5D_2 - y^5P_2^\circ$
U	2742.256	20	III	36455.56	−01	$a^5F_3 - w^5D_2^\circ$
U	2742.017	2	III	36458.75	−02	$a^5D_2 - y^3F_3^\circ$
U	2741.578	(2)		36464.57	−04	$a^3F_2 - w^3F_2^\circ$
W	2741.10	(3)		36470.94	+02	$c^3P_2 - q^3G_3^\circ$
G	2738.210	(2)		36509.42	+01	$a^5F_1 - v^5D_2^\circ$
V	§2737.643	(2)		36516.99	−03	$a^3H_6 - s^3G_5^\circ$
G	2737.310	20r	II	36521.43	−01	$a^5D_1 - y^5P_1^\circ$
G	§2736.960	(3)		36526.10	+04	$a^5F_2 - y^5S_2^\circ$
U	2735.614	8	III	36544.07	−02	$a^5P_2 - t^5P_1^\circ$
A	2735.475	8	III	36545.92	00	$a^5F_4 - w^5D_3^\circ$
G	2734.613	(1)		36557.44	+03	$a^5F_3 - w^5F_2^\circ$
G	2734.266	2	III	36562.08	−01	$a^5P_3 - t^5P_2^\circ$
G	2734.002	2	III	36565.62	−02	$a^5F_2 - v^5D_3^\circ$
G	2733.581	15	II	36571.25	−01	$a^5F_5 - w^5D_4^\circ$
U	2731.281	(2)		36602.04	00	$b^3F_2 - t^3F_3^\circ$
G	2730.981	2	III	36606.06	−01	$a^5F_1 - v^5D_1^\circ$
U	2728.973	(2)		36632.99	−06	$a^5D_1 - y^3F_2^\circ$
G	2728.819	2	III	36635.07	+02	$a^3H_4 - u^3H_4^\circ$
G	2728.020	3	III	36645.79	+02	$a^5F_4 - w^5F_4^\circ$
G	2726.237	(2)		36669.75	+01	$b^3F_2 - t^3F_2^\circ$
G	2726.054	6	III	36672.21	−01	$a^5F_1 - v^5D_0^\circ$
U	2725.805	(1)		36675.57	00	$b^3F_3 - t^3F_4^\circ$
U	2725.606	(2)		36678.24	−09	$a^5F_2 - v^5D_2^\circ$
G	§2724.951	10	III	36687.06	00	$a^5F_3 - v^5D_4^\circ$
A	2723.577	15	II	36705.57	00	$a^5D_2 - y^5P_1^\circ$
U	2722.032	(2)		36726.40	+09	$a^3F_4 - y^1G_4$
G	2720.902	40r	II	36741.65	−01	$a^5D_3 - y^5P_2^\circ$
U	2720.516	(1)		36746.86	+02	$a^5D_3 - y^3F_3^\circ$
G	2720.194	(3)		36751.21	+06	$a^5P_3 - 13_4^\circ$
G	2719.418	3	III?	36761.71	+06	$a^3H_5 - u^3H_5^\circ$
G	2719.027	60R	II	36766.98	−02	$a^5D_4 - y^5P_3^\circ$
					+44	$(b^3F_3 - t^3F_3^\circ)$
C	2718.435	6	III?	36774.99	00	$a^5F_2 - v^5D_1^\circ$
G	2717.786	2	III	36783.78	−01	$a^5F_3 - y^5S_2^\circ$
G	2717.368	(1)		36789.43	−03	$a^5F_4 - w^5F_3^\circ$
U	§2716.41	(1)		36802.41	+12	$a^3H_5 - u^3H_4^\circ$
V	2716.259	(2)		36804.45	−03	$a^3H_4 - u^3F_4^\circ$
U	2715.323	1	III	36817.14	−04	$a^5D_2 - y^3F_2^\circ$
G	2714.868	1	III	36823.32	−05	$a^5F_3 - v^5D_3^\circ$

Ref	λ I A	Int	T C	Wave Number Observed	o−c	Desig
U	2714.062	(2)		36834.25	+01	$b^3F_3 - t^3F_2^\circ$
C	2711.655	4	III	36866.94	+05	$a^5F_4 - w^5F_5^\circ$
G	2710.543	2	III	36882.07	+03	$a^3F_2 - v^3G_3^\circ$
G	2709.989	(2)		36889.60	+02	$z^7D_4^\circ - 2$
W	2709.7	(1)		36893.5	−1	$b^3G_5 - q^3G_5^\circ$
G	2708.570	4	IV	36908.93	−02	$b^3F_4 - t^3F_4^\circ$
C	2706.581	8	III	36936.04	−02	$a^5F_3 - v^5D_2^\circ$
G	2706.012	4	IV	36943.81	00	$a^3H_6 - u^3H_6^\circ$
U	2702.453	(2)		36992.47	−02	$a^3H_6 - u^3H_5^\circ$
G	2701.908	(2)		36999.92	00	$b^3F_4 - t^3F_3^\circ$
A	2699.107	6	III	37038.32	−03	$a^5F_4 - v^5D_4^\circ$
G	2697.019	2	III	37066.99	+02	$a^3F_3 - v^3G_4^\circ$
G	2696.284	(5)		37077.10	−07	$z^7D_5^\circ - 1$
U	2695.662	(2gn)		37085.65	−01	$z^7D_3^\circ - 3$
G	2695.032	1	III	37094.32	+05	$a^5F_5 - w^5F_4^\circ$
G	2694.536	(5)		37101.15	00	$z^7D_5^\circ - 2$
G	2694.222	(1)		37105.47	+22	$a^5D_3 - y^3F_2^\circ$
U	2692.658	(3)		37127.02	−14	$a^5F_1 - x^3D_2^\circ$
G	2692.247	(2)		37132.69	+01	$a^3F_4 - w^3F_4^\circ$
G	2690.067	2	III	37162.78	−01	$a^5D_4 - y^3F_3^\circ$
G	2689.827	2	III	37166.10	00	$a^3F_3 - y^3H_4^\circ$
A	2689.212	8	III	37174.60	+06	$a^5F_4 - v^5D_3^\circ$
U	2684.857	(2)		37234.89	−05	$a^5F_2 - x^3D_3^\circ$
G	2681.586	(2)		37280.31	+07	$z^7D_4^\circ - 3$
U	2680.91	(1)		37289.71	+05	$a^3F_3 - v^3G_3^\circ$
G	2680.452	2	III	37296.08	−01	$a^5F_2 - x^3D_2^\circ$
A	2679.062	10	III	37315.43	+04	$a^5F_5 - w^5F_5^\circ$
U	2674.71	(1)		37376.14	+07	$a^3P_2 - w^1D_2^\circ$
C	2673.213	1	III	37397.07	−04	$a^5F_1 - x^3D_1^\circ$
G	2669.492	2	IV	37449.20	−01	$a^3H_5 - x^3I_6^\circ$
G	2667.912	1	IIIA	37471.37	−01	$a^5D_2 - y^3D_3^\circ$
U	2666.970	3	III	37484.60	−06	$a^3F_4 - v^3G_5^\circ$
G	2666.811	8	III	37486.84	−01	$a^5F_5 - v^5D_4^\circ$
G	2666.398	2	III	37492.65	−02	$a^5F_3 - x^3D_3^\circ$
C	2662.056	3	III	37553.79	−03	$a^5F_3 - x^3D_2^\circ$
U	2661.196	(2)		37565.93	−11	$a^5F_2 - x^3D_1^\circ$
G	2660.396	1	III	37577.22	−01	$a^5F_2 - y^3G_3^\circ$
G	2656.792	(2)		37628.19	00	$a^3F_4 - y^3H_5^\circ$
G	2656.145	3	III	37637.36	00	$a^3H_6 - x^3I_7^\circ$
U	2655.14	(1)		37651.61	−05	$a^3F_4 - v^3G_4^\circ$
C	2651.706	2	III	37700.36	−02	$a^5F_3 - y^3G_4^\circ$
C	2647.558	3	III	37759.43	−02	$a^5D_3 - y^3D_3^\circ$
G	2645.422	1	IIIA	37789.92	−02	$a^5D_1 - y^3D_2^\circ$
C	2643.997	8	III	37810.27	+01	$a^5F_1 - x^5G_2^\circ$
G	2641.645	4	III	37843.94	−02	$a^5F_4 - x^3D_3^\circ$
G	2636.477	1	III	37918.11	+03	$a^5F_4 - y^3G_5^\circ$
A	2635.808	8	III	37927.73	00	$a^5F_2 - x^5G_3^\circ$
G	2632.593	2	IIIA	37974.05	−01	$a^5D_2 - y^3D_2^\circ$
G	2632.238	4	III	37979.18	00	$a^5F_2 - x^5G_2^\circ$
G	§2629.579	2	III	38017.58	−11	$a^5D_0 - y^3D_1^\circ$
G	2623.532	5	III	38105.19	+02	$a^5F_3 - x^5G_4^\circ$
U	2623.366	2	III	38107.62	−01	$a^5D_1 - y^3D_1^\circ$
G	2618.708	2	III	38175.40	+02	$a^5D_4 - y^3D_3^\circ$
G	2618.018	5	III	38185.46	00	$a^5F_3 - x^5G_3^\circ$
G	2614.494	1	III	38236.91	00	$a^5F_3 - x^5G_2^\circ$
G	2612.771	2	III	38262.13	00	$a^5D_3 - y^3D_2^\circ$
G	2610.750	1	III	38291.74	−02	$a^5D_2 - y^3D_1^\circ$
G	2606.826	6	III	38349.38	−02	$a^5F_4 - x^5G_5^\circ$
G	2605.656	6	III	38366.60	+02	$a^5F_5 - y^3G_5^\circ$
U	2599.565	6	III	38456.49	+03	$a^5F_4 - x^5G_4^\circ$
U	2598.855	(1)		38466.99	−34	$a^3F_2 - 5^\circ?$
G	2595.422	(2)		38517.86	+01	$a^5F_1 - y^3P_0^\circ$
G	2594.150	1	III	38536.76	+01	$a^5F_4 - x^5G_3^\circ$
G	2593.510	(3)		38546.27	−01	$z^7D_5^\circ - h^7D_5$
U	2586.557	(1)		38649.88	−05	$a^3G_5 - t^3H_6^\circ$

TABLE B—(*Continued*)

Ref	λ I A	Int	T C	Wave Number Observed	o−c	Desig
A	2584.536	8	III	38680.10	+03	$a^5F_5 - x^5G_6^\circ$
G	2580.450	(2)		38741.34	00	$a^5F_2 - y^3P_2^\circ$
G	2580.062	(2)		38747.17	00	$a^5F_1 - y^3P_1^\circ$
G	*2579.266	(4)		38759.12	−11	$a^5F_2 - u^5D_3^\circ$
					−02	$a^5F_4 - z^3I_5^\circ$
G	2576.688	4	III	38797.91	+01	$a^5F_5 - x^5G_5^\circ$
G	2572.752	(4)		38857.26	+03	$a^3F_2 - u^3G_3^\circ$
W	§2571.57	(3)		38875.11	+16	$a^3F_3 - 5^\circ$
G	§2569.742	(4)		38902.77	−02	$a^5F_2 - u^5D_2^\circ$
G	2569.595	(6)		38904.99	+03	$a^5F_5 - x^5G_4^\circ$
G	§2568.862	(5)		38916.09	−01	$a^5F_2 - y^3P_1^\circ$
W	2567.86	(3)		38931.27	+03	$a^5P_1 - u^3F_2^\circ$
G	2564.555	(4)		38981.44	+02	$a^5F_1 - w^3D_2^\circ$
V	2563.820	(2)		38992.62	−16	$a^5F_3 - u^5D_4^\circ$
G	*2562.224	(5)		39016.91	+11	$a^5F_1 - u^5D_0^\circ$
					−05	$a^5F_3 - u^5D_3^\circ$
G	2561.852	(3)		39022.57	+04	$a^5F_1 - u^5D_1^\circ$
U	2561.262	(2)		39031.56	+12	$a^5F_2 - w^3D_3^\circ$
G	2560.556	(4)		39042.32	−03	$a^5F_1 - x^3F_2^\circ$
G	2557.268	(1)		39092.51	00	$a^3F_4 - x^3H_5^\circ$
G	2556.862	1	III	39098.72	+02	$a^5F_5 - z^3I_6^\circ$
U	2556.298	(4)		39107.35	+08	$a^3F_3 - u^3G_4^\circ$
G	2555.648	(1)		39117.29	−08	$a^5F_1 - w^3D_1^\circ$
G	2552.827	(4)		39160.51	00	$a^5F_3 - u^5D_2^\circ$
G	2552.604	2	III	39163.94	−01	$a^5D_1 - y^7P_2^\circ$
G	2549.612	10r	III	39209.89	−01	$a^5D_3 - x^5D_4^\circ$
G	2545.977	10r	III	39265.87	−01	$a^5D_2 - x^5D_3^\circ$
G	2544.706	6	IV	39285.48	00	$b^3F_4 - r^3G_5^\circ$
G	2543.920	6	IV	39297.61	+07	$b^3F_3 - r^3G_4^\circ$
C	2542.101	6	IV	39325.74	00	$b^3F_2 - r^3G_3^\circ$
G	2540.971	10R	III	39343.22	−01	$a^5D_1 - x^5D_2^\circ$
U	2539.575	(1)		39364.85	+15	$a^5F_3 - x^3F_3^\circ$
G	2539.355	(7)		39368.26	+01	$a^5F_4 - u^5D_3^\circ$
G	2537.454	(5)		39397.75	+05	$a^3F_4 - u^3G_5^\circ$
U	§2536.738	(5)		39408.87	−20	$a^5F_3 - w^3D_2^\circ$
G	2535.604	8r	III	39426.48	+01	$a^5D_0 - x^5D_1^\circ$
G	2535.128	(5)		39433.90	−02	$a^5F_2 - 1_2^\circ$
G	2532.874	(2)		39468.98	−02	$a^5F_3 - x^3F_2^\circ$
W	2531.5	(1)		39490.4	+2	$b^3F_3 - r^3G_3^\circ$
C	2530.694	3	III	39502.98	−14	$a^5D_2 - y^7P_3^\circ$
G	2529.833	3	III	39516.42	+01	$a^5D_1 - x^5D_1^\circ$
G	2529.134	10r	III	39527.35	−01	$a^5D_2 - x^5D_2^\circ$
W	2528.91	(3)		39530.85	−07	$b^3F_4 - r^3G_4^\circ$
G	2527.433	15r	II	39553.94	−01	$a^5D_3 - x^5D_3^\circ$
G	2524.290	8r	II	39603.20	+02	$a^5D_1 - x^5D_0^\circ$
G	2522.848	40R	II	39625.83	00	$a^5D_4 - x^5D_4^\circ$
G	2522.488	(6)		39631.48	−17	$a^5F_4 - z^3H_5^\circ$
G	2521.917	(7)		39640.46	00	$a^5F_4 - w^3D_3^\circ$
C	2519.628	(10)		39676.46	−02	$a^5F_1 - w^5G_2^\circ$
G	2518.100	12r	II	39700.53	−01	$a^5D_2 - x^5D_1^\circ$
G	2517.658	(8)		39707.50	+01	$a^5F_2 - w^5G_3^\circ$
G	2516.569	(5)		39724.68	−02	$a^5F_3 - z^1G_4^\circ$
G	2516.249	(2)		39729.74	−03	$a^5F_4 - z^3H_4^\circ$
G	2515.848	(2)		39736.07	+06	$a^3F_3 - u^3D_2^\circ$
G	2513.847	(3)		39767.70	00	$b^3F_2 - q^3G_3^\circ$
G	2512.361	5r	III	39791.22	+03	$a^5D_3 - y^7P_3^\circ$
G	2510.833	15R	II	39815.43	00	$a^5D_3 - x^5D_2^\circ$
G	2508.751	(5)		39848.47	+01	$a^5F_2 - x^3G_3^\circ$
C	2507.899	6	III	39862.01	+01	$a^5F_3 - w^5G_4^\circ$
G	2506.569	(4)		39883.16	00	$b^3F_3 - t^3H_4^\circ$
G	2505.004	(3)		39908.08	+04	$b^3F_4 - t^3H_5^\circ$
G	2503.491	(3)		39932.19	−01	$b^3F_3 - q^3G_3^\circ$
G	2501.692	(6)		39960.91	−02	$a^5F_5 - x^3F_4^\circ$
G	2501.130	20R	II	39969.88	00	$a^5D_4 - x^5D_3^\circ$
G	§2498.895	10	IV	40005.62	−34	$a^5D_3 - y^7P_4^\circ$

Ref	λ I A	Int	T C	Wave Number Observed	o−c	Desig
G	2496.992	(4)		40036.12	−07	$b^3F_4 - q^3G_5^\circ$
C	2496.532	6	III	40043.48	+03	$a^5F_4 - w^5G_5^\circ$
G	§2495.869	(5)		40054.12	+02	$a^5F_5 - z^3H_6^\circ$
G	2494.250	(5)		40080.13	−02	$a^5F_5 - z^3H_5^\circ$
G	*2493.998	(6)		40084.17	+06	$a^5F_3 - x^3G_4^\circ$
					−01	$a^5F_1 - v^5F_2^\circ$
W	2492.64	(2)		40105.95	−24	$a^5F_3 - x^3G_3^\circ$
W	2492.17	(2)		40113.57	00	$b^3F_4 - q^3G_4^\circ$
G	2491.983	(8)		40116.58	+04	$b^3F_4 - t^3H_4^\circ$
G	2491.155	20R	II	40129.92	−01	$a^5D_1 - x^5F_2^\circ$
G	2490.642	30R	II	40138.18	+05	$a^5D_2 - x^5F_3^\circ$
G	2489.751	15r	II	40152.55	00	$a^5D_0 - x^5F_1^\circ$
G	2488.942	(6)		40165.58	00	$b^3F_4 - q^3G_3^\circ$
G	2488.143	40R	II	40178.49	−05	$a^5D_3 - x^5F_4^\circ$
G	2487.368	(4)		40191.01	−01	$a^5D_2 - z^5S_2^\circ$
C	2487.064	(12)		40195.92	+02	$a^5F_1 - v^5F_1^\circ$
G	2486.690	(10)		40201.97	00	$a^5F_3 - v^5F_4^\circ$
G	§2486.372	(10)		40207.11	−01	$a^5D_4 - y^7P_3^\circ$
G	2485.989	(10)		40213.31	+02	$a^5F_4 - w^5G_4^\circ$
G	§2484.186	15R	II	40242.49	00	$a^5D_1 - x^5F_1^\circ$
G	2483.531	10	II?	40253.10	−01	$a^5F_2 - v^5F_2^\circ$
G	2483.270	60R	II	40257.32	−05	$a^5D_4 - x^5F_5^\circ$
G	2479.775	20R	II	40314.06	00	$a^5D_2 - x^5F_2^\circ$
G	2479.478	6	III	40318.90	−01	$a^5F_2 - x^3P_2^\circ$
G	2476.861	(2)		40361.49	+06	$a^5F_1 - x^3P_1^\circ$
G	2476.654	3	III	40364.86	+04	$a^5F_2 - v^5F_1^\circ$
C	2474.813	(8)		40394.88	−02	$a^5F_3 - v^5F_3^\circ$
G	2473.156	(3)	II	40421.96	+07	$a^5D_4 - y^7P_4^\circ$
V	2472.910	12R	II	40425.97	−23	$a^5D_3 - x^5F_3^\circ$
V	2472.875	(5)		40426.54	−08	$a^5D_2 - x^5F_1^\circ$
G	*2472.343	5	III	40435.24	−16	$a^5F_4 - x^3G_4^\circ$
					+13	$a^5F_5 - w^5G_6^\circ$
G	*2470.961	(4)		40457.86	+38	$a^5F_4 - x^3G_3^\circ$
					+01	$a^5F_4 - x^3G_5^\circ$
C	2468.878	4	III	40491.98	+03	$a^5F_5 - w^5G_5^\circ$
G	2467.730	(5)		40510.83	00	$a^5F_3 - v^5F_2^\circ$
C	2465.148	6	III	40553.25	−01	$a^5F_4 - v^5F_4^\circ$
G	§2463.728	(6)		40576.62	−02	$a^5F_3 - x^3P_2^\circ$
G	2462.645	10r	II	40594.46	−01	$a^5D_4 - x^5F_4^\circ$
G	2462.178	4	III	40602.15	+02	$a^5D_3 - x^5F_2^\circ$
G	2458.564	(4)		40661.83	+04	$a^5F_5 - w^5G_4^\circ$
C	2457.596	6	II	40677.86	+04	$a^5F_5 - v^5F_5^\circ$
C	2453.475	5	III	40746.18	−01	$a^5F_4 - v^5F_3^\circ$
G	2452.590	(2)		40760.88	−02	$a^3H_4 - t^3H_5^\circ$
A	§2447.708	4	II?	40842.17	+04	$a^5D_4 - x^5F_3^\circ$
G	2445.210	(6)		40883.89	−01	$a^5F_5 - x^3G_4^\circ$
C	2443.871	(20)		40906.29	−05	$a^5F_5 - x^3G_5^\circ$
C	2442.567	(20)		40928.12	−02	$a^3H_5 - t^3H_5^\circ$
G	2440.106	(15)		40969.40	00	$a^3H_4 - t^3H_4^\circ$
G	2439.743	(25)		40975.49	−01	$a^3H_6 - t^3H_6^\circ$
C	2438.181	2	III	41001.74	−02	$a^5F_5 - v^5F_4^\circ$
G	2429.810	(4)		41142.98	+04	$a^5F_1 - v^3D_1^\circ$
G	*2423.094	(4)		41257.01	+13	$a^5F_2 - v^3D_2^\circ$
					−15	$a^5F_2 - w^3F_3^\circ$
G	2420.390	(2)		41303.10	+05	$a^5F_5 - w^3G_5^\circ$
G	2419.879	(2)		41311.82	−04	$a^5F_2 - v^3D_1^\circ$
G	2419.058	(2)		41325.84	+05	$a^5F_4 - y^1G_4^\circ$
G	2417.490	(2)		41352.64	+03	$a^3F_4 - 9_4^\circ$
G	*2408.045	(3)		41514.83	+22	$a^5F_3 - v^3D_2^\circ$
					−05	$a^5F_3 - w^3F_3^\circ$
U	2398.215	(1)		41684.98	+11	$a^3F_4 - y^1F_3^\circ$
C	2389.971	(25)		41828.75	−01	$a^5D_2 - x^5P_3^\circ$
W	2385.9	(1)		41900.1	+3	$a^5F_3 - v^3G_4^\circ$
U	2381.831	(1)		41971.69	−01	$a^5D_1 - x^5P_2^\circ$
U	2377.991	(2)		42039.46	+06	$a^3F_3 - t^3G_3^\circ$

TABLE B—(*Continued*)

Ref	λ I A	Int	T C	Wave Number Observed	o−c	Desig
C	2374.517	(10)		42100.96	−02	$a^5D_0 - x^5P_1^\circ$
G	2373.618	(20)		42116.90	+07	$a^5D_3 - x^5P_3^\circ$
C	2371.428	(15)		42155.79	−04	$a^5D_2 - x^5P_2^\circ$
G	2369.454	(8)		42190.91	−01	$a^5D_1 - x^5P_1^\circ$
U	2365.509	(1n)		42261.27	+07	$a^3F_4 - t^3G_4^\circ$
U	2355.915	(1)		42433.35	−16	$a^5D_2 - y^5G_3^\circ$
G	§2355.327	(2)		42443.95	+05	$a^5D_3 - x^5P_2^\circ$
G	*2350.408	(5)		42532.77	+13	$a^5F_5 - v^3G_5^\circ$
					+01	$a^5D_4 - x^5P_3^\circ$
U	2341.575	(1n)		42693.20	+19	$a^5D_3 - z^5H_4^\circ$
G	2329.637	(2)		42911.95	+03	$a^5D_4 - y^5G_5^\circ$
C	2320.356	(40)	III	43083.58	−03	$a^5D_3 - w^5D_4^\circ$
G	2317.892	(2)		43129.37	+09	$a^3F_2 - s^3G_3^\circ$
C	2313.102	(40)	III	43218.67	−03	$a^5D_2 - w^5D_3^\circ$
C	2308.997	(30)	III	43295.50	−01	$a^5D_1 - w^5D_2^\circ$
G	2306.378	(4)		43344.66	+05	$a^3F_3 - s^3G_4^\circ$
G	2306.164	(2)		43348.68	+07	$a^5F_3 - t^5D_4^\circ$
G	2304.727	(5)		43375.71	+05	$a^5F_2 - t^5D_3^\circ$
C	2303.579	(20)	II	43397.32	−03	$a^5D_1 - w^5F_2^\circ$
C	2303.422	(15)		43400.27	−08	$a^5D_0 - w^5F_1^\circ$
C	2301.682	(20)		43433.09	−02	$a^5D_0 - w^5D_1^\circ$
U	2300.599	(1)		43453.53	−10	$a^3F_4 - v^3H_5^\circ$
C	2300.140	(30)		43462.21	−03	$a^5D_2 - w^5F_3^\circ$
U	2299.453	(1)		43475.19	−16	$a^5F_1 - t^5D_2^\circ$?
C	2299.218	(25)	III	43479.63	−01	$a^5D_2 - w^5D_2^\circ$
G	2298.657	(6)		43490.24	−05	$a^5D_1 - w^5F_1^\circ$
U	2298.175	10r	II	43499.37	−17	$a^5D_4 - w^5D_4^\circ$
C	2297.785	(35d)		43506.74	−03	$a^5D_3 - w^5D_3^\circ$
C	2296.925	(15d)		43523.04	−01	$a^5D_1 - w^5D_1^\circ$
U	2295.535	(1n)		43549.37	+05	$a^3F_4 - x^1H_5^\circ$
C	2294.406	(25)		43570.81	−02	$a^5D_1 - w^5D_0^\circ$
C	2293.845	(25)		43581.46	−02	$a^5D_2 - w^5F_2^\circ$
X	2292.79	(1)		43601.52	+08	$a^5F_1 - 7_2^\circ$
C	2292.523	(30)		43606.60	−02	$a^5D_3 - w^5F_4^\circ$
G	2291.624	(4)		43623.70	−03	$a^5D_1 - y^5S_2^\circ$
C	*2291.122	(15)		43633.26	−13	$a^5F_3 - t^5D_3^\circ$
					−08	$a^5F_2 - u^5F_3^\circ$
G	2290.771	(3)		43639.93	−01	$a^5F_4 - u^5F_5^\circ$
G	2290.546	(9)		43644.23	−04	$a^5F_2 - t^5D_2^\circ$
G	2290.064	(3)Ni?		43653.41	00	$a^5F_3 - u^5F_4^\circ$
G	2289.032	(10)		43673.09	+22	$a^5F_1 - u^5F_2^\circ$
C	2287.632	(15)		43699.82	−08	$a^5F_4 - t^5D_4^\circ$
C	2287.248	(30)		43707.16	−02	$a^5D_2 - w^5D_1^\circ$
C	2284.087	(40)		43767.63	−08	$a^5D_3 - w^5D_2^\circ$
C	2283.653	(12)		43775.95	−05	$a^5D_1 - v^5D_2^\circ$
G	2283.299	(9)		43782.74	+02	$a^5D_0 - v^5D_1^\circ$
G	2283.079	(9)		43786.96	−08	$a^5F_1 - t^5D_0^\circ$
G	2282.861	(4)		43791.14	00	$a^5F_1 - u^5F_1^\circ$
U	2281.986	(1)		43807.93	+07	$a^5D_2 - y^5S_2^\circ$
X	2281.66	(1)		43814.2	−3	$a^3F_4 - w^1F_3^\circ$
G	2280.222	(8)		43841.82	+03	$a^5F_2 - u^5F_2^\circ$
C	§2279.922	(10)		43847.58	+14	$a^5D_2 - v^5D_3^\circ$
U	2278.614	(2)		43872.75	+09	$a^5D_1 - v^5D_1^\circ$
G	2277.663	(12)		43891.07	00	$a^5F_3 - u^5F_3^\circ$
C	2277.098	(9)		43901.96	−04	$a^5F_3 - t^5D_2^\circ$
C	2276.025	(12)		43922.65	−05	$a^5D_4 - w^5D_3^\circ$
G	2275.593	(2)		43930.99	+03	$a^3F_4 - s^3G_5^\circ$
G	2275.189	(6)		43938.79	00	$a^5D_1 - v^5D_0^\circ$
C	*2274.088	(9)		43960.05	−01	$a^5F_2 - u^5F_1^\circ$
					−08	$a^5D_2 - v^5D_2^\circ$
G	2272.816	(8)		43984.66	−02	$a^5F_4 - t^5D_3^\circ$
C	2272.067	(15)		43999.16	−04	$a^5D_3 - v^5D_4^\circ$
C	2271.781	(40)		44004.69	−01	$a^5F_4 - u^5F_4^\circ$
C	2270.860	(18)		44022.54	−01	$a^5D_4 - w^5F_4^\circ$
G	2269.093	(18)		44056.82	+03	$a^5D_2 - v^5D_1^\circ$
G	2267.465	(15)		44088.45	+01	$a^5F_5 - u^5F_5^\circ$
G	2267.080	(9)		44095.93	00	$a^5D_3 - y^5S_2^\circ$
G	2266.903	(10)		44099.38	−14	$a^5F_3 - u^5F_2^\circ$
X	2265.61	(1)		44124.5	0	$a^5F_2 - u^5P_1^\circ$
C	2265.053	(20)		44135.39	−12	$a^5D_3 - v^5D_3^\circ$
C	2264.389	(45)		44148.32	−08	$a^5F_5 - t^5D_4^\circ$
U	2263.476	(6)		44166.14	−10	$a^5D_4 - w^5F_3^\circ$
G	2260.860	12	Fe II	44217.24	00	$(a^5F_3 - u^5P_2^\circ)$
U	2260.594	(2)		44222.44	+06	$a^3F_3 - u^3F_3^\circ$
C	2259.511	15		44243.63	−04	$a^5D_4 - w^5F_5^\circ$
U	2259.279	(1)		44248.17	−03	$a^5D_3 - v^5D_2^\circ$
U	2256.750	(1)		44297.76	+06	$a^3F_3 - u^3F_2^\circ$
C	2255.861	(45)		44315.21	+01	$a^5F_4 - u^5P_3^\circ$
G	2251.865	(12)		44393.84	+08	$a^5D_1 - x^3D_2^\circ$
G	2250.784	(10)		44415.16	+03	$a^5D_4 - v^5D_4^\circ$
C	2248.858	(25)		44453.20	00	$a^5F_5 - u^5F_4^\circ$
U	2247.461	(1)		44480.82	−08	$a^5F_5 - 4_4^\circ$
C	2245.651	(15)		44516.67	−07	$a^5D_2 - x^3D_3^\circ$
X	2245.14	(1)		44526.8	+1	$a^5F_2 - u^3D_1^\circ$
U	2243.911	(1)		44551.19	−25	$a^5D_4 - v^5D_3^\circ$
U	2242.579	(15)		44577.65	−24	$a^5D_2 - x^3D_2^\circ$
X	2241.85	(1)		44592.1	+3	$a^5F_4 - u^3D_3^\circ$
C	2240.627	(4)		44616.48	−02	$a^3F_4 - u^3F_4^\circ$
U	2238.259	(2)		44663.68	−02	$a^5D_1 - x^3D_1^\circ$
U	2237.814	(2n)		44672.56	+07	$a^3F_2 - t^3F_3^\circ$
U	2234.432	(2)		44740.16	−03	$a^3F_2 - t^3F_2^\circ$
C	2231.211	(15)		44804.75	−05	$a^5D_3 - x^3D_3^\circ$
U	2229.066	(5)		44847.85	+02	$a^5D_2 - x^3D_1^\circ$
U	2228.489	(1)		44859.46	+44	$a^5D_2 - y^3G_3^\circ$
C	2228.170	(10)		44865.88	−08	$a^5D_3 - x^3D_2^\circ$
G	2222.75	(7)		44975.1	+1	$a^3F_4 - v^1G_4^\circ$
U	2220.912	(2)		45012.50	−02	$a^5D_3 - y^3G_4^\circ$
U	2217.744	(1)		45076.78	−07	$a^5D_1 - x^5G_2^\circ$
U	2217.578	(1n)		45080.17	+06	$a^3F_3 - t^3F_3^\circ$
C	2211.234	(7)		45209.48	−05	$a^5D_2 - x^5G_3^\circ$
C	2210.686	(9)		45220.69	−05	$a^5D_4 - x^3D_3^\circ$
C	2207.068	(6)		45294.81	−05	$a^5D_4 - y^3G_5^\circ$
C	2201.117	(4)		45417.26	−05	$a^5D_3 - x^5G_4^\circ$
C	2200.723	(15)		45425.39	−09	$a^5D_1 - w^5P_2^\circ$
U	2200.370	(10r) (5)		45432.67	+30	$a^5D_0 - w^5P_1^\circ$
U	2197.230	(1)		45497.59	−01	$a^5D_3 - x^5G_3^\circ$
C	2196.040	(50)		45522.24	−07	$a^5D_1 - w^5P_1^\circ$
U	2193.564	(2)		45573.62	−21	$a^3F_4 - t^3F_4^\circ$
U	2193.411	(2)		45576.80	−10	$a^5F_4 - s^3D_3^\circ$
G	2191.836	(60)		45609.55	−06	$a^5D_2 - w^5P_2^\circ$
C	2191.202	(10)		45622.74	−07	$a^5D_0 - z^3S_1^\circ$
U	2189.393	(1n)		45660.44	−17	$a^5F_3 - t^5P_3^\circ$
U	2189.183	(1)		45664.82	+02	$a^3F_4 - t^3F_3^\circ$
C	2187.192	(40)		45706.38	−06	$a^5D_2 - w^5P_1^\circ$
C	2186.890	(5)		45712.69	−06	$a^5D_1 - z^3S_1^\circ$
G	2186.483	(40)		45721.19	−02	$a^5D_3 - w^5P_3^\circ$
U	2186.241	(3)		45726.26	+08	$a^5D_4 - x^5G_5^\circ$
U	2183.465	(1)		45784.39	−05	$a^5D_1 - y^3P_0^\circ$
U	2181.133	(1n)		45833.33	+09	$a^5D_4 - x^5G_4^\circ$
C	2180.866	(4)		45838.94	−06	$a^5D_1 - y^3P_2^\circ$
U	*2178.073	(35)		45897.72	+04	$a^5D_3 - w^5P_2^\circ$
					−16	$a^5D_2 - z^3S_1^\circ$
C	2176.837	(6)		45923.77	−05	$a^5D_0 - y^3P_1^\circ$
U	2176.396	(1)		45933.08	+02	$a^5F_3 - y^1F_3^\circ$
C	2173.212	(8)		46000.37	−08	$a^5D_1 - u^5D_2^\circ$
C	2172.581	(6)		46013.72	−04	$a^5D_1 - y^3P_1^\circ$
U	2172.137	(2)		46023.12	−01	$a^5D_2 - y^3P_2^\circ$
G	2171.292	(40)		46041.04	+01	$a^5D_2 - u^5D_3^\circ$
G	2166.769	(100)		46137.13	−01	$a^5D_4 - w^5P_3^\circ$

TABLE B—(*Continued*)

Ref	λ I A	Int	T C	Wave Number Observed	o−c	Desig
U	2165.537	(1n)		46163.38	−09	$a^5F_3 - 10_3^{\circ}$
C	2164.547	(7)		46184.49	−09	$a^5D_2 - u^5D_2^{\circ}$
C	2163.860	{(6) (1)}		46199.15	−03	$a^5D_0 - u^5D_1^{\circ}$
C	2161.577	(5)		46247.94	−07	$a^5D_1 - w^3D_2^{\circ}$
U	2160.236	(1)		46276.65	−10	$a^5F_3 - 11_3^{\circ}$
X	2159.92	(3)		46283.4	0	$a^5D_1 - u^5D_0^{\circ}$
U	2159.645	(4)		46289.30	+18	$a^5D_1 - u^5D_1^{\circ}$
U	2159.425	(2)		46294.02	00	$a^5D_0 - w^3D_1^{\circ}$
U	2158.922	(4)		46304.81	−11	$a^5D_3 - u^5D_4^{\circ}$
U	2158.622	{(1) (1)}		46311.25	+05	$a^5D_3 - y^3P_2^{\circ}$
C	2157.792	(5)		46329.06	−04	$a^5D_3 - u^5D_3^{\circ}$
U	*2155.238	(2)		46383.95	−28	$a^5F_3 - t^5P_2^{\circ}$
					−01	$a^5D_1 - w^3D_1^{\circ}$
U	2155.012	(3)		46388.82	+05	$a^5D_2 - x^3F_3^{\circ}$
C	2154.458	(2)		46400.74	+15	$a^5F_5 - 9_4^{\circ}$
C	2153.004	(5)		46432.07	−07	$a^5D_2 - w^3D_2^{\circ}$
C	*2151.099	{(3) (2)}		46473.19	−06	$a^5D_2 - u^5D_1^{\circ}$
					−08	$a^5D_3 - x^3F_4^{\circ}$
C	2150.182	(3)		46493.01	−06	$a^5D_2 - x^3F_2^{\circ}$
U	2149.416	(1)		46509.57	+18	$a^5F_3 - t^3G_4^{\circ}$
U	2149.170	(1)		46514.89	+13	$a^5F_4 - 10_3^{\circ}$
U	2148.394	(1n)		46531.69	+10	$a^5D_1 - 1_2^{\circ}$
U	2146.710	(2n)		46568.19	+10	$a^5D_2 - w^3D_1^{\circ}$
C	2145.188	(3)		46601.23	−08	$a^5D_3 - w^3D_3^{\circ}$
U	2144.576	(1)		46614.52	−03	$a^5F_2 - t^3G_3^{\circ}$
U	2142.141	(1n)		46667.51	+01	$a^5D_1 - y^3S_1^{\circ}$
C	2141.715	(1)		46676.79	−05	$a^5D_3 - x^3F_3^{\circ}$
U	2141.083	(1)		46690.56	−05	$a^5D_3 - z^3H_4^{\circ}$
U	2139.929	(2)		46715.74	+02	$a^5D_2 - 1_2^{\circ}$
C	*2139.695	{(3) (2)}		46720.85	00	$a^5D_4 - u^5D_4^{\circ}$
					+64	$a^5D_3 - w^3D_2^{\circ}$
C	2138.589	(3)		46745.01	−02	$a^5D_4 - u^5D_3^{\circ}$
U	2133.311	(1)		46860.64	−04	$a^5F_4 - t^3G_4^{\circ}$
C	2132.015	(4)		46889.13	−08	$a^5D_4 - x^3F_4^{\circ}$
U	2130.417	(1)		46924.29	−29	$a^5F_4 - 13_4^{\circ}$
U	2126.212	(1)		47017.08	−16	$a^5D_4 - w^3D_3^{\circ}$
N	2122.188	1		47106.2	−3	$a^5D_4 - z^3H_4^{\circ}$
N	2119.125	5		47174.3	+2	$a^5D_3 - w^5G_4^{\circ}$
C	2115.168	20		47262.54	−09	$a^5D_2 - v^5P_3^{\circ}$
N	2114.588	25		47275.5	+1	$a^5D_1 - v^5P_2^{\circ}$
N	2113.08	20		47309.1	−1	$a^5F_5 - t^3G_4^{\circ}$
C	2112.966	25		47311.79	−03	$a^5D_0 - v^5P_1^{\circ}$
C	2110.233	30		47373.06	+51	$a^5D_0 - v^5F_1^{\circ}$
					−02	$(a^5F_5 - 13_4^{\circ})$
C	2108.955	30		47401.76	00	$a^5D_1 - v^5P_1^{\circ}$
N	2108.302	12		47416.4	−2	$a^5D_1 - x^3P_2^{\circ}$
N	2108.188	1p		47419.0	+7	$a^5D_3 - x^3G_3^{\circ}$
N	2108.139	12		47420.1	−1	$a^5D_4 - w^5G_5^{\circ}$
N	2106.380	25		47459.7	+2	$a^5D_2 - v^5P_2^{\circ}$
N	2106.260	20		47462.4	−1	$a^5D_1 - v^5F_1^{\circ}$
N	2103.964	1		47514.2	+1	$a^5D_3 - v^5F_4^{\circ}$
N	2103.048	25		47534.9	0	$a^5D_2 - v^5F_2^{\circ}$
N	2102.910	20		47538.0	+1	$a^5D_0 - x^3P_1^{\circ}$
C	2102.349	30		47550.69	−01	$a^5D_3 - v^5P_3^{\circ}$
C	2100.795	30		47585.86	−03	$a^5D_2 - v^5P_1^{\circ}$
N	2100.144	10		47600.6	−1	$a^5D_2 - x^3P_2^{\circ}$

Ref	λ I A	Int	T C	Wave Number Observed	o−c	Desig
N	2098.953	25		47627.6	−4	$a^5D_1 - x^3P_1^{\circ}$
N	2098.081	15p		47647.4	+8	$a^5D_2 - v^5F_1^{\circ}$
N	2095.451	1		47707.2	+2	$a^5D_3 - v^5F_3^{\circ}$
N	2093.660	40		47748.0	+4	$a^5D_3 - v^5P_2^{\circ}$
N	2090.862	20		47811.9	−3	$a^5D_2 - x^3P_1^{\circ}$
N	2090.380	30		47822.9	−1	$a^5D_3 - v^5F_2^{\circ}$
N	2087.525	25		47888.3	−5	$a^5D_3 - x^3P_2^{\circ}$
N	2084.117	50		47966.6	0	$a^5D_4 - v^5P_3^{\circ}$
N	2058.100	1		48572.9	0	$a^3F_4 - t^3H_5^{\circ}$
N	2047.241	2		48830.5	0	$a^3F_4 - q^3G_3^{\circ}$?
N	2016.512	5		49574.5	0	$a^5F_4 - v^1G_4^{\circ}$
	λ *Vacuum*					
N	1974.059	1		50657.0	−5	$a^5D_2 - t^5D_3^{\circ}$
N	1973.911	1		50660.8	0	$a^5D_3 - t^5D_4^{\circ}$
N	1970.771	0		50741.6	−3	$a^5D_1 - t^5D_2^{\circ}$
N	1964.043	20		50915.4	+3	$a^5D_2 - u^5F_3^{\circ}$
N	1963.629	15		50926.1	0	$a^5D_2 - t^5D_2^{\circ}$
N	1963.110	25		50939.6	+1	$a^5D_1 - u^5F_2^{\circ}$
N	1962.871	20		50945.8	+3	$a^5D_3 - t^5D_3^{\circ}$
N	1962.746	15		50949.0	+3	$a^5D_1 - t^5D_1^{\circ}$
N	1962.100	30		50965.8	−8	$a^5D_3 - u^5F_4^{\circ}$
N	1962.031	25		50967.6	−2	$a^5D_0 - u^5F_1^{\circ}$
N	1961.236	20		50988.2	+2	$a^5D_2 - u^5P_3^{\circ}$
N	1960.129	25		51017.0	+3	$a^5D_4 - u^5F_5^{\circ}$
N	1958.739	15		51053.2	−4	$a^5D_1 - t^5D_0^{\circ}$
N	*1958.598	30		51056.9	−8	$a^5D_1 - u^5F_1^{\circ}$
					−3	$a^5D_1 - u^5P_2^{\circ}$
N	1957.831	25		51076.9	+2	$a^5D_4 - t^5D_4^{\circ}$
N	1956.026	30		51124.1	+5	$a^5D_2 - u^5F_2^{\circ}$
N	*1955.690	20		51132.8	−1	$a^5D_2 - t^5D_1^{\circ}$
					+6	$a^5D_0 - u^5P_1^{\circ}$
N	1952.997	20		51203.4	+2	$a^5D_3 - u^5F_3^{\circ}$
N	1952.596	30		51213.9	−2	$a^5D_3 - t^5D_2^{\circ}$
N	1952.262	20		51222.6	+4	$a^5D_1 - u^5P_1^{\circ}$
N	*1951.556	25		51241.2	−7	$a^5D_2 - u^5F_1^{\circ}$
					−1	$a^5D_2 - u^5P_2^{\circ}$
N	1950.223	20		51276.2	+2	$a^5D_3 - u^5P_3^{\circ}$
N	1946.978	25		51361.6	+1	$a^5D_4 - t^5D_3^{\circ}$
N	1946.219	10		51381.7	+2	$a^5D_4 - u^5F_4^{\circ}$
N	1945.294	25		51406.1	−2	$a^5D_2 - u^5P_1^{\circ}$
N	1945.070	20		51412.0	+3	$a^5D_3 - u^5F_2^{\circ}$
N	1940.649	25		51529.2	−2	$a^5D_3 - u^5P_2^{\circ}$
N	1937.274	35		51618.9	−2	$a^5D_4 - u^5F_3^{\circ}$
N	1934.528	25		51692.2	+2	$a^5D_4 - u^5P_3^{\circ}$
N	1903.37	1		52538.4	+6	$a^5D_3 - s^3D_3^{\circ}$
N	1888.32	12n		52957.2	+1	$a^5D_2 - y^1F_3^{\circ}$
N	1887.761	14		52972.8	0	$a^5D_3 - t^5P_3^{\circ}$
G	1880.14	5		53187.6	+1	$a^5D_2 - 10_3^{\circ}$
N	1878.849	2		53224.1	−1	$a^5D_1 - t^5P_2^{\circ}$
N	1876.421	10		53292.9	−1	$a^5D_0 - t^5P_1^{\circ}$
N	1873.259	15		53382.9	−1	$a^5D_1 - t^5P_1^{\circ}$
N	1873.052	12		53388.8	+1	$a^5D_4 - t^5P_3^{\circ}$
N	1872.359	15		53408.6	+3	$a^5D_2 - t^5P_2^{\circ}$
N	1866.815	10		53567.2	+1	$a^5D_2 - t^5P_1^{\circ}$
N	1866.07	12		53588.6	−3	$a^5D_3 - 11_3^{\circ}$
N	1863.54	0p		53661.2?	+1	$a^5D_4 - y^1F_3^{\circ}$
N	1862.318	15		53696.5	+1	$a^5D_3 - t^5P_2^{\circ}$
N	1855.58	15		53891.5	0	$a^5D_4 - 10_3^{\circ}$

NOTES TO TABLE B

* Blend.
§ Blend with *Fe* II.
( ) Masked.
** Notes by A. S. King as follows:

5204.582	Blend with *Cr*
4058.227	Blend with *Co*
3998.054	May be partly *Co*
3940.882	May be partly *Sr* and *Co*
3902.948	Blend with *Cr*
3786.176	Probably double
3753.610	Blend with *Ti*
3703.556	Blend with *V*
3605.450	Blend with *Cr* to violet
3369.549	Blend with *Ni*
3196.930	Blend with *Ni* to red
3134.111	Blend with *Ni*, but chiefly *Fe*
3092.778	Blend with *Al*

References for Wave-length and Intensity:

A International Standards, *Trans. Internat. Astron. Union* **6**: 80, 1938.
B International Standards, *Trans. Internat. Astron. Union* **3**: 86, 1928.
C Meggers and Humphreys, *Bur. Standards Jour. Research* **18**: 543 (RP 992), 1937.
D Kiess, *Bur. Standards Jour. Research* **20**: 33 (RP 1062), 1938.
E Meggers, *Bur. Standards Jour. Research* **14**: 33 (RP 755), 1935.
F Meggers and Kiess, *Bur. Standards Jour. Research* **9**: 309 (RP 473), 1932.
G Burns and Walters, *Publ. Allegheny Observ.* **8**: 39 (No. 4), 1931.
H Jackson, *Proc. Royal Soc.* **A 133**: 553, 1931.
I Babcock, *Mount Wilson Contr. No. 343*, 1927; *Astrophys. Jour.* **66**: 256, 1927.
J St. John and Babcock, *Mount Wilson Contr. No. 202*; *Astrophys. Jour.* **53**: 260, 1921. (Corrected by values in Table IV, Ref. I: mean of grating and interf. meas. used.)
K Burns and Walters, *Publ. Allegheny Observ.* **6**: 159 (No. 11), 1929.
L Meggers and Kiess, *Sci. Papers Bur. Standards* **19**: 273 (No. 479), 1924.
M King, *Mount Wilson Contr. No. 496*, 1934; *Astrophys. Jour.* **80**: 124, 1934.
N Green, *Phys. Rev.* **55**: 1209, 1939; and unpublished material, 1937.
O Meggers and Kiess, *Sci. Papers Bur. Standards* **14**: 642 (No. 324), 1918.
Q Babcock, unpublished material.
R King, A. S., unpublished material.
S Dobbie, unpublished material.
SS Solar Spectrum Wave-length (SSb denotes that the line is blended in the solar spectrum).
T Smith, Sinclair, unpublished material.
U Harrison, unpublished material, June 1942.
V Burns, *Lick Observ. Bull.* **8**: 27 (No. 247), 1913.
W Kayser, *Handbuch der Spectroscopie* **6**: 896, 1912.
X Schumacher, *Zeitschr. f. Wissen. Photographie* **19**: 149, 1919. (Corrected to agree with meas. in Ref. G and K.)
Y Dingle, *Monthly Notices Royal Astron. Soc.* **94**: 866, 1934.

For Intensity see also:

King, A. S., *Mount Wilson Contr. No. 247*, 1922; *Astrophys. Jour.* **56**: 318, 1922.
King, A. S., *Mount Wilson Contr. No. 66*, 1913; *Astrophys. Jour.* **37**: 239, 1913.

TABLE C

PREDICTED LINES OF FE I PRESENT IN THE SOLAR SPECTRUM

Solar λ	Solar Int	Grade	Wave Number		Desig	Solar λ	Solar Int	Grade	Wave Number		Desig
			Solar	o−c					Solar	o−c	
10987.02	1	fb	9099.16	+17	$b^3P_2 - z^3D_1^\circ$	8950.217	−1	g	11169.85	−02	$y^5D_3^\circ - e^5D_4$
10780.69	−2N	g	9273.30	+01	$b^3H_6 - z^3G_5^\circ$	8931.76	−2N	g	11192.93	+03	$a^1G_4 - z^3G_4^\circ$
10725.20	0	g	9321.28	−01	$b^3D_2 - y^3D_2^\circ$	8922.643	−2	g	11204.37	+03	$x^5F_5^\circ - f^5F_4$
10616.73	1	g	9416.52	+02	$b^3H_5 - z^3G_4^\circ$	8905.989	0	f	11225.32	00	$x^5F_3^\circ - e^5P_2$
10577.15	1	g	9451.75	−01	$b^3H_4 - z^3G_3^\circ$	8902.926	−3	g	11229.18	+01	$x^5D_2^\circ - e^7G_3$
10555.70	0	f	9470.95	−07	$w^5D_3^\circ - g^5F_4$	8895.98	−3	g	11237.95	+03	$z^3G_4^\circ - e^5F_5$
10388.77	−2	g	9623.14	−04	$w^5D_3^\circ - h^5D_3$	8892.11	−3	f	11242.84	+03	$x^5F_4^\circ - e^5P_3$
10379.04	−1	g	9632.16	−03	$a^5P_1 - z^5F_2^\circ$	8887.07	−3N	f	11249.22	+04	$x^5D_3^\circ - e^5G_3$
10364.05	0	f	9646.10	+08	$w^5D_3^\circ - f^5P_2$	8878.775	−2	g	11259.73	−01	$y^5D_2^\circ - e^5D_3$
10362.72	−1	f	9647.33	+01	$w^5D_2^\circ - g^5F_3$	8878.271	−1	g	11260.36	−02	$b^3G_4 - z^3G_5^\circ$
10333.21	−1	f	9674.88	+03	$d^3F_4 - u^5D_4^\circ$	8834.025	−2	g	11316.76	+01	$y^5D_1^\circ - e^5D_2$
10332.36	−1	g	9675.69	−02	$b^3D_1 - y^3D_1^\circ$	8828.103	−3	g	11324.36	−02	$x^5D_3^\circ - e^3D_3$
10311.96	−1	g	9694.82	−07	$a^3P_0 - z^5P_1^\circ$	*8819.51	−3	g	11335.38	−04	$x^5D_1^\circ - e^3D_2$
10307.46	−3N	f	9699.05	+02	$d^3F_4 - u^5D_3^\circ$	8816.876	−2	f	11338.78	−02	$x^5D_2^\circ - e^7S_3$
10283.87	−2	g	9721.30	00	$w^5D_1^\circ - h^5D_1$	8805.19	−2	f	11353.82	+02	$x^5D_4^\circ - e^5G_4$
10265.22	−2	g	9738.96	+01	$a^5P_1 - z^5F_1^\circ$	8798.07	−3	g	11363.01	−03	$y^7P_3^\circ - e^7S_3$
10156.56	0	g	9843.16	−05	$d^3F_4 - x^3F_4^\circ$	8779.08	−2	g	11387.59	+06	$y^5D_0^\circ - e^5D_1$
10155.19	1	g	9844.49	00	$a^5P_3 - z^5F_3^\circ$	8767.68	−3Nd?	f	11402.40	−03	$z^5P_2^\circ - X_3$
10149.13	−3	g	9850.37	−03	$x^5F_1^\circ - f^5D_0$	8727.19	−3	g	11455.30	−11	$c^3F_2 - x^3D_3^\circ$
10143.48	−3	f	9855.84	+12	$z^3D_3^\circ - X_2$	8700.314	−2	g	11490.69	+04	$x^5D_3^\circ - e^7G_3$
10137.14	−1	g	9862.01	−08	$x^5F_2^\circ - f^5D_1$	8698.717	0	g	11492.80	00	$b^3G_4 - z^5G_3^\circ$
10084.41	0	g	9913.58	+01	$d^3F_3 - x^3F_4^\circ$	*8689.788	−1	g	11504.61	−09	$a^1G_4 - z^3G_3^\circ$
10081.43	0	g	9916.51	−03	$a^3P_1 - z^5P_2^\circ$	*8686.75	3	g	11508.63	+03	$x^5D_2^\circ - e^3D_2$
10080.43	−1	f	9917.49	+01	$x^5F_1^\circ - f^7D_1$	8680.82	−3	g	11516.49	−07	$c^3F_2 - x^3D_2^\circ$
10070.58	−2	f	9927.20	00	$w^5D_0^\circ - g^5F_1$	8679.646	2	gb	11518.05	−03	$y^7P_2^\circ - e^7S_3$
10058.36	−3	f	9939.26	−08	$a^5P_2 - z^5F_1^\circ$	8671.879	0	g	11528.36	−03	$x^5D_0^\circ - e^5P_1$
10032.89	0	f	9964.49	−05	$w^5D_1^\circ - f^5G_2$	8663.723	−1	g	11539.22	+01	$x^5D_0^\circ - g^5D_3$
10019.81	0	f	9977.49	−05	$w^5D_2^\circ - f^5G_3$	8656.672	−1	g	11548.61	−02	$x^5D_0^\circ - e^3D_1$
10016.76	2	f	9980.54	−09	$x^5F_2^\circ - f^7D_2$	8654.436	−1	g	11551.60	−05	$a^3D_2 - y^3D_3^\circ$
9970.23	−2	g	10027.11	+02	$c^3P_2 - z^3P_1^\circ$	8652.475	−1	g	11554.21	+02	$y^5D_3^\circ - e^5D_3$
9967.30	3	fb	10030.06	+02	$x^5F_2^\circ - f^7D_1$	8632.424	0	g	11581.06	+01	$y^5D_4^\circ - e^5D_4$
9953.51	−1	g	10043.95	−07	$w^5D_3^\circ - h^5D_2$	8616.284	3	g	11602.75	−02	$x^5D_4^\circ - e^7G_5$
9951.19	0	g	10046.30	−04	$w^5D_4^\circ - h^5D_3$	8613.946	2	g	11605.89	−02	$x^5D_2^\circ - e^5P_3$
9950.62	0	g	10046.87	+09	$d^3F_4 - x^3F_3^\circ$	8610.609	2	g	11610.40	+02	$z^3G_4^\circ - e^5F_4$
9924.41	0	f	10073.40	−06	$a^1D_2 - y^3D_2^\circ$	8607.075	1	g	11615.16	00	$x^5D_1^\circ - e^5P_1$
9920.54	−1	g	10077.33	−09	$x^5F_1^\circ - e^7F_1$	8592.119	−2	g	11635.38	−02	$x^5D_1^\circ - e^3D_1$
9913.19	2	g	10084.81	00	$x^5F_5^\circ - e^7F_6$	8584.791	−2	g	11645.31	+03	$x^5D_1^\circ - g^5D_2$
9878.200	1	g	10120.53	−02	$x^5F_5^\circ - f^7D_5$	8576.48	−3	f	11656.60	+03	$d^3F_4 - y^1G_4^\circ$
9800.80	−3	g	10200.45	−02	$x^5F_1^\circ - e^7F_2$	8571.807	2	g	11662.95	+04	$x^5D_1^\circ - e^5P_2$
9771.07	−3	f	10231.49	−01	$d^3F_2 - w^3D_1^\circ$	8567.776	−1	g	11668.44	+01	$x^5D_4^\circ - e^3D_3$
9764.37	−3	f	10238.51	+03	$w^5D_3^\circ - f^5G_3$	8562.109	0	g	11676.16	+02	$z^3G_3^\circ - e^5F_3$
9608.93	−3	fb	10404.14	−04	$y^7P_3^\circ - e^7P_3$	8538.021	2	g	11709.10	−01	$x^5D_4^\circ - e^7G_4$
9573.65	−3N	g	10442.47	00	$x^5F_2^\circ - e^7G_3$	8527.847	0	g	11723.07	+05	$x^5D_0^\circ - g^5D_1$
9531.226	2	fb	10488.95	−01	$x^5F_3^\circ - e^7F_2$	8525.008	−1	f	11726.98	+05	$d^3F_3 - y^1G_4^\circ$
9433.34	−3	g	10597.80	−05	$x^5F_4^\circ - e^7F_4$	8519.10	−3	g	11735.11	−06	$x^5D_3^\circ - f^5F_2$
9409.59	−3NN	f	10624.54	−05	$x^5F_4^\circ - e^5G_3$	8509.65	−1	g	11748.15	−01	$z^5G_4^\circ - e^5F_5$
9383.423	−2	f	10654.17	−03	$y^7P_3^\circ - e^7P_2$	8496.483	−1	g	11766.35	+03	$z^5G_3^\circ - e^5F_4$
9297.14	0	g	10753.05	+01	$y^3D_3^\circ - e^3F_2$	8493.796	1	gb	11770.07	−01	$x^5D_3^\circ - e^3D_2$
9289.44	−2	f	10761.97	−05	$x^5F_3^\circ - f^5F_3$	8481.986	1	g	11786.46	−04	$c^3F_2 - x^3D_1^\circ$
9248.76	−3	f	10809.30	+06	$y^7P_2^\circ - e^7P_2$	8480.636	0	g	11788.33	−01	$x^5D_2^\circ - e^5P_1$
9203.21	−3d?	f	10862.80	−12	$x^5F_3^\circ - f^5F_2$	8466.510	−3	g	11808.00	+04	$c^3F_3 - x^3D_3^\circ$
9173.12	−3	g	10898.43	+10	$b^3F_2 - z^3D_1^\circ$	8466.102	−3	g	11808.57	00	$x^5D_2^\circ - e^3D_1$
9156.26	0	g	10918.50	−04	$b^3G_3 - z^5G_4^\circ$	8465.173	−2	g	11809.86	+07	$x^5D_1^\circ - g^5D_1$
9140.12	−3	f	10937.78	+04	$a^3D_3 - y^3F_3^\circ$	8461.472	−3	f	11815.03	−09	$z^5P_3^\circ - X_3$
9116.940	2N	fb	10965.58	−06	$x^5D_1^\circ - e^5G_2$	8458.99	−3N	g	11818.50	+04	$x^5D_2^\circ - g^5D_2$
9112.19	−3	g	10971.30	+07	$x^5F_5^\circ - e^7G_5$	8447.678	−3	g	11834.32	−07	$a^5F_3 - z^7D_4^\circ$
9038.79	−3	g	11060.39	+06	$b^3G_5 - z^5G_6^\circ$	8447.34	−3	f	11834.80	+10	$x^5D_4^\circ - e^7G_3$
9024 70	−2	g	11077.66	+09	$x^5F_5^\circ - e^7G_4$	8434.509	0d?	g	11852.80	+01	$x^5D_1^\circ - g^5D_0$
8994.66	−3d	f	11114.67	−03	$a^3D_1 - y^3F_2^\circ$	8425.889	−3	g	11864.93	+01	$a^5F_1 - z^7D_1^\circ$
*8978.16	−3	g	11135.08	−15	$x^5D_1^\circ - e^7G_2$	8414.084	0	gb	11881.57	−01	$z^3G_3^\circ - e^3F_4$
		g		+01	$a^1P_1 - y^3D_2^\circ$	8401.695	−2	g	11899.08	−03	$z^5G_2^\circ - e^5F_3$
8967.59	−3	g	11148.21	−06	$y^7P_4^\circ - e^7S_3$	8382.217	−3	g	11926.74	+02	$a^5F_2 - z^7D_2^\circ$
8956.30	−3	g	11162.26	−06	$x^5D_1^\circ - e^7G_1$	8369.858	−1?	f	11944.35	+02	$x^5D_4^\circ - e^7S_3$

TABLE C—(*Continued*)

Solar λ	Solar Int	Grade	Wave Number Solar	Wave Number o−c	Desig
8358.504	2	g	11960.58	+04	$b^3G_4 - z^3G_3^\circ$
8356.02	−3	f	11964.14	+08	$b^1D_2 - z^3S_1^\circ$
8355.15	−3	g	11965.38	+01	$y^5D_4^\circ - e^5D_3$
8349.02	3	gb	11974.17	+05	$a^5F_4 - z^7D_5^\circ$
8345.19	−3	g	11979.66	+01	$a^3G_5 - y^5F_5^\circ$
8342.290	3	fb	11983.82	−12	$b^3G_5 - z^3G_4^\circ$
8307.603	0	fb	12033.87	+02	$a^5F_2 - z^7D_1^\circ$
8303.17	−3	g	12040.29	−08	$a^3G_4 - y^5F_4^\circ$
8299.985	−1	g	12044.91	+04	$X_3 - v^3P_2^\circ$
8269.644	0	g	12089.09	+01	$d^3F_4 - v^3D_3^\circ$
8263.850	0	gb	12097.58	+01	$x^5D_3^\circ - e^5P_2$
8254.32	−3	f	12111.54	+03	$a^1G_4 - y^3F_4^\circ$
8204.95	0N	g	12184.42	−02	$a^5F_3 - z^7D_2^\circ$
8204.09	−2	g	12185.69	+01	$a^5F_4 - z^7D_4^\circ$
8196.51	−2	f	12196.96	+01	$d^3F_4 - w^3F_3^\circ$
8171.239	0	f	12234.69	+09	$a^1F_3 - w^3H_4^\circ$
8146.67	−3	g	12271.58	00	$a^3D_1 - y^3D_2^\circ$
8129.35	−3	g	12297.72	−05	$a^3G_3 - y^5F_2^\circ$
8112.179	−2	g	12323.76	−01	$a^3G_5 - y^5F_4^\circ$
8108.312	−2	g	12329.64	+03	$a^3G_4 - y^5F_3^\circ$
8072.162	−1	g	12384.85	−01	$a^3P_1 - z^3D_1^\circ$
8027.93	−1	g	12453.09	+05	$a^3D_3 - y^3D_2^\circ$
8016.523	−1	fb	12470.81	−01	$y^3D_2^\circ - e^5S_2$
8002.56	−2	f	12492.57	−01	$d^3F_2 - w^3F_2^\circ$
7965.55	−1N	f	12550.61	−05	$x^5F_2^\circ - f^5P_2$
7964.970	3	gb	12551.53	−06	$x^5F_3^\circ - g^5F_4$
7954.97	−1N	g	12567.30	−05	$b^3G_4 - y^3F_4^\circ$
7941.79	1N	fb	12588.16	+08	$a^1G_4 - y^3F_3^\circ$
7924.169	−2	g	12616.15	−05	$y^3D_2^\circ - e^3D_3$
7820.81	−1	f	12782.89	−01	$b^1D_2 - 1_2^\circ$
7813.67	−3NN	f	12794.57	−07	$x^5F_1^\circ - f^5P_1$
7810.815	1	g	12799.24	−01	$x^5F_4^\circ - g^5F_4$
7807.916	6	g	12804.00	+09	$x^5F_5^\circ - g^5F_5$
7802.51	1	g	12812.86	−04	$x^5F_2^\circ - g^5F_3$
7766.62	−3	f	12872.10	+19	$z^3F_3^\circ - e^5D_4$
7746.605	1	g	12905.33	−09	$x^5F_3^\circ - f^3D_3$
7745.521	1	g	12907.14	−06	$x^5F_2^\circ - f^5P_1$
7737.65	−1NN	g	12920.26	+01	$z^5G_3^\circ - e^3F_3$
7733.738	−2	gb	12926.81	−08	$x^5F_3^\circ - f^5G_4$
7720.72	−1	g	12948.60	−06	$x^5F_2^\circ - h^5D_2$
7719.046	3	g	12951.41	00	$x^5F_4^\circ - h^5D_3$
7689.04	−3	f	13001.95	+09	$x^5F_1^\circ - h^5D_1$
7664.18	−3	g	13044.12	−06	$y^3D_1^\circ - e^3D_1$
7661.48	−3	g	13048.72	−04	$x^5F_2^\circ - f^3D_2$
7647.84	−2NN	g	13072.00	00	$z^5G_2^\circ - e^3F_2$
7617.985	−2	g	13123.22	−03	$c^3F_2 - u^5D_2^\circ$
7617.245	−3	g	13124.49	−10	$x^5F_3^\circ - h^5D_2$
7588.310	3	g	13174.54	−01	$x^5F_4^\circ - f^5G_4$
7582.120	0	g	13185.30	+05	$x^5D_3^\circ - h^5D_4$
7573.72	−3	f	13199.92	+05	$z^3F_2^\circ - e^5D_2$
7552.795	−2	g	13236.49	00	$x^5F_4^\circ - g^5F_3$
7551.108	0	g	13239.45	−01	$x^5F_2^\circ - g^5F_2$
7547.904	0	g	13245.07	−03	$x^5F_1^\circ - f^5G_2$
7540.444	0	g	13258.17	−01	$a^3G_4 - z^5G_4^\circ$
7537.96	−3N	f	13262.54	+01	$v^5D_3^\circ - i^5D_3$
7528.18	−2N	f	13279.77	−06	$x^5F_4^\circ - e^5H_5$
7526.67	−3N	f	13282.43	−03	$v^5D_4^\circ - i^5D_4$
*7512.166	−2	g	13308.08	−08	$a^3P_2 - z^3D_2^\circ$
		g		+01	$c^3F_3 - u^5D_4^\circ$
7508.60	−3N	g	13314.40	−12	$x^5D_2^\circ - h^5D_3$
7506.030	1	g	13318.96	−09	$x^5F_3^\circ - f^5G_3$
7501.280	−2	g	13327.39	−06	$c^3F_2 - x^3F_3^\circ$
7495.66	−3	f	13337.39	+03	$x^5D_2^\circ - f^5P_2$
7494.74	−3	f	13339.02	−03	$a^3F_3 - z^5D_4^\circ$
7486.118	−3	g	13354.39	+02	$z^3D_3^\circ - e^5D_4$
7484.308	−1	g	13357.62	−04	$x^5F_2^\circ - f^5G_2$
7482.213	−1	g	13361.36	−02	$x^5F_2^\circ - e^3G_3$
7481.736	−3	g	13362.21	+02	$a^3G_3 - z^5G_3^\circ$
7477.595	0N	f	13369.60	−14	$z^3F_4^\circ - e^5D_4$
7476.87	−3	g	13370.90	+09	$c^3F_2 - w^3D_2^\circ$
*7474.513	−3N	f	13375.12	−02	$z^3F_2^\circ - e^5D_1$
		f		+16	$z^3D_2^\circ - e^5D_3$
7471.757	−2	g	13380.05	−02	$a^3G_4 - z^3G_5^\circ$
7463.395	−1	f	13395.04	−03	$x^5F_3^\circ - e^5H_4$
7461.25	−3	f	13398.89	+05	$v^5D_4^\circ - i^5D_3$
7452.110	−1	g	13415.33	−06	$x^5F_3^\circ - g^5F_2$
7447.912	0	fb	13422.88	+16	$v^5D_3^\circ - i^5D_2$
7443.25	−2	f	13431.30	+03	$x^5F_2^\circ - f^3D_1$
7431.97	−3N	gb	13451.68	−06	$y^5P_1^\circ - e^7P_2$
7420.241	−3N	f	13472.94	−08	$x^5F_2^\circ - e^5H_3$
7418.330	−3	g	13476.42	−01	$c^3F_3 - x^3F_4^\circ$
7415.193	−1	g	13482.12	00	$x^5F_5^\circ - e^3G_5$
7400.851	−3	g	13508.24	+03	$b^3F_2 - y^5F_2^\circ$
7398.96	−3N	g	13511.70	+05	$x^5F_5^\circ - f^5G_4$
7396.526	−1	g	13516.14	−04	$x^5D_2^\circ - f^3D_3$
7385.51	−3N	f	13536.30	+04	$y^3D_2^\circ - g^5D_1$
7385.00	−3	f	13537.24	−07	$x^5F_3^\circ - e^3G_3$
7382.614	0	g	13541.62	+04	$a^3G_5 - z^5G_4^\circ$
7373.011	2	fb	13559.08	−06	$a^3P_2 - z^3D_1^\circ$
7359.983	−3	f	13583.25	−06	$x^5F_5^\circ - e^3H_6$
7356.76	−3	f	13589.20	+09	$y^5P_1^\circ - f^7D_2$
7348.51	−2N	g	13604.46	00	$c^3F_3 - w^3D_3^\circ$
7347.15	−3	g	13606.98	+02	$a^3G_3 - z^5G_2^\circ$
7344.200	0	gb	13612.44	−04	$a^3G_4 - z^5G_3^\circ$
7341.78	−3	f	13616.93	00	$x^5F_5^\circ - e^5H_5$
7330.150	0	f	13638.53	+01	$y^5P_1^\circ - f^7D_1$
7325.28	−3	f	13647.60	+10	$z^3D_2^\circ - e^5D_2$
7317.43	−1	g	13662.24	−04	$x^5D_1^\circ - f^3D_2$
7316.739	1	g	13663.53	+06	$a^3G_5 - z^3G_5^\circ$
7312.08	−2N	f	13672.24	−05	$x^5F_4^\circ - e^3H_5$
7311.265	2	g	13673.76	−02	$z^3P_1^\circ - e^5F_2$
7300.548	−1	f	13693.83	+07	$c^3F_3 - z^3H_4^\circ$
7295.28	−3	g	13703.73	00	$y^5P_2^\circ - e^7P_2$
7278.526	−3	f	13735.27	−09	$x^5D_2^\circ - h^5D_2$
7268.566	1	f	13754.09	+03	$z^3F_4^\circ - e^5D_3$
7266.96	−3	f	13757.13	+07	$a^5P_3 - z^3F_4^\circ$
7261.30	−3	g	13767.85	−04	$x^5D_4^\circ - g^5F_4$
7261.016	1	g	13768.39	−03	$a^3G_4 - z^3G_4^\circ$
7225.79	−3N	g	13835.51	+05	$x^5D_2^\circ - f^3D_2$
7216.63	−1	g	13853.07	+09	$x^5D_1^\circ - g^5F_2$
7213.847	0	g	13858.42	−01	$z^3P_1^\circ - e^5F_1$
7205.536	−2	fb	13874.40	−04	$y^3D_3^\circ - g^5D_2$
7190.128	1	g	13904.13	−01	$c^3P_0 - y^3D_1^\circ$
7162.34	0	g	13958.08	+06	$x^5D_0^\circ - f^3D_1$
7160.859	−2	fb	13960.96	−02	$x^5F_4^\circ - e^3H_4$
7127.573	3	g	14026.16	00	$x^5D_2^\circ - g^5F_2$
7125.33	−2N	g	14030.58	−10	$d^3F_4 - t^3D_4^\circ$
7120.58	−3	g	14039.94	−05	$c^3F_3 - z^1G_4^\circ$
7120.03	1	fb	14041.02	−04	$y^5P_3^\circ - f^7D_4$
7118.105	0	g	14044.82	+03	$x^5D_1^\circ - f^3D_1$
7114.574	0	g	14051.79	−03	$a^3G_5 - z^3G_4^\circ$
7109.70	−3	f	14061.42	−04	$y^5P_2^\circ - e^5G_3$
7107.25	−3	f	14066.26	+09	$a^1F_3 - t^3G_3^\circ$
7105.87	−3	f	14069.00	+07	$c^3F_2 - x^3G_3^\circ$
7103.150	0	gb	14074.39	+01	$a^3H_5 - y^5F_5^\circ$
7101.31	−3	f	14078.04	−04	$a^5P_2 - z^3F_3^\circ$
7093.09	0	g	14094.35	+03	$y^5P_3^\circ - e^7P_2$
7089.71	−2N	gb	14101.07	+03	$d^3F_3 - t^5D_4^\circ$
7079.27	−1N	g	14121.86	+14	$x^5D_4^\circ - f^3D_3$
7074.485	−3N	g	14131.41	−08	$y^3F_3^\circ - e^3D_3$

TABLE C—(*Continued*)

Solar λ	Solar Int	Grade	Wave Number		Desig
			Solar	o−c	
7072.80	−2	f	14134.78	+03	$c^3F_4 - z^3H_5^\circ$
7068.64	0	f	14143.10	−09	$x^5D_4^\circ - f^5G_4$
7068.07	−1	f	14144.24	−11	$x^5D_2^\circ - f^5G_2$
7062.79	−3	f	14154.80	00	$x^5D_2^\circ - g^5F_1$
7057.92	−2	f	14164.58	+08	$z^5P_2^\circ - e^7D_2$
*7034.090	−2	{f f}	14212.57	{−06 −02	$y^5P_3^\circ - e^5G_4$ $y^5P_2^\circ - e^5G_2$
7031.40	−3N	g	14218.01	+04	$x^5D_2^\circ - f^3D_1$
7031.09	−2	g	14218.63	−15	$y^3F_2^\circ - e^3D_2$
7028.59	−1	gb	14223.69	−01	$c^3P_1 - y^3D_1^\circ$
7022.385	−1	g	14236.27	+02	$y^5F_1^\circ - e^3F_2$
6997.080	−1	f	14287.74	+10	$x^5D_3^\circ - g^5F_2$
6983.52	−3N	g	14315.49	+03	$d^3F_4 - t^5D_3^\circ$
6979.156	−3	f	14324.44	+03	$b^3P_2 - y^3F_3^\circ$
6970.495	1	g	14342.23	−03	$c^3P_2 - y^3D_2^\circ$
6963.01	−3N	f	14357.65	+01	$c^3F_2 - v^5F_3^\circ$
6953.057	1N	fb	14378.20	−09	$z^5P_3^\circ - e^7D_3$
6942.84	−2NN	f	14399.36	−04	$c^3F_3 - x^3G_4^\circ$
6936.496	−1	f	14412.53	−04	$y^5P_2^\circ - e^7S_3$
6932.498	−3	g	14420.84	−02	$d^3F_2 - t^5D_3^\circ$
6930.384	−3NN	f	14425.24	−08	$y^5P_3^\circ - e^7F_4$
6926.385	−3	f	14433.57	+03	$d^3F_3 - 4_4^\circ$
6920.168	−2	f	14446.54	−01	$y^5P_2^\circ - f^5F_3$
6881.054	−3N	fb	14528.65	+02	$y^3F_2^\circ - g^5D_2$
*6879.55	−3	g	14531.83	+08	$z^5D_2^\circ - X_3$
6864:324	−2N	f	14564.07	−02	$y^5P_3^\circ - e^7F_2$
6860.099	−3	g	14573.04	+08	$y^3D_1^\circ - f^5P_2$
6859.493	−3	f	14574.32	−01	$b^3P_1 - y^3F_2^\circ$
6848.87	−3	f	14596.93	−01	$y^5P_2^\circ - f^5F_1$
6845.98	−3	fb	14603.09	−10	$y^5P_3^\circ - e^5G_2$
6841.642	−3	f	14612.35	−01	$X_2 - w^1F_3^\circ$
6824.857	−3	f	14648.28	−08	$x^5D_2^\circ - e^3P_2$
6819.49	−3	f	14659.81	−15	$c^3P_2 - y^3D_1^\circ$
6808.769	−3N?	f	14682.90	+07	$b^3P_2 - y^3F_2^\circ$
6805.752	−3	g	14689.42	−05	$d^3F_2 - t^5D_2^\circ$
6803.854	−3	f	14693.50	−03	$y^5P_3^\circ - e^7G_3$
6803.27	−3	f	14694.77	+06	$y^5P_3^\circ - f^5F_4$
6801.849	−3	g	14697.83	+04	$a^3F_2 - z^5F_1^\circ$
6801.202	−3	f	14699.23	+26	$z^5D_1^\circ - X_2$?
6794.623	−2	f	14713.47	−04	$x^5D_3^\circ - f^3F_4$
6786.460	−3	g	14731.16	−11	$z^5D_3^\circ - X_3$
6785.88	−3N	f	14732.42	00	$c^3F_4 - v^5F_5^\circ$
6785.764	−2	f	14732.68	−01	$d^3F_3 - y^1D_2^\circ$
6783.28	−3	g	14738.07	−02	$b^3F_4 - z^3G_5^\circ$
6769.682	−3	f	14767.68	−05	$d^3F_2 - y^1D_2^\circ$
6764.19	−3	g	14779.66	−14	$d^3F_4 - u^3G_3^\circ$
6756.568	−3	fb	14796.34	−02	$b^1D_2 - w^3F_2^\circ$?
6753.470	−2	f	14803.12	−04	$y^5P_3^\circ - e^7S_3$
6746.975	−2	g	14817.37	−03	$b^3F_2 - z^5G_2^\circ$
6745.984	−1	g	14819.55	−06	$c^3F_4 - w^5G_3^\circ$
6737.28	−3	g	14838.70	+05	$z^5D_2^\circ - X_2$
6736.546	−3	g	14840.31	+03	$b^1D_2 - z^1D_2^\circ$
6712.467	−3	f	14893.55	−06	$x^5D_2^\circ - f^3F_3$
6711.282	−3	g	14896.18	−09	$d^3F_2 - t^5D_1^\circ$
*6700.919	−3	f	14919.22	−04	$X_3 - w^1F_3^\circ$
6696.322	0	g	14929.46	−04	$y^3D_1^\circ - f^5P_1$
6682.24	−3	f	14960.92	−02	$c^3F_4 - x^3G_5^\circ$
6681.30	−3	f	14963.02	+08	$z^3G_5^\circ - e^7F_6$
*6677.54	−3	{g f}	14971.45	{−01 −12	$z^5D_4^\circ - X_3$ $x^5D_1^\circ - e^3P_1$
6676.89	−3	g	14972.90	−07	$y^5P_3^\circ - e^3D_2$
6647.856	−3	g	15038.30	+13	$z^5D_3^\circ - X_2$
6635.702	−1	f	15065.84	−05	$z^3G_4^\circ - e^7F_5$
6615.01	−3	f	15112.97	+05	$z^3G_3^\circ - e^7F_4$
6609.693	−2	g	15125.13	−02	$a^5F_2 - z^7F_3^\circ$
6601.14	−3	f	15144.72	−03	$x^5D_2^\circ - e^3P_1$
6555.864	−3N	f	15249.31	+02	$c^3F_4 - v^5F_3^\circ$
6551.714	−2	fb	15258.98	−07	$a^5F_2 - z^7F_1^\circ$
6524.749	−3	f	15322.03	+03	$x^5D_1^\circ - e^3P_0$
6494.510	1	g	15393.38	+04	$y^3D_3^\circ - f^5P_3$
6483.954	−1N	g	15418.44	00	$a^3F_4 - z^5F_3^\circ$
6472.152	−3N	f	15446.55	−02	$z^5G_4^\circ - e^5G_5$?
6468.842	−3	f	15454.45	+04	$y^3D_2^\circ - h^5D_1$
6456.874	−1N	f	15483.10	−02	$y^3D_2^\circ - f^5G_3$
6419.676	−1	f	15572.81	−06	$z^3F_3^\circ - e^5F_4$
6396.395	−3N	f	15629.49	−01	$b^1G_4 - y^3G_4^\circ$
6388.424	−2	f	15649.00	−02	$z^5F_4^\circ - e^7D_5$
6385.744	0	f	15655.56	−02	$y^3D_3^\circ - g^5F_3$
6376.180	−3	f	15679.05	+09	$z^5G_6^\circ - e^5G_6$
6353.856	−3	f	15734.13	−04	$a^5F_4 - z^7F_3^\circ$
6351.305	−3	f	15740.45	−04	$z^5G_5^\circ - e^5G_6$
6339.982	−3	f	15768.56	−06	$z^5F_3^\circ - e^7D_4$
6315.425	−1	g	15829.88	−02	$c^3F_3 - v^3D_2^\circ$
6307.885	−2N	f	15848.80	−08	$b^3D_3 - x^3D_3^\circ$
6293.952	−1d?	g	15883.88	−08	$y^3D_1^\circ - e^3P_2$
6290.547	−2N	g	15892.48	00	$b^3F_3 - y^5P_3^\circ$
6284.007	−3	f	15909.02	−01	$a^3D_2 - x^5P_3^\circ$
6253.847	−1	f	15985.74	−06	$y^3D_3^\circ - f^5G_3$
6251.292	−3	f	15992.28	−08	$y^3F_3^\circ - h^5D_4$
6249.657	−2	f	15996.46	−03	$z^5F_4^\circ - e^7D_4$
6219.528	−3	f	16073.95	+01	$z^5F_2^\circ - e^7D_2$
6209.760	−3N	f	16099.24	−08	$z^3D_1^\circ - e^5F_2$
6187.413	−1	g	16157.39	−01	$b^3P_2 - y^3D_1^\circ$
6171.010	−3	f	16200.33	−01	$y^3D_3^\circ - f^5G_2$
6157.427	−2	f	16236.07	−03	$a^3D_2 - x^5P_2^\circ$
6148.668	−2	f	16259.19	−07	$z^5G_6^\circ - f^5F_5$
6145.415	−2N	f	16267.80	00	$z^5F_4^\circ - e^7D_3$
6139.663	−2N	g	16283.04	−03	$b^3F_3 - y^5P_2^\circ$
6137.509	−2	f	16288.76	00	$z^5F_5^\circ - e^7D_4$
6124.084	−3	f	16324.46	−01	$a^1F_3 - u^3F_2^\circ$
6120.257	−1	f	16334.67	−02	$a^5F_4 - z^7P_4^\circ$
6114.396	−3N	f	16350.33	+04	$z^3D_2^\circ - e^5F_2$
6106.865	−3	g	16370.49	−06	$b^3F_2 - y^5P_1^\circ$
6105.144	0	g	16375.11	+03	$y^3F_4^\circ - g^5F_5$
6098.259	0	f	16393.59	+06	$y^5P_3^\circ - f^5P_3$
6097.106	−3	f	16396.70	−06	$a^5P_3 - z^3P_2^\circ$
6091.738	−2	f	16411.14	+01	$y^5P_2^\circ - f^5P_2$
6083.708	−3N	fb	16432.80	−10	$z^3D_3^\circ - e^5F_3$
6081.843	−3	f	16437.84	+01	$c^3F_3 - v^3G_3^\circ$
6081.723	−3	g	16438.17	00	$z^5G_3^\circ - g^5D_2$
6065.813	−3	f	16481.28	−02	$b^3H_4 - z^5H_4^\circ$
6060.829	−3	f	16494.84	−06	$y^5F_4^\circ - f^5D_3$
6051.037	−3N	f	16521.52	−11	$b^3F_4 - y^3F_3^\circ$
6019.386	−2	f	16608.40	−08	$a^1H_5 - y^3G_4^\circ$
6018.314	0	fb	16611.36	+07	$y^3F_2^\circ - h^5D_1$
6016.930	−3	f	16615.18	+05	$d^3F_4 - y^1F_3^\circ$
6015.264	−2	g	16619.78	−04	$a^5P_1 - y^5F_2^\circ$
6012.784	−2N	fb	16626.64	−08	$y^5P_3^\circ - g^5F_4$
6009.853	−3	fb	16634.74	−06	$a^3D_3 - x^5P_2^\circ$
6007.722	−2	f	16640.64	+07	$b^3H_5 - z^5H_5^\circ$
5996.510	−3	f	16671.76	−06	$y^5F_2^\circ - e^5G_3$
5996.230	−3	f	16672.54	−02	$a^3D_1 - x^5P_1^\circ$
5995.949	−2N	f	16673.32	−05	$y^5P_2^\circ - g^5F_3$
5991.575	−3	f	16685.49	00	$d^3F_3 - y^1F_3^\circ$
5981.398	−3	f	16713.88	−06	$a^1I_6 - z^3I_6^\circ$
5978.149	−2	g	16722.96	+06	$y^5P_1^\circ - h^5D_1$
5976.171	−3	f	16728.50	+02	$b^1D_2 - v^3F_3^\circ$
5974.596	−3N	f	16732.91	+07	$y^5D_2^\circ - f^5D_3$
5973.362	−3	g	16736.36	+03	$y^3F_2^\circ - g^5F_2$
5958.351	−1	g	16778.53	−03	$a^5P_3 - y^5F_3^\circ$

TABLE C—(*Continued*)

Solar λ	Solar Int	Grade	Wave Number		Desig
			Solar	o−c	
5955.117	−3	f	16787.64	00	$d^3F_3 - x^1F_3^\circ$
5952.192	−3	g	16795.89	−02	$x^5F_2^\circ - i^5D_3$
5950.142	−2	f	16801.68	−04	$y^5P_3^\circ - f^5P_2$
*5943.602	0	g	16820.16	−06	$a^5P_2 - y^5F_2^\circ$
5943.117	−3	f	16821.54	−02	$c^3F_2 - z^1F_3^\circ$
5942.737	−3	f	16822.61	−07	$d^3F_2 - x^1F_3^\circ$
5933.811	−2	f	16847.92	−02	$y^5P_1^\circ - g^5F_2$
5931.905	−3	f	16853.33	−05	$c^3F_4 - y^3H_4^\circ$
5928.527	−3	f	16862.93	−09	$y^5D_1^\circ - f^5D_1$
5901.533	−1	f	16940.07	−02	$y^5F_4^\circ - e^5G_4$
5893.243	−2	f	16963.89	00	$y^5D_1^\circ - f^5D_0$
5892.478	−2	f	16966.10	−04	$y^5P_1^\circ - f^5G_2$
5891.186	1	fb	16969.82	−07	$y^3F_2^\circ - e^5H_3$
5890.508	−1d?	gb	16971.77	−07	$x^5F_3^\circ - i^5D_3$
5887.478	−2	f	16980.51	−04	$y^5P_3^\circ - f^3D_3$
5881.728	−2	g	16997.11	+09	$a^5P_3 - y^5F_2^\circ$
5881.288	1	g	16998.38	−03	$y^3F_3^\circ - f^5G_3$
5879.503	0	f	17003.54	−05	$y^5P_2^\circ - f^5G_3$
5876.302	−1	g	17012.80	−08	$y^5F_1^\circ - f^5F_2$
5867.010	−3	f	17039.74	−01	$y^5P_1^\circ - f^3D_1$
5861.120	0	g	17056.87	−04	$y^5F_2^\circ - f^5F_3$
5859.965	−3	f	17060.24	−01	$y^5D_2^\circ - f^7D_3$
5858.790	0	g	17063.65	−05	$y^5F_4^\circ - f^5F_5$
5858.284	−2	f	17065.12	−05	$a^3H_5 - y^3F_4^\circ$
5849.698	0	f	17090.17	−08	$b^1G_4 - x^3F_4^\circ$
5845.298	0	g	17103.04	−08	$x^5F_4^\circ - i^5D_4$
5835.589	−1	f	17131.49	−03	$b^3P_2 - x^5D_3^\circ$
5835.434	−1	g	17131.95	−08	$x^5F_3^\circ - i^5D_2$
5835.114	0	g	17132.89	−04	$y^5F_3^\circ - f^5F_4$
5827.886	0	f	17154.14	00	$z^5D_1^\circ - e^7D_2$
5826.649	−2N	g	17157.78	−04	$y^5F_2^\circ - f^5F_2$
5815.229	(1)	g	17191.47	+01	$d^3F_4 - t^3G_4^\circ$
5813.341	−3N	f	17197.06	−03	$y^5D_2^\circ - f^7D_2$
5809.878	−2N	g	17207.31	+01	$y^5F_2^\circ - f^5F_1$
5807.995	−1	g	17212.88	−07	$y^3F_3^\circ - f^5G_2$
5807.804	−1	f	17213.45	−04	$z^5D_0^\circ - e^7D_1$
5807.253	−3	f	17215.08	−09	$b^3H_6 - z^5H_6^\circ$
5796.674	−2	f	17246.50	00	$y^5D_2^\circ - f^7D_1$
5791.539	0	g	17261.79	−03	$d^3F_3 - t^3G_4^\circ$
5787.277	−2	f	17274.50	−01	$a^3D_3 - w^5D_4^\circ$
5787.024	−1	gb	17275.26	−11	$y^5F_3^\circ - f^5F_3$
5778.814	−2	fb	17299.80	−02	$y^5P_3^\circ - f^3D_2$
5762.847	0	gb	17347.73	−04	$y^5F_1^\circ - e^3D_1$
5761.094	−2	f	17353.01	−05	$y^5D_1^\circ - e^5G_2$
5760.538	−2	f	17354.69	−02	$y^5D_3^\circ - f^7D_3$
5754.923	−2N	g	17371.62	−10	$a^3P_0 - y^5P_1^\circ$
5753.983	−2N	fb	17374.46	−03	$a^3H_4 - y^3F_3^\circ$
5753.400	−2N	g	17376.22	−06	$y^5F_3^\circ - f^5F_2$
5749.640	−2	f	17387.58	+04	$z^3G_4^\circ - h^5D_4$
5747.865	−2	f	17392.95	−05	$b^3P_2 - x^5D_2^\circ$
5739.810	−2	fb	17417.36	−10	$y^5D_2^\circ - e^5G_3$
5738.242	0	g	17422.12	−05	$y^5F_4^\circ - f^5F_4$
5732.886	−2	f	17438.40	−07	$y^5D_4^\circ - f^5D_3$
5732.311	0	g	17440.14	−08	$x^5F_5^\circ - i^5D_4$
*5721.717	−1	f f	17472.44	−03 −04	$y^5F_2^\circ - e^5P_1$ $y^5D_3^\circ - e^5G_4$
5720.902	0	g	17474.92	−06	$y^3F_4^\circ - f^5G_3$
5715.476	−2	f	17491.51	−03	$y^5D_3^\circ - f^7D_2$
5714.903	−3	f	17493.27	−07	$z^5D_3^\circ - e^7D_2$
5709.931	−1	f	17508.50	00	$y^5F_3^\circ - e^5P_3$
5707.249	−2	f	17516.73	+01	$b^3D_3 - u^5D_2^\circ$
5706.116	0	f	17520.21	00	$y^5F_2^\circ - e^5P_2$
5696.108	0	fb	17550.99	−01	$y^3F_4^\circ - e^5H_4$
5691.707	−1	g	17564.56	−05	$y^5F_4^\circ - f^5F_3$
5690.074	−3	f	17569.60	−02	$x^5D_1^\circ - i^5D_2$
5685.887	−2	f	17582.54	−07	$x^5D_2^\circ - i^5D_3$
5678.616	−2	g	17605.05	−05	$a^3P_1 - y^5P_2^\circ$
5678.407	−2	g	17605.70	−08	$z^3D_3^\circ - e^3F_2$
5677.705	−1	f	17607.87	−08	$y^5D_4^\circ - e^5G_5$
5661.988	−2	f	17656.75	−07	$z^3P_1^\circ - e^5P_1$
5661.028	−2	f	17659.75	00	$d^3F_2 - t^3G_3^\circ$
5658.672	0	g	17667.10	+01	$y^5F_2^\circ - g^5D_1$
5652.026	−3	g	17687.88	−04	$y^5D_1^\circ - f^5F_2$
5651.477	0	g	17689.59	−03	$z^3G_3^\circ - f^5G_4$
5650.287	−3	g	17693.32	+08	$y^3F_4^\circ - e^3G_3$
5648.917	−3	f	17697.61	−06	$a^3D_3 - w^5D_3^\circ$
5646.690	−1	f	17704.59	+03	$z^3P_1^\circ - e^5P_2$
5644.352	−2	f	17711.92	00	$y^5D_3^\circ - e^5G_3$
5643.945	−2	f	17713.20	−01	$c^3F_4 - z^1F_3^\circ$
5642.764	−1	f	17716.91	−04	$y^3F_3^\circ - e^3P_2$
5636.004	−3	f	17738.16	−01	$y^5D_2^\circ - e^7G_2$
5634.526	−3	fb	17742.81	+01	$x^5D_2^\circ - i^5D_2$
5627.100	−2	g	17766.22	−07	$y^5F_5^\circ - f^5F_4$
5623.644	−3	f	17777.14	−01	$a^3D_1 - w^5D_2^\circ$
5617.152	−1	f	17797.69	−05	$y^5F_4^\circ - e^5P_3$
5615.169	−3	g	17803.98	+04	$z^5G_4^\circ - g^5F_5$
5614.284	−1N	g	17806.78	+03	$x^5F_2^\circ - g^5G_2$
5613.716	−2N	f	17808.58	−05	$x^5D_1^\circ - 4_2$
5609.991	−2	f	17820.41	−06	$b^3D_2 - u^5D_1^\circ$
5608.982	−1	g	17823.61	00	$z^3P_2^\circ - g^5D_3$
5607.673	0	f	17827.77	−03	$y^5D_3^\circ - e^7G_4$
5602.569	−2	f	17844.01	−08	$x^5D_3^\circ - i^5D_3$
5595.069	−2	g	17867.93	−03	$x^5F_3^\circ - g^5G_3$
5583.992	−1N	g	17903.38	−08	$y^5D_2^\circ - f^5F_2$
5579.357	−1	f	17918.25	−07	$y^5D_0^\circ - e^3D_1$
5577.035	−1	g	17925.71	00	$x^5F_4^\circ - g^5G_4$
5570.070	−2	g	17948.13	−03	$b^3P_1 - z^5S_2^\circ$
5568.709	−3	f	17952.51	−01	$c^3F_3 - v^3F_3^\circ$
5568.466	−3	f	17953.30	−09	$y^5D_3^\circ - e^7G_3$
5568.081	−2	g	17954.53	−03	$y^5D_3^\circ - f^5F_4$
5566.819	−2	f	17958.61	00	$a^3D_3 - w^5D_2^\circ$
*5563.700	0	f f	17968.67	−02 −03	$a^3P_1 - y^3F_2^\circ$ $c^3F_3 - u^5F_4^\circ$
5562.128	−2	f	17973.75	−03	$z^3G_5^\circ - e^5H_6$
5559.652	−1	f	17981.76	−05	$x^5D_2^\circ - 4_2$
*5557.921	0	fb	17987.36	−07	$c^3P_0 - x^5P_1^\circ$
5555.180	−3	f	17996.23	−04	$a^1D_2 - z^3S_1^\circ$
5553.238	−2N	g	18002.53	−05	$y^5D_1^\circ - e^5P_1$
5552.854	−3	f	18003.77	00	$b^3P_2 - x^5F_3^\circ$
5552.702	−1	f	18004.26	−02	$x^5D_3^\circ - i^5D_2$
5551.781	−3N	g	18007.25	−03	$y^5D_4^\circ - f^5F_5$
5551.310	−3N	f	18008.78	−05	$a^1P_1 - x^3D_1^\circ$
5549.661	−1	g	18014.13	00	$x^5F_5^\circ - g^5G_5$
5549.534	−3	f	18014.54	+06	$z^3G_5^\circ - g^5F_4$
*5543.049	−2	g g	18035.62	−04 −06	$b^1G_4 - x^3G_5^\circ$ $y^5D_2^\circ - e^5P_3$
5541.592	−3	f	18040.36	−04	$a^3D_2 - v^5D_2^\circ$
5536.599	−1	g	18056.63	−03	$b^3P_2 - z^5S_2^\circ$
5529.791	−3	f	18078.86	+03	$b^3P_0 - x^5F_1^\circ$
5528.905	−1	g	18081.76	−03	$z^3G_3^\circ - f^5G_3$
5521.303	−1	f	18106.65	−06	$z^3G_4^\circ - e^5H_5$
5518.546	−3	gb	18115.70	+08	$x^5F_4^\circ - g^5G_3$
5516.306	−2	f	18123.05	−05	$y^5D_4^\circ - e^5G_3$
5512.414	−1	f	18135.85	−05	$z^5G_3^\circ - f^3D_3$
5510.243	−3	f	18142.99	−05	$c^3F_4 - u^5F_5^\circ$
5505.734	−3	f	18157.85	+05	$z^3G_3^\circ - e^5H_4$
5499.600	−1	f	18178.10	−02	$z^3G_3^\circ - g^5F_2$
5496.577	−1	f	18188.10	−04	$x^5D_4^\circ - i^5D_3$
5493.356	−3	f	18198.77	−08	$b^3D_2 - y^3S_1^\circ$
5489.872	−1	f	18210.31	−09	$z^5G_2^\circ - f^3D_2$

TABLE C—(*Continued*)

Solar λ	Solar Int	Grade	Wave Number Solar	Wave Number o−c	Desig	Solar λ	Solar Int	Grade	Wave Number Solar	Wave Number o−c	Desig
5488.173	−1d?	gb	18215.95	−12	$y^3F_3^\circ - f^3F_2$	5257.648	0	fb	19014.64	00	$a^1H_5 - x^3G_5^\circ$
*5487.524	−1N	g	18218.11	00	$y^5D_2^\circ - e^5P_1$	5255.747	−2	f	19021.51	+03	$y^5F_2^\circ - f^5P_2$
5482.268	−2N	f	18235.57	−01	$b^3D_1 - y^3S_1^\circ$	5255.666	−2	g	19021.80	+06	$y^5F_4^\circ - g^5F_5$
5479.984	−2	f	18243.17	−12	$x^5D_3^\circ - 4_2$	5253.256	−3	f	19030.53	−04	$b^3D_1 - v^5F_1^\circ$
5474.098	−3	g	18262.79	−02	$x^5F_5^\circ - g^5G_4$	5253.040	−1	f	19031.31	−05	$a^3P_2 - y^5P_1^\circ$
5473.172	0	gb	18265.88	+03	$y^5D_2^\circ - e^5P_2$	5246.007	−3	fb	19056.83	−03	$a^3D_3 - x^3D_2^\circ$
5469.283	−1N	gb	18278.87	+04	$z^5G_5^\circ - g^5F_5$	5245.738	−3	f	19057.80	−08	$a^1P_1 - z^3S_1^\circ$
5469.076	−3	f	18279.56	+05	$b^1D_2 - v^3P_2^\circ$	5245.638	−3	g	19058.17	−06	$z^5G_5^\circ - e^3H_6$
5461.824	−2N	f	18303.83	−09	$z^5P_1^\circ - e^5F_2$	5240.360	−2	g	19077.36	−01	$b^3H_5 - y^3G_4^\circ$
5455.094	−3	f	18326.41	00	$a^3D_3 - v^5D_3^\circ$	5238.253	−3N	f	19085.04	00	$z^3F_2^\circ - e^5G_3$
5453.996	−2	g	18330.10	−04	$y^5D_3^\circ - e^5P_3$	5236.386	−2	g	19091.84	−01	$z^5G_5^\circ - e^5H_5$
5443.425	−2	g	18365.70	−04	$y^5D_4^\circ - f^5F_4$	5226.388	−3	f	19128.36	+10	$b^3G_5 - y^5G_5^\circ$
5438.055	−2	f	18383.83	−06	$d^3F_4 - v^3H_5^\circ$	5218.516	−3	f	19157.22	−03	$d^3F_2 - s^3G_3^\circ$
5437.209	0	g	18386.69	−07	$z^5G_5^\circ - f^5G_6$	5217.677	−3N	g	19160.30	+05	$z^3F_2^\circ - e^3D_3$
5435.184	−1	f	18393.54	−05	$z^3G_4^\circ - f^5G_3$	5213.816	−2	f	19174.49	−10	$z^3F_3^\circ - e^5G_4$
5429.858	1	f	18411.58	−10	$z^3G_3^\circ - e^5H_3$	5213.353	−2	g	19176.19	−02	$z^3G_5^\circ - e^3H_4$
5429.513	1	g	18412.75	+02	$y^5D_2^\circ - g^5D_1$	5209.896	−2	g	19188.91	00	$b^3H_6 - y^3G_5^\circ$
5429.434	−2	g	18413.02	00	$c^3F_3 - u^3G_3^\circ$	5206.821	−2	g	19200.25	−06	$y^5F_2^\circ - f^3D_3$
5428.715	−2	f	18415.46	−03	$c^3F_2 - t^3D_1^\circ$?	5197.944	0	f	19233.04	−05	$y^5F_1^\circ - f^5P_1$
5423.760	−2	f	18432.29	−08	$b^1G_4 - w^3G_5^\circ$	5196.270	−3N	f	19239.23	−11	$b^3G_5 - y^5G_4^\circ$
5422.167	−1	g	18437.70	−06	$z^5G_6^\circ - f^5G_5$	5184.199	−2	f	19284.03	−09	$z^5G_5^\circ - e^3G_4$
5421.846	−1	g	18438.79	+02	$y^3F_4^\circ - f^3F_3$	5172.219	−1	f	19328.69	−05	$b^3F_4 - x^5D_3^\circ$
5412.795	−1	f	18469.62	+01	$z^3G_4^\circ - e^5H_4$	5169.302	−1	gb	19339.60	−01	$c^3F_4 - t^3D_3^\circ$
5412.577	−2	f	18470.37	−05	$d^3F_3 - v^3H_4^\circ$	*5168.193	−2d?	f	19343.75	−02	$z^3F_3^\circ - e^7F_3$
5406.781	1	fb	18490.16	−04	$z^5G_4^\circ - f^3D_3$	5167.718	−1	f	19345.53	−05	$a^1P_1 - u^5D_2^\circ$
5406.342	−1	f	18491.67	+05	$c^3F_4 - v^3F_3^\circ$	5164.687	−2	f	19356.88	+04	$b^3F_3 - x^5D_2^\circ$
5405.360	1	f	18495.03	−03	$z^3G_5^\circ - e^5H_5$	5159.971	−2	g	19374.57	−08	$y^5F_1^\circ - f^3D_2$
5401.272	0	g	18508.99	−05	$z^5G_6^\circ - e^5H_6$	5150.196	−1	g	19411.34	−01	$a^1H_5 - w^3G_5^\circ$
5396.908	−2	f	18523.99	−03	$c^3P_2 - x^5P_2^\circ$	5146.319	−2	g	19425.97	−07	$z^5G_4^\circ - f^3F_4$
5391.793	−3	f	18541.56	−04	$a^3G_5 - x^5F_5^\circ$	5145.740	−3N	f	19428.15	−05	$b^1G_4 - 3_3^\circ$
5385.591	−1	f	18562.92	−04	$b^1G_4 - w^3G_4^\circ$	5141.542	−3	f	19444.01	+02	$b^1G_4 - w^3F_3^\circ$
5384.204	−3	f	18567.70	+07	$z^5P_2^\circ - e^5F_2$	5130.936	−3	g	19484.21	−10	$z^5G_5^\circ - e^3H_5$
5374.771	−2	f	18600.28	+03	$a^1H_5 - w^5G_5^\circ$	5127.690	−1	g	19496.54	−04	$a^5D_3 - z^7D_2^\circ$
5358.120	−1	f	18658.09	−07	$a^3D_2 - x^3D_2^\circ$	5124.617	−1	g	19508.23	−05	$b^3H_4 - z^3I_5^\circ$
5346.341	−2	f	18699.19	−01	$z^5P_3^\circ - e^5F_3$	5123.290	−2N	g	19513.29	−03	$z^5G_3^\circ - f^3F_3$
5339.428	−1	f	18723.40	−09	$z^3G_4^\circ - e^5H_3$	5120.888	−3	g	19522.44	+02	$z^5G_2^\circ - f^3F_2$
5334.339	−2	g	18741.27	−05	$y^5D_4^\circ - e^5P_3$	5119.917	−3	f	19526.14	−07	$z^3F_4^\circ - e^7F_5$
5331.197	−3	f	18752.31	+02	$z^5P_2^\circ - e^5F_1$	5114.516	−2d?	fb	19546.76	00	$d^3F_4 - u^3F_4^\circ$
5327.895	−1	f	18763.93	−14	$z^5G_3^\circ - f^5G_2$	5096.187	−2	f	19617.06	−06	$d^3F_3 - u^3F_4^\circ$
5327.266	−2	f	18766.15	−04	$b^3D_2 - v^5F_3^\circ$	*5091.726	−2	f	19634.25	00	$a^1P_1 - u^5D_1^\circ$
5319.216	−2	g	18794.55	+01	$c^3F_4 - u^3G_4^\circ$	5086.776	−2	f	19653.35	−04	$y^5D_3^\circ - f^5P_3$
5318.040	−3	fb	18798.71	00	$b^3G_3 - y^5G_3^\circ$	5085.907	−3N	f	19656.71	+10	$z^3F_3^\circ - f^5F_4$
5317.549	−1d?	g	18800.44	−07	$c^3F_3 - t^3D_3^\circ$	5085.685	−2	fb	19657.58	−02	$y^5F_5^\circ - e^5H_6$
5315.785	−2N	f	18806.68	−03	$b^3D_2 - v^5P_2^\circ$	5084.563	−2	f	19661.91	−05	$b^1G_4 - v^3G_5^\circ$
5308.690	−1	f	18831.81	+06	$y^5F_3^\circ - f^5P_3$	5082.656	−3	f	19669.29	+12	$c^3P_0 - v^5D_1^\circ$
5301.314	−2	f	18858.02	+06	$z^3G_5^\circ - e^5H_4$	5081.845	−2	fb	19672.43	+07	$z^3F_4^\circ - e^5G_4$
5300.407	−2	f	18861.24	+02	$d^3F_4 - s^3G_5^\circ$	5080.937	−3	g	19675.94	+05	$b^3H_6 - z^3I_6^\circ$
5293.041	−1	g	18887.49	−03	$z^3G_5^\circ - e^3H_5$	5075.167	−3	gb	19698.31	+01	$y^5F_5^\circ - g^5F_4$
5288.379	−3	f	18904.14	00	$b^3G_4 - y^5G_4^\circ$	5069.627	−3	f	19719.82	−13	$b^3F_3 - x^5F_4^\circ$
5285.131	0	g	18915.76	−04	$z^3G_4^\circ - f^3F_4$	5064.975	1	g	19737.95	−10	$y^5F_3^\circ - f^3D_2$
5284.618	−1	g	18917.59	−01	$c^3F_2 - t^3D_2^\circ$	5052.993	−1	g	19784.75	−08	$b^3H_5 - z^3I_5^\circ$
5284.284	−3	f	18918.79	−06	$b^3D_1 - v^5F_2^\circ$	5051.311	−2N	g	19791.34	−08	$y^5F_4^\circ - g^5F_3$
5281.165	−2	f	18929.96	+04	$d^3F_3 - s^3G_4^\circ$	5050.139	−2	f	19795.93	−03	$z^3F_4^\circ - f^5F_5$
5279.675	−2	g	18935.31	−08	$b^3H_4 - y^3G_3^\circ$	5047.125	−2	f	19807.75	+06	$d^3F_3 - u^3F_3^\circ$
*5277.312	−1	g g	18943.78	+01 −02	$b^3H_5 - y^3G_5^\circ$ $z^5G_3^\circ - e^3H_4$	*5041.325	−1	f f	19830.54	+02 −04	$z^3P_0^\circ - g^5F_1$ $a^1F_3 - r^3G_3^\circ$
5275.286	1	f	18951.06	+04	$a^1D_2 - y^3S_1^\circ$	5040.248	−2	f	19834.77	+01	$y^5F_4^\circ - e^5H_5$
5273.602	−3N	g	18957.11	+07	$z^5G_5^\circ - e^3G_5$	5027.531	−3	f	19884.95	−10	$z^3F_4^\circ - e^7F_4$
5270.067	−2	f	18969.83	−01	$b^3D_1 - v^5P_1^\circ$	5027.355	−1	f	19885.64	−05	$z^3F_2^\circ - e^5P_1$
5267.280	0	g	18979.86	00	$z^5G_4^\circ - e^5H_4$	5025.081	−1	f	19894.64	+01	$z^3P_1^\circ - g^5F_2$
5265.424	−3	g	18986.56	−01	$z^5G_5^\circ - f^5G_4$	5021.694	−2	f	19908.06	−07	$y^5D_1^\circ - f^5P_1$
5262.889	−1	f	18995.70	−01	$a^3D_3 - x^3D_3^\circ$	5021.603	0	fb	19908.42	00	$y^5F_3^\circ - e^5H_4$
5262.624	−2	g	18996.65	−05	$z^5G_6^\circ - e^3H_6$	5019.737	−1	f	19915.82	+01	$z^3F_2^\circ - g^5D_2$
5261.503	−3	f	19000.70	−03	$b^3G_5 - y^5G_6^\circ$	5019.189	−2	f	19918.00	−05	$d^3F_2 - u^3F_2^\circ$
5259.095	−3N	g	19009.40	−02	$z^5G_4^\circ - e^3H_5$	5016.484	0	g	19928.74	00	$y^5F_3^\circ - g^5F_2$

TABLE C—(*Continued*)

Solar λ	Solar Int	Grade	Wave Number		Desig
			Solar	o−c	
5015.301	−2	f	19933.44	00	$z^3F_2^\circ - e^5P_2$
5012.700	0	f	19943.78	−06	$y^5F_2^\circ - e^5H_3$
5012.160	1	g	19945.93	−02	$y^5D_2^\circ - f^3D_3$
5011.209	−2	f	19949.72	+13	$y^5D_1^\circ - h^5D_2$
5010.331	−2	f	19953.21	−12	$b^3F_4 - x^5F_4^\circ$
5006.695	−2	f	19967.70	+09	$b^3F_3 - x^5F_3^\circ$
5003.881	−3	f	19978.93	−11	$b^3F_2 - x^5F_2^\circ$
4995.411	−1	f	20012.81	−02	$z^3P_1^\circ - f^5G_2$
4992.787	−2	f	20023.32	+05	$z^3P_1^\circ - g^5F_1$
4991.862	−1	fb	20027.03	00	$y^5F_4^\circ - e^3G_4$
4987.857	−3	fb	20043.11	−12	$z^3F_4^\circ - g^5D_4$
4987.654	−3	f	20043.95	−12	$y^5F_5^\circ - e^3G_5$
4986.915	−1	g	20046.90	−04	$y^5F_3^\circ - f^5G_2$
4985.992	−1	f	20050.61	−05	$y^5F_3^\circ - e^3G_3$
4979.840	−3	f	20075.38	00	$c^3P_2 - w^5D_1^\circ$
4978.116	−2	f	20082.34	−02	$z^3D_1^\circ - e^5P_1$
4972.914	−3	f	20103.34	−07	$a^3D_2 - y^3P_2^\circ$
4970.653	0	fb	20112.48	+01	$z^3D_1^\circ - g^5D_2$
4966.286	−2	f	20130.17	+07	$z^3D_1^\circ - e^5P_2$
4957.705	−2	g	20165.01	−11	$y^5D_2^\circ - h^5D_2$
4954.298	−3	f	20178.88	00	$y^5F_5^\circ - e^5H_5$
4945.284	−3	f	20215.66	+02	$c^3P_2 - v^5D_3^\circ$
4942.602	−2	f	20226.63	−04	$y^5F_3^\circ - e^3H_4$
4939.481	−3	f	20239.41	−08	$c^3F_2 - 11_3^\circ$
4933.193	0	g	20265.21	−02	$y^5D_2^\circ - f^3D_2$
4930.067	−2	f	20278.06	−10	$a^3D_2 - y^3P_1^\circ$
4926.848	−2	f	20291.31	−10	$a^1I_6 - y^3H_5^\circ$
4922.162	−2	f	20310.62	+06	$z^3P_2^\circ - g^5F_2$
4919.749	−2	f	20320.59	−06	$a^3D_1 - y^3P_2^\circ$
4916.678	−2	f	20333.28	−05	$z^3D_2^\circ - e^5P_1$
4911.541	0	f	20354.54	−07	$y^5F_3^\circ - f^3F_4$
4908.608	−2	g	20366.70	+02	$a^3P_0 - x^5D_1^\circ$
4906.775	−2	f	20374.31	+11	$y^5F_4^\circ - g^7D_3$
4893.707	−2	f	20428.72	−03	$z^3P_2^\circ - f^5G_2$
4893.572	−3	f	20429.29	+09	$y^5F_5^\circ - g^7D_4$
4887.369	−1	f	20455.21	+02	$c^3F_4 - 9_4^\circ$
4886.179	−3	g	20460.19	−02	$c^3P_0 - x^3D_1^\circ$
4877.592	0	g	20496.21	+08	$z^7P_3^\circ - e^5D_4$
4876.204	−2	f	20502.05	−06	$a^3D_3 - y^3P_2^\circ$
4875.741	−3	f	20503.99	−10	$d^3F_4 - t^3F_4^\circ$
4874.363	0	g	20509.79	−04	$c^3P_1 - x^3D_2^\circ$
4873.754	−1	g	20512.36	−05	$a^3D_2 - w^3D_2^\circ$
4872.910	−2	f	20515.90	−01	$y^5F_4^\circ - e^3H_4$
4872.703	−3	f	20516.78	−05	$z^3P_1^\circ - e^3P_2$
4871.937	1	f	20520.00	00	$a^3D_3 - u^5D_3^\circ$
4870.049	−1	f	20527.96	+01	$z^3D_2^\circ - g^5D_1$
4869.471	0	f	20530.39	−08	$a^1D_2 - v^3D_3^\circ$
4868.38?	−1	g	20535.00	−01	$a^3F_3 - y^5D_4^\circ$
4867.641	−3	f	20538.11	−01	$b^3H_5 - x^3F_4^\circ$
4867.544	−1	g	20538.52	−05	$a^3F_2 - y^5D_3^\circ$
4863.782	−3	g	20554.41	00	$z^7P_2^\circ - e^5D_3$
4862.553	−2	g	20559.60	−08	$y^5D_3^\circ - f^3D_2$
4859.306	−2	f	20573.34	00	$a^3D_2 - x^3F_2^\circ$
*4858.264	−2	f	20577.75	+01	$y^5F_2^\circ - f^3F_3$
4849.662	−1	g	20614.26	+04	$a^1H_5 - y^3H_6^\circ$
4843.370	−2	f	20641.03	+09	$a^1H_5 - v^3G_5^\circ$
4842.734	−3	f	20643.74	−11	$y^5F_4^\circ - f^3F_4$
4841.675	−2	gb	20648.26	−11	$a^3D_2 - w^3D_1^\circ$
4839.790	−3	g	20656.30	−08	$y^5P_2^\circ - i^5D_3$
4838.094	−1	f	20663.54	−01	$a^3D_3 - u^5D_2^\circ$
4837.668	−3	fb	20665.36	−06	$d^3F_3 - t^3F_3^\circ$
4822.676	−2	g	20729.60	−05	$a^3D_1 - w^3D_2^\circ$
4816.684	−3	g	20755.39	−06	$b^3H_5 - z^3H_4^\circ$
4815.231	−1	f	20761.65	−06	$a^1P_1 - x^3P_2^\circ$
4813.727	−3N	f	20768.14	−02	$d^3F_2 - t^3F_2^\circ$

Solar λ	Solar Int	Grade	Wave Number		Desig
			Solar	o−c	
4802.525	−1	g	20816.58	+01	$y^5P_2^\circ - i^5D_2$
4801.622	−2	f	20820.49	+01	$z^3P_0^\circ - e^3P_1$
4800.544	−3	g	20825.17	+04	$b^3H_4 - z^1G_4^\circ$
4799.071	−3	f	20831.56	−05	$y^5F_2^\circ - f^3F_2$
4794.365	−1	g	20852.01	−04	$a^3P_1 - x^5D_1^\circ$
4790.752	−1	f	20867.73	−02	$a^3D_3 - x^3F_3^\circ$
4790.570	−1	f	20868.53	−05	$y^5D_3^\circ - f^5G_2$
4782.813	−2	g	20902.37	−11	$b^3H_6 - z^3H_6^\circ$
4780.822	−1	g	20911.08	−03	$a^3D_3 - w^3D_2^\circ$
4766.879	−1	g	20972.24	−06	$z^5F_2^\circ - e^3F_3$
4760.076	−1	g	21002.21	−02	$z^7P_2^\circ - e^5D_1$
4749.260	−3N	f	21050.05	−02	$y^5F_3^\circ - f^3F_2$
4744.644	−2	f	21070.52	−03	$a^5F_2 - z^5P_3^\circ$
4742.939	−2	f	21078.10	−02	$y^5D_2^\circ - e^3P_2$
4727.003	−2	f	21149.16	+02	$a^3D_1 - y^3S_1^\circ$
4716.838	−1	f	21194.73	+04	$a^3D_3 - 1_2^\circ$
4701.910	0N	g	21262.02	−04	$z^5F_1^\circ - e^3F_2$
4700.441	−1	f	21268.67	−08	$a^5P_2 - y^3D_1^\circ$
4690.382	0	f	21314.28	−03	$a^5F_1 - z^5P_2^\circ$
4688.382	0	fb	21323.37	−01	$y^5D_2^\circ - f^3F_3$
4687.678	−1	f	21326.57	−06	$b^3P_0 - w^5F_1^\circ$
4687.313	0	f	21328.23	−04	$a^5F_3 - z^5P_3^\circ$
4685.036	−1	fb	21338.60	−02	$b^3P_1 - w^5F_2^\circ$
4679.985	−2	f	21361.63	−09	$y^5D_1^\circ - f^3F_2$
4678.422	−1	g	21368.77	−05	$z^5F_2^\circ - e^3F_3$
4677.604	0	f	21372.50	−08	$y^5D_3^\circ - e^3P_2$
4674.658	0	f	21385.97	−04	$a^3F_3 - z^3P_2^\circ$
4673.280	1	g	21392.28	−01	$z^5P_2^\circ - e^7P_2$
4672.839	1	f	21394.30	−02	$a^3F_2 - z^3P_1^\circ$
4672.038	−1	f	21397.96	−08	$c^3F_3 - w^1G_4^\circ$
4668.072	1	g	21416.14	00	$z^5P_1^\circ - e^5S_2$
4665.551	−1	f	21427.72	+04	$c^3F_4 - 13_4^\circ$
4665.259	−2	f	21429.05	−10	$z^3P_2^\circ - e^3P_1$
4661.336	−1N	f	21447.09	−03	$b^3P_2 - w^5F_2^\circ$
4653.505	−1	f	21483.18	−06	$a^5F_2 - z^5P_2^\circ$
4643.217	−1	g	21530.78	−10	$a^3F_4 - y^5D_3^\circ$
4642.593	−1N	g	21533.68	−04	$z^5F_3^\circ - e^3F_2$
4641.218	0	f	21540.05	−01	$b^3P_2 - w^5F_1^\circ$
4636.678	−1	f	21561.14	−09	$a^1G_4 - z^3I_5^\circ$
4635.630	0	g	21566.02	−04	$z^7F_3^\circ - e^5D_4$
4632.818	1	g	21579.11	+04	$z^5P_2^\circ - f^7D_1$
4632.147	0	fb	21582.24	−02	$a^1D_2 - w^3P_2^\circ$
4631.039	0N	f	21587.39	−04	$y^5D_4^\circ - f^3F_4$
4638.687	−1	f	21598.37	+03	$z^5P_1^\circ - e^7F_2$
4625.440	−1N	g	21613.53	00	$z^3F_2^\circ - f^3D_3$
4621.622	−1	f	21631.39	+02	$z^3D_1^\circ - f^5P_2$
4620.140	−1	f	21638.32	−06	$c^3P_1 - w^5P_1^\circ$?
4612.620	−1	g	21673.60	+10	$b^3P_2 - y^5S_2^\circ$
4611.194	0	g	21680.30	−02	$z^7F_4^\circ - e^5D_4$
4611.075	−1	f	21680.86	−12	$a^3D_2 - x^3P_2^\circ$
4607.100	−1N	f	21699.57	−11	$a^1P_1 - v^3D_2^\circ$
4606.015	−2N	f	21704.68	−14	$b^3D_3 - t^5D_4^\circ$
4605.105	−2	f	21708.97	−03	$b^3P_0 - v^5D_1^\circ$
4604.852	−1	g	21710.17	+02	$a^1I_6 - x^3H_6^\circ$
4604.247	−1N	f	21713.02	−06	$b^3P_2 - v^5D_3^\circ$
4603.352	0	f	21717.24	−03	$b^3P_1 - v^5D_2^\circ$
4598.745	0	f	21738.99	−02	$z^5P_2^\circ - e^7F_1$
*4598.374	−1	f	21740.74	−22	$a^5F_3 - z^5P_2^\circ$
		f		−04	$z^3F_3^\circ - h^5D_3$
4597.038	−2	f	21747.06	+11	$a^5F_2 - z^5P_1^\circ$
4595.213	−1	g	21755.70	−03	$a^1I_6 - x^3H_5^\circ$
4587.726	−1	f	21791.21	−04	$z^3F_2^\circ - f^5P_1$
4585.601	−2	f	21801.30	−04	$c^3P_2 - w^5P_3^\circ$
4583.721	−1	f	21810.24	−03	$c^3P_0 - y^3P_1^\circ$
4579.692	−1	f	21829.43	−04	$b^3D_3 - v^3F_2^\circ$

TABLE C—(*Continued*)

Solar λ	Solar Int	Grade	Wave Number Solar	Wave Number o−c	Desig
*4579.061	−1N	f f	21832.44	+03 −05	$a^3D_1 - v^5F_2^\circ$ $z^3D_3^\circ - h^5D_4$
4571.448	0	g	21868.80	−03	$z^7F_2^\circ - e^5D_3$
4569.073	−2	g	21880.17	−07	$b^3H_5 - w^3G_5^\circ$
4568.607	−1	f	21882.40	+06	$z^3D_2^\circ - f^5P_2$
4555.740	−2	f	21944.20	+07	$a^3D_1 - v^5F_1^\circ$
4546.682	−1	f	21987.92	+01	$z^3D_1^\circ - f^5P_1$
4546.479	−1	f	21988.90	−06	$c^3F_2 - w^1D_2^\circ$
4545.547	−2	g	21993.41	−03	$b^3D_3 - v^3F_3^\circ$
4544.490	−1N	f	21998.52	+05	$z^3F_2^\circ - h^5D_1$
4543.237	−1	f	22004.59	−09	$b^3D_2 - t^5D_3^\circ$
4538.958	−2	f	22025.34	−05	$c^3F_2 - w^1F_3^\circ$
4538.604	−1N	f	22027.05	−12	$z^3F_2^\circ - f^5G_3$
4538.185	−2N	f	22029.08	+07	$y^5D_4^\circ - f^3F_3$
4528.825	0	f	22074.61	−03	$c^3P_2 - w^5P_1^\circ$?
4528.764	0	g	22074.91	−02	$b^3H_4 - y^1G_4^\circ$
4526.414	1	f	22086.37	−07	$z^3F_4^\circ - g^5F_4$
4521.670	−2N	f	22109.54	−12	$a^3D_1 - x^3P_1^\circ$
4518.589	0	fb	22124.62	−05	$a^5P_1 - y^7P_2^\circ$
4517.600	−2	g	22129.46	−01	$z^3D_1^\circ - f^3D_2$
4516.464	−2	f	22135.03	−08	$z^5P_2^\circ - f^5F_3$
4516.273	1	f	22135.96	−02	$z^5P_3^\circ - e^7F_4$
4516.091	−2N	f	22136.86	−03	$a^3D_3 - w^3G_4^\circ$
4511.069	−1	f	22161.50	−12	$z^3F_3^\circ - h^5D_2$
4510.836	0	f	22162.64	−08	$z^5P_3^\circ - e^5G_3$
4504.208	−2N	f	22195.26	+13	$z^3D_1^\circ - h^5D_1$
4498.562	−1N	f	22223.11	−13	$z^3D_3^\circ - h^5D_3$
4494.064	0	f	22245.36	−07	$z^3F_2^\circ - e^3G_3$
4492.970	−2	f	22250.77	+06	$a^3D_3 - w^3G_3^\circ$
4490.236	−1	g	22264.32	+01	$z^7F_1^\circ - e^5D_1$
4487.754	0	g	22276.63	−05	$b^3H_6 - z^1H_5^\circ$
4487.006	−1	f	22280.35	+01	$z^3D_2^\circ - h^5D_2$
4485.978	0	f	22285.45	−05	$z^5P_2^\circ - f^5F_1$
4483.780	0	f	22296.37	+01	$b^3D_3 - u^3G_4^\circ$
4481.033	−1	fb	22310.04	+02	$b^3D_1 - t^5D_2^\circ$
4480.278	−1	fb	22313.80	−04	$z^5P_3^\circ - e^5G_2$
4479.971	1	g	22315.33	+01	$z^3F_2^\circ - f^3D_1$
*4479.001	−1N	f f	22320.16	+04 −01	$b^3D_3 - u^5P_3^\circ$ $z^3D_1^\circ - g^5F_2$
*4472.544	−2	g	22352.39	−13	$a^3F_4 - y^5F_3^\circ$
*4463.139	−1	f	22399.49	+03	$c^3P_1 - u^5D_0^\circ$
4461.822	−2N	f	22406.10	−13	$b^3G_3 - u^5D_3^\circ$
4452.618	0	f	22452.42	00	$z^3F_3^\circ - g^5F_2$
4450.764	0	fb	22461.77	+02	$z^3F_4^\circ - f^5G_4$
4441.557	−1	f	22508.33	+01	$z^3D_3^\circ - g^5F_3$
4438.526	−1N	f	22523.70	+01	$z^3F_4^\circ - g^5F_3$
4433.394	−1	f	22549.77	−01	$b^3G_3 - u^5D_2^\circ$
4429.207	−1	fb	22571.09	−05	$z^3D_2^\circ - g^5F_2$
4428.713	−1	fb	22573.61	+16	$b^3D_3 - u^5P_2^\circ$
4428.551	1	f	22574.43	+09	$z^3F_3^\circ - e^3G_3$
*4425.771	−1N	f f	22588.61	−11 +08	$z^5D_2^\circ - e^3F_2$ $b^3D_2 - u^5P_2^\circ$
4419.785	0	fb	22619.20	−02	$a^3D_2 - w^3F_3^\circ$
4418.576	−1d?	f	22625.39	+13	$b^3D_1 - u^5P_2^\circ$
4414.234	−1	fb	22647.65	−01	$c^3P_1 - 1_2^\circ$
4414.050	−2	f	22648.59	−12	$z^5P_3^\circ - f^5F_2$
4412.424	−1N	fb	22656.94	+03	$a^5P_3 - y^7P_3^\circ$
4405.420	−1	f	22692.96	−10	$z^3D_2^\circ - e^3G_3$
4405.035	1	gb	22694.94	−08	$a^5D_3 - z^7F_3^\circ$
4404.105	−1	f	22699.73	−05	$z^3D_2^\circ - g^5F_1$
4394.306	−2	f	22750.35	00	$z^3F_3^\circ - e^3H_4$
4393.701	0	f	22753.48	0	$b^3D_2 - u^5P_1^\circ$
4393.039	0	f	22756.91	−06	$c^3P_2 - x^3F_3^\circ$
4392.313	−1N	fb	22760.68	−01	$a^1D_2 - v^3F_3^\circ$
4391.877	0	gb	22762.94	−01	$z^3D_2^\circ - f^3D_1$
4382.003	−1	f	22814.22	+10	$b^1G_4 - w^3H_5^\circ$
4375.487	−1	g	22848.20	−04	$a^1H_5 - u^3G_4^\circ$
4373.899	0	f	22856.50	−01	$b^3D_2 - t^3D_3^\circ$
4369.716	0	gb	22878.38	+09	$z^3F_3^\circ - f^3F_4$
4368.644	0	fb	22883.98	+07	$a^3D_3 - w^3F_4^\circ$
4367.065	−2	f	22892.26	00	$z^3G_5^\circ - g^5G_5$
4358.926	−1	f	22935.01	+13	$z^3D_3^\circ - g^5F_2$
*4357.519	0Nd?	fb	22942.41	+04	$z^3D_1^\circ - e^3P_2$
4354.267	0Nd?	fb	22959.55	+06	$z^3F_4^\circ - e^3H_5$
4351.392	−1	f	22974.72	−12	$z^5F_2^\circ - f^5D_3$
4343.214	2	fb	23017.98	+06	$a^3D_3 - w^3F_3^\circ$
4341.565	−1	f	23026.72	+03	$a^3D_1 - w^3F_2^\circ$
4341.252	−1	fb	23028.38	−10	$z^5F_3^\circ - f^5D_4$
4338.835	0d	fb	23041.21	+02	$a^3P_0 - x^5P_1^\circ$
4333.053	−1	f	23071.95	+01	$b^1D_2 - t^3F_2^\circ$
4330.824	−1	f	23083.83	−09	$c^3P_2 - 1_2^\circ$
4329.552	−1	f	23090.61	−04	$a^5P_1 - x^5F_2^\circ$
4325.960	1	f	23109.78	−05	$b^3H_5 - v^3G_5^\circ$
4323.372	0	f	23123.61	−03	$a^3H_4 - y^5G_5^\circ$
4322.703	−1N	f	23127.19	−03	$b^3F_2 - w^5F_3^\circ$
4319.456	0	f	23144.58	−04	$b^3F_2 - w^5D_2^\circ$
4318.801	−1	f	23148.09	+06	$b^3F_3 - w^5F_4^\circ$
4315.956	−1	f	23163.35	00	$a^3H_5 - y^5G_5^\circ$
4313.037	1N	gb	23179.02	+02	$a^3G_3 - y^3G_4^\circ$
4310.381	1	g	23193.30	−04	$z^3D_2^\circ - e^3P_2$
4309.463	1	f	23198.24	−03	$c^3P_0 - v^5P_1^\circ$
4307.058	−2d?	f	23211.20	+12	$z^5F_4^\circ - f^7D_5$
4300.220	−1	f	23248.11	−07	$z^3F_4^\circ - e^3H_4$
4299.486	0	f	23252.08	+01	$a^3D_3 - z^1D_2^\circ$
4298.199	1	f	23259.04	+04	$c^3P_0 - v^5F_1^\circ$
4292.136	2	f	23291.89	−03	$a^5P_3 - x^5F_3^\circ$
4283.414	−1	f	23339.32	−08	$b^3F_2 - w^5F_1^\circ$
4281.601	−1	f	23349.20	−03	$a^3H_4 - y^5G_3^\circ$
4280.638	0	f	23354.45	−03	$b^3G_3 - w^5G_3^\circ$
4278.002	−3	f	23368.85	+04	$y^5F_3^\circ - i^5D_4$
4277.392	0	f	23372.18	+02	$b^3F_2 - w^5D_1^\circ$
4271.961	1N	fb	23401.89	−07	$a^3H_5 - y^5G_4^\circ$
4271.637	0Nd?	f	23403.66	+06	$a^5P_2 - x^5F_1^\circ$
4270.331	−1	f	23410.82	−14	$b^3F_3 - w^5F_2^\circ$
4269.862	2N	fb	23413.39	+04	$z^5F_3^\circ - f^7D_4$
4260.735	1	fb	23463.55	−03	$b^3P_1 - w^5P_1^\circ$
4259.309	1Nd?	fb	23471.40	+18	$b^3G_4 - w^5G_4^\circ$
4256.317	−1	f	23487.90	−01	$a^3H_5 - z^5H_4^\circ$
4253.914	1d	fb	23501.17	+11	$b^3D_2 - 8_1^\circ$
*4253.542	−1	f	23503.23	−13	$z^5F_5^\circ - f^7D_5$
4250.916	1	f	23517.74	−09	$c^3P_1 - v^5P_1^\circ$
4249.352	−1	f	23526.40	−16	$a^3P_1 - x^5P_1^\circ$
4248.416	1	f	23531.58	−07	$a^5F_1 - z^3D_2^\circ$
*4247.317	2	f f	23537.67	−03 −12	$a^3H_4 - z^5H_3^\circ$ $b^3D_1 - 8_1^\circ$
4246.023	0	g	23544.84	−03	$a^3D_1 - w^3P_0^\circ$
4239.958	1	fb	23578.52	−04	$c^3P_1 - v^5F_1^\circ$
4239.368	2	f	23581.81	−01	$b^3D_3 - s^3D_3^\circ$
4238.622	−1N	gb	23585.95	−07	$a^1I_6 - y^3I_5^\circ$
4237.680	1	f	23591.20	−04	$b^3G_3 - v^5F_4^\circ$
4236.645	−1	f	23596.97	+07	$b^3D_2 - s^3D_3^\circ$
4235.838	0	f	23601.45	−01	$a^3H_6 - z^5H_5^\circ$
4235.640	−1	f	23602.56	+03	$b^3F_4 - w^5F_5^\circ$
4225.717	1	fb	23657.98	−07	$y^5F_4^\circ - i^5D_4$
4224.634	−1	f	23664.05	−02	$a^3G_3 - x^5G_3^\circ$
4223.733	0	g	23669.10	−02	$b^3G_5 - z^1G_4^\circ$
4220.053	1	gb	23689.73	−01	$z^3D_2^\circ - e^3P_1$
4219.736	−1	f	23691.52	+02	$z^5P_2^\circ - f^5P_3$
4219.421	3	g	23693.28	−05	$b^3G_4 - x^3G_4^\circ$
4218.226	1	f	23700.00	−08	$a^3H_5 - z^5H_6^\circ$

TABLE C—(*Continued*)

Solar λ	Solar Int	Grade	Wave Number Solar	Wave Number o−c	Desig
4213.422	−1	f	23727.02	+01	$a^3G_4 - x^5G_5^\circ$
4212.043	−1N	f	23734.79	+13	$z^5F_2^\circ - e^3D_3$
4210.404	3	g	23744.02	−07	$c^3P_1 - x^3P_1^\circ$
4200.789	1	gb	23798.37	−05	$a^3F_2 - y^5P_3^\circ$
4200.104	−1	g	23802.25	−09	$z^3D_3^\circ - f^3F_3$
4199.379	−1	f	23806.36	−06	$b^3G_5 - w^5G_4^\circ$
4197.362	−1	g	23817.80	+10	$z^3F_4^\circ - f^3F_3$
4197.102	2	gb	23819.28	−02	$a^5F_2 - z^3F_3^\circ$
4194.491	0	f	23834.10	+03	$a^3G_4 - x^5G_4^\circ$
4181.549	1	f	23907.87	+02	$a^1D_2 - u^3D_1^\circ$
4181.194	0	f	23909.90	+01	$b^3D_1 - z^1P_1^\circ$
4180.404	1	f	23914.42	+06	$a^3G_4 - x^5G_3^\circ$
4175.914	1	g	23940.13	−11	$z^5F_3^\circ - e^7G_4$
4173.151	−1d?	g	23955.98	+18	$z^5F_3^\circ - g^5D_4$
*4172.978	1	f f	23956.97	−04 −06	$b^3D_3 - 9_4^\circ$ $y^5D_1^\circ - i^5D_2$
4169.096	−1	g	23979.28	−01	$a^5F_1 - z^3F_2^\circ$
4164.265	0	fb	24007.10	−16	$z^5F_2^\circ - e^7G_1$
4163.358	−2	f	24012.32	−05	$y^5D_2^\circ - i^5D_3$
4162.910	−1	f	24014.91	+09	$c^3P_2 - v^5F_1^\circ$
4152.085	1	f	24077.52	−07	$c^3F_4 - v^1G_4^\circ$
4149.501	0	f	24092.51	−07	$b^1G_4 - 10_3^\circ$
4148.260	−1	f	24099.72	+03	$z^5P_2^\circ - f^5P_2$
4147.490	1	f	24104.20	+01	$z^5P_3^\circ - f^5P_3$
4147.347	2	g	24105.03	−03	$z^5F_5^\circ - e^5G_4$
4143.510	2	g	24127.35	−07	$z^5F_4^\circ - e^3D_3$
4137.984	0	f	24159.57	−06	$z^7F_5^\circ - e^5F_5$
4137.417	2	g	24162.88	+03	$y^5F_2^\circ - g^5G_3$
4134.196	0	g	24181.70	−02	$b^3F_2 - x^3D_3^\circ$
4132.540	3	g	24191.40	00	$y^5F_3^\circ - g^5G_4$
*4131.959	2	g g b	24194.79	+06 −15	$z^5D_2^\circ - f^5D_3$ $z^5F_2^\circ - f^5F_1$
4131.758	1	f	24195.97	−07	$y^5D_1^\circ - 4_2$
4129.466	2	g	24209.40	−04	$z^5F_3^\circ - f^5F_3$
4125.234	−2	f	24234.24	−03	$a^3H_4 - w^5F_4^\circ$
4121.991	−1	f	24253.30	+07	$a^1D_2 - 8_1^\circ$
4119.672	0	f	24266.95	−07	$z^7F_3^\circ - e^5F_4$
4115.896	−1	f	24289.22	−05	$b^3D_3 - y^1F_3^\circ$
4112.176	−2	f	24311.19	−02	$a^3G_5 - z^3I_6^\circ$
4112.083	−1	f	24311.74	+02	$a^1D_2 - v^3P_2^\circ$
4108.303	−2	f	24334.11	+04	$z^5P_1^\circ - f^3D_2$
4108.138	1	g	24335.09	−04	$z^5D_3^\circ - e^7P_4$
4105.065	−1	f	24353.30	−04	$z^5F_1^\circ - e^5P_1$
4104.472	0	g	24356.82	−07	$b^3G_4 - w^3G_3^\circ$
*4103.623	0N	g	24361.86	00	$a^3D_3 - z^1F_3^\circ$
4100.916	0	fb	24377.94	−02	$a^3H_4 - w^5F_3^\circ$
*4100.350	0	f g b	24381.31	+03 00	$z^7F_4^\circ - e^5F_4$ $y^5F_3^\circ - g^5G_3$
4099.996	0	g	24383.41	−05	$z^5F_1^\circ - g^5D_2$
4097.018	−1	f	24401.13	+04	$z^5F_1^\circ - e^5P_2$
4096.951	−1	f	24401.53	+02	$a^3H_5 - w^5F_4^\circ$
4096.217	1	g	24405.91	−03	$a^5F_3 - z^3F_2^\circ$
4095.646	−2	g	24409.30	−10	$a^1I_6 - y^1H_5^\circ$
4095.274	0	f	24411.53	−04	$y^5D_2^\circ - 4_2$
4090.773	−1	f	24438.38	−12	$b^1G_4 - t^3G_4^\circ$
4090.326	0N	gb	24441.05	+05	$a^3F_2 - y^5P_1^\circ$
4087.801	−1	f	24456.15	−08	$z^5P_2^\circ - f^5P_1$
4084.152	−2	gb	24478.00	+08	$z^5D_4^\circ - f^7D_5$
4079.186	2	f	24507.80	−05	$z^5F_2^\circ - e^5P_2$
4074.686	2	f	24534.86	+10	$b^3D_2 - 10_3^\circ$
4067.856	−1	g	24576.06	−02	$y^5F_5^\circ - g^5G_5$
4067.604	0	g	24577.59	−01	$a^3D_2 - v^3F_2^\circ$
4067.493	0	g	24578.25	−02	$b^3G_5 - w^3G_4^\circ$
4066.006	−1	g	24587.24	+10	$z^5F_5^\circ - f^5F_4$
4064.054	1	gb	24599.05	+07	$b^3G_5 - z^1H_5^\circ$

Solar λ	Solar Int	Grade	Wave Number Solar	Wave Number o−c	Desig
4058.469	−1	f	24632.90	−06	$b^3D_3 - 11_3^\circ$
4057.671	0	f	24637.75	−07	$a^1P_1 - t^3D_1^\circ$?
4046.465	1	f	24705.97	−06	$y^5D_3^\circ - 4_2$
4046.083	2	gb	24708.31	−08	$z^5D_2^\circ - f^7D_1$
4045.601	2	gb	24711.25	−05	$z^5D_4^\circ - e^7P_3$
4044.497	1	f	24717.99	−02	$y^5D_4^\circ - i^5D_3$
4043.993	2	gb	24721.08	−05	$z^5D_3^\circ - e^7P_2$
4043.692	0	fb	24722.92	−01	$a^3P_0 - v^5D_1^\circ$
4042.763	−1	f	24728.60	−05	$z^5D_1^\circ - e^7F_1$
4036.377	1	f	24767.72	−06	$a^3G_3 - w^3D_3^\circ$
4035.986	−1	fb	24770.12	−02	$b^3G_3 - w^3F_4^\circ$
4031.718	2	fb	24796.34	+06	$b^3G_3 - v^3D_3^\circ$
4030.901	2	fb	24801.37	−02	$b^1G_4 - t^3G_3^\circ$
4020.024	0	f	24868.47	+14	$z^5D_2^\circ - e^7F_1$
4011.901	−1	g	24918.82	−10	$b^3G_5 - y^1G_4^\circ$
4009.549	1	f	24933.44	−04	$z^5D_4^\circ - e^7F_5$
4006.159	1	g	24954.53	−02	$z^5D_2^\circ - e^3D_3$
4005.484	1	f	24958.74	+02	$b^3F_3 - x^5G_4^\circ$
4005.390	1	f	24959.33	−04	$a^3P_1 - y^5S_2^\circ$
3998.475	0	f	25002.49	−10	$b^3H_4 - 6_5^\circ$
*3997.493	2	f f	25008.63	−05 −07	$z^5D_3^\circ - e^7F_3$ $z^5D_3^\circ - e^5S_2$
3996.790	1	f	25013.03	−01	$y^5D_3^\circ - g^5G_4$
*3996.265	1	f g	25016.32	+09 −04	$b^3G_4 - v^3D_3^\circ$ $z^5D_0^\circ - e^7G_1$
3994.272	−2	f	25028.80	−02	$z^7F_2^\circ - e^5F_1$
3992.646	0	f	25038.99	−02	$b^3F_3 - x^5G_3^\circ$
3990.569	0d?	fb	25052.02	−11	$z^5D_3^\circ - e^7F_4$
3989.611	−1N	fb	25058.04	−05	$b^3H_5 - 4_4^\circ$
3989.262	−1N	gb	25060.23	−15	$z^5D_1^\circ - e^7G_2$
3986.298	0	g	25078.86	−01	$z^5D_3^\circ - e^5G_3$
3985.322	1	f	25085.01	−03	$b^3F_4 - x^5G_5^\circ$
3984.943	1	f	25087.39	−08	$z^5D_1^\circ - e^7G_1$
3984.451	−1	f	25090.49	+03	$b^3F_3 - x^5G_2^\circ$
3983.818	0	g	25094.48	+10	$b^3G_3 - w^3F_2^\circ$
3979.117	−1	g	25124.12	+02	$b^3G_4 - w^3F_3^\circ$
3974.636	2	fb	25152.45	+08	$a^1G_4 - y^3H_4^\circ$
3971.010	1N	f	25175.42	−12	$y^5D_4^\circ - g^5G_5$
3965.844	−1	f	25208.20	−10	$a^3P_1 - v^5D_1^\circ$
3962.651	−1	f	25228.52	+03	$b^3D_3 - t^3G_3^\circ$
3962.400	0	g	25230.12	+13	$z^5D_3^\circ - e^5G_2$
3959.454	−1	f	25248.89	+02	$z^5D_4^\circ - e^7F_3$
3955.768	0	f	25272.42	+03	$b^3F_4 - x^5G_3^\circ$
3955.221	1	g	25275.91	−01	$a^1G_4 - v^3G_3^\circ$
3950.801	−2	g	25304.19	−10	$z^7D_3^\circ - e^5D_3$
3949.235	1	g	25314.22	−01	$z^7D_1^\circ - e^5D_2$
3948.476	−1N	g	25319.09	+03	$z^5D_4^\circ - e^5G_3$
3948.284	1	g	25320.32	−02	$z^5D_3^\circ - e^7G_3$
3947.980	−2	f	25322.27	+14	$a^3D_2 - u^5F_1^\circ$
3936.772	0d?	gb	25394.36	+10	$z^5D_4^\circ - e^3D_3$
3927.610	0	f	25453.60	00	$a^3G_4 - z^1G_4^\circ$
3925.540	1	f	25467.02	+07	$a^3D_3 - u^5P_3^\circ$
3923.043	1	g	25483.23	−07	$a^3D_3 - y^1D_2^\circ$
*3922.086	1	g g	25489.44	00 −07	$z^5D_0^\circ - e^3D_1$ $z^7D_1^\circ - e^5D_1$
3918.575	0	fb	25512.28	+02	$b^3P_2 - v^5F_1^\circ$
3914.428	1	f	25539.31	−06	$a^3D_1 - u^5F_1^\circ$
3910.536	2	g	25564.72	−14	$z^5D_3^\circ - f^5F_2$
3908.687	−1	g	25576.82	−02	$z^7D_3^\circ - e^5D_2$
3906.965	1	gb	25588.09	+03	$z^5D_1^\circ - e^5P_2$
3905.681	2N	gb	25596.50	−14	$z^7D_2^\circ - e^5D_1$
3905.191	1	g	25599.72	−05	$z^5D_3^\circ - e^3D_2$
3905.011	1	f	25600.90	−03	$z^5F_2^\circ - f^5P_3$
3889.360	1d?	fb	25703.91	+ 1	$a^3D_1 - u^5P_1^\circ$
3885.935	0	gb	25726.57	−05	$b^1G_4 - x^1H_5^\circ$

TABLE C—(*Continued*)

Solar λ	Solar Int	Grade	Wave Number Solar	Wave Number o−c	Desig
3885.758	1	g	25727.74	00	$z^5D_2^\circ - e^5P_2$
3878.196	1	g	25777.91	−03	$z^5D_1^\circ - g^5D_0$
3867.442	0	g	25849.58	+02	$b^3F_2 - u^5D_2^\circ$
*3864.307	3	{g g} b	25870.55	{+04 −03	$b^3F_3 - u^5D_3^\circ$ $z^5D_4^\circ - g^5D_3$
3863.703	1	g	25874.60	−02	$z^5D_2^\circ - g^5D_1$
3860.730	−2	f	25894.52	+08	$z^5F_4^\circ - g^5F_5$
3858.474	0	gb	25909.69	+06	$z^5D_3^\circ - g^5D_2$
3848.299	2	g	25978.17	−05	$b^3F_2 - w^3D_3^\circ$
3846.290	1	fb	25991.74	−02	$b^1G_4 - w^1F_3^\circ$
3845.224	1	fb	25998.94	−07	$z^5F_3^\circ - g^5F_4$
3843.717	2N	fb	26009.13	+01	$z^5F_2^\circ - f^5P_2$
3842.905	1	g	26014.63	−06	$b^3F_3 - x^3F_4^\circ$
3840.203	0	g	26032.93	−03	$a^3P_2 - w^5D_1^\circ$
3834.476	0	g	26071.81	−12	$a^3D_3 - u^3D_2^\circ$
3826.627	1N	fb	26125.29	+04	$a^3H_4 - x^5G_3^\circ$
3824.752	−1	f	26138.10	−13	$b^3F_2 - u^5D_1^\circ$
3819.496	2	f	26174.06	+04	$z^5F_3^\circ - f^5P_2$
3816.924	1	g	26191.73	−01	$z^7P_2^\circ - f^5D_2$
3813.930	1	f	26212.26	+06	$a^3H_5 - x^5G_4^\circ$
3811.809	2	fb	26226.85	−03	$z^5F_4^\circ - g^5F_4$
3810.902	0	gb	26233.09	+02	$b^3F_2 - w^3D_1^\circ$
3803.260	1	f	26285.80	−11	$a^3P_2 - v^5D_2^\circ$
3789.824	1	g	26378.98	−06	$z^5F_4^\circ - h^5D_3$
3785.792	1	f	26407.08	−09	$z^5F_5^\circ - f^5G_5$
3784.254	0	fb	26417.81	+12	$b^3H_5 - w^3H_4^\circ$
3771.499	1	g	26507.15	+02	$b^3H_6 - w^3H_5^\circ$
3764.223	1	g	26558.39	−07	$a^5P_2 - w^5F_2^\circ$
3763.573	1	g	26562.97	−05	$a^3P_0 - z^3S_1^\circ$
3761.069	1	f	26580.66	−05	$z^5F_4^\circ - f^3D_3$
3758.131	−1	f	26601.44	−14	$z^5F_2^\circ - f^5G_3$
3754.876	−1N	f	26624.50	+13	$b^1G_4 - u^3H_4^\circ$
3751.092	1	gb	26651.36	−04	$a^5P_2 - w^5F_1^\circ$
3748.907	0	fb	26666.89	+03	$a^3G_5 - z^1H_5^\circ$
3747.006	2	gb	26680.42	−02	$z^7P_3^\circ - e^7P_2$
3742.569	2	gb	26712.05	−08	$z^7P_2^\circ - e^5G_3$
3742.148	1	f	26715.06	−02	$z^3F_3^\circ - g^5G_4$
3735.702	0Nd?	fb	26761.15	+03	$a^3P_1 - w^5P_2^\circ$
3733.197	1	g	26779.10	+01	$b^3F_4 - w^5G_5^\circ$
3731.161	−1	f	26793.72	−08	$b^1G_4 - u^3F_4^\circ$
3729.341	0	f	26806.79	00	$a^1G_4 - u^5F_4^\circ$
3727.687	1	gb	26818.69	−08	$b^3F_3 - w^5G_3^\circ$
3727.533	1	f	26819.80	−04	$z^5F_2^\circ - e^3G_3$
3727.028	0	fb	26823.43	−02	$a^3D_1 - z^1P_1^\circ$
3726.067	−1	f	26830.35	−02	$b^3G_5 - x^1G_4^\circ$
3722.760	0	fb	26854.18	+04	$z^5F_2^\circ - g^7D_3$
*3722.238	1	{f f}	26857.95	{00 −06	$a^3P_1 - w^5P_1^\circ$ $c^3P_1 - t^5D_2^\circ$
3717.837	0	f	26889.74	+01	$z^5F_2^\circ - f^3D_1$
3717.188	−1	f	26894.43	−03	$z^5F_5^\circ - f^5G_4$
3709.032	1	g	26953.57	−03	$z^7P_2^\circ - e^7G_3$
3708.189	−2	g	26959.70	−04	$b^3F_3 - x^3G_3^\circ$
*3705.710	2	fb	26977.73	+03	$a^3G_3 - 3_3^\circ$
3705.264	−1N	f	26980.98	−04	$z^5F_3^\circ - f^5G_2$
3704.798	−2	f	26984.38	+01	$b^1G_4 - u^3F_3^\circ$
3703.449	−2	f	26994.20	−14	$z^5F_4^\circ - f^5G_3$
3700.601	1	g	27014.98	+04	$z^5D_3^\circ - h^5D_4$
3699.397	−2	f	27023.77	+06	$z^3D_2^\circ - g^5G_3$
3698.018	1N	gb	27033.85	+08	$a^5P_2 - v^5D_1^\circ$
*3693.784	0	{g f}	27064.83	{−05 +02	$a^3F_3 - x^5D_4^\circ$ $c^3P_1 - t^5D_1^\circ$
3691.536	−3	f	27081.32	−06	$z^5F_1^\circ - g^7D_1$
3691.177	−2N	g	27083.95	00	$b^3F_2 - v^5F_3^\circ$
3689.375	3	fb	27097.18	−04	$z^7P_2^\circ - f^5F_3$
3685.663	−3	f	27124.47	00	$b^3F_2 - v^5P_2^\circ$

Solar λ	Solar Int	Grade	Wave Number Solar	Wave Number o−c	Desig
3683.757	−3	f	27138.50	+07	$z^3D_2^\circ - g^5G_2$
*3682.174	2	{g g}	27150.17	{−04 −21	$z^7P_3^\circ - e^7F_2$ $z^7P_4^\circ - f^7D_3$
3675.767	0	f	27197.49	−05	$z^3D_3^\circ - g^5G_4$
3675.450	−1	g	27199.84	−04	$b^3F_2 - v^5F_2^\circ$
3673.684	−1	f	27212.91	00	$z^3F_4^\circ - g^5G_4$
3670.221	−3	f	27238.59	+04	$a^3F_2 - y^7P_3^\circ$
3666.850	−2	f	27263.63	−02	$z^7P_2^\circ - g^5D_3$
3666.285	2	f	27267.83	+02	$a^3D_2 - 10_3^\circ$
3662.738	−2	f	27294.24	−03	$c^3P_2 - t^5D_2^\circ$
3660.412	1	fb	27311.58	−02	$b^3F_2 - v^5F_1^\circ$
3658.025	1	g	27329.40	−02	$b^3G_3 - u^3G_4^\circ$
3656.358	1	g	27341.86	−07	$z^7F_2^\circ - f^5D_3$
3655.356	1	g	27349.32	−08	$a^3P_1 - y^3P_1^\circ$
*3653.353	1	{g f}	27364.35	{−03 −02	$b^3F_3 - v^5F_2^\circ$ $z^7F_3^\circ - e^7P_4$
3652.261	−1	g	27372.53	00	$c^3P_2 - y^1D_2^\circ$
3651.040	−1	f	27381.68	−05	$z^5D_4^\circ - f^5G_5$
3649.699	−1	f	27391.74	−03	$z^7P_4^\circ - f^5F_5$
3647.563	0	f	27407.78	−06	$z^5D_2^\circ - f^3D_3$
3646.098	−1	f	27418.80	00	$z^7F_2^\circ - e^7P_3$
3641.460	1	fb	27453.72	−10	$z^7F_1^\circ - f^5D_2$
3638.169	1	f	27478.55	−08	$z^7F_4^\circ - e^7P_4$
3637.059	1	g	27486.93	−07	$b^3G_3 - u^3G_3^\circ$
*3636.486	1	{f g}	27491.27	{+14 +02	$a^3F_3 - y^7P_2^\circ$ $z^5D_2^\circ - g^5F_3$
3635.829	−1	g	27496.23	−07	$z^7F_5^\circ - e^7F_6$
3634.536	−2	g	27506.02	−14	$z^7F_2^\circ - f^5D_2$
3633.653	−2	f	27512.70	−08	$z^7P_2^\circ - e^5P_1$
3628.829	2	gb	27549.27	−10	$b^3G_4 - u^3G_4^\circ$
3627.360	−2N	f	27560.43	−09	$z^7P_2^\circ - e^5P_2$
3624.065	2	g	27585.49	−06	$z^5D_2^\circ - f^5P_1$
3623.511	−1	f	27589.70	00	$z^7P_3^\circ - g^5D_3$
3621.202	2	fb	27607.30	−05	$z^5D_3^\circ - f^3D_3$
*3620.881	1	{f g}	27609.74	{−02 −08	$z^7F_0^\circ - f^5D_1$ $b^3H_4 - t^3G_4^\circ$
3619.671	−1	g	27618.97	−06	$a^3P_1 - u^5D_0^\circ$
3618.924	0	g	27624.68	−08	$a^3P_1 - u^5D_1^\circ$
3618.616	1N	g	27627.03	+02	$z^5D_2^\circ - h^5D_2$
3618.305	3	fb	27629.40	−04	$z^7F_6^\circ - e^7P_4$
3617.961	2N	fb	27632.03	+08	$a^3H_4 - w^5G_5^\circ$
3615.005	−3	g	27654.62	00	$z^7D_5^\circ - e^5F_5$
3613.953	−3	fb	27662.68	−01	$b^3H_5 - 12_5^\circ$
3613.450	2	f	27666.52	+01	$a^3D_3 - 10_3^\circ$
3613.110	2	fb	27669.13	−21	$z^7F_2^\circ - f^7D_3$
3612.520	−2	f	27673.64	−08	$b^3H_4 - 13_4^\circ$
3609.473	2	fb	27697.00	−10	$z^7F_3^\circ - f^7D_4$
3606.539	2	g	27719.54	−06	$a^3P_1 - w^3D_1^\circ$
3606.379	0	g	27720.76	−02	$b^3F_4 - w^3G_4^\circ$
3602.709	−2	f	27749.00	−06	$z^7P_4^\circ - e^7G_3$
3601.429	−2	f	27758.87	−05	$a^3P_2 - w^5P_3^\circ$
3597.253	−3d?	f	27791.09	−13	$a^1I_6 - x^3I_6^\circ$
3594.105	−2	g	27815.43	−08	$z^7D_4^\circ - e^5F_4$
3593.795	−2	f	27817.83	+01	$a^3H_4 - v^5F_5^\circ$
3588.535	3	g	27858.61	−08	$z^7P_4^\circ - e^7S_3$
3588.247	0	f	27860.84	−10	$a^3F_3 - y^7P_4^\circ$
3582.332	2	f	27906.84	+07	$z^5D_1^\circ - g^5F_1$
3579.836	2	f	27926.30	−08	$z^5D_3^\circ - e^3G_4$
3572.322	−3	f	27985.04	−02	$a^3H_5 - v^5F_6^\circ$
3570.598	−3	g	27998.55	+05	$z^7D_3^\circ - e^5F_3$
3566.316	0N	f	28032.16	−06	$a^3P_2 - w^5P_1^\circ$
3565.839	0	f	28035.91	−10	$z^5D_2^\circ - f^5G_2$
3564.566	2	g	28045.93	−05	$a^3H_4 - x^3G_3^\circ$
3564.525	3	g	28046.25	−09	$a^3H_4 - x^3G_5^\circ$
3563.612	−1	g	28053.43	−05	$z^7F_6^\circ - e^5G_5$

TABLE C—(*Continued*)

Solar λ	Solar Int	Grade	Wave Number: Solar	Wave Number: o−c	Desig
3562.608	−3	f	28061.34	−09	$b^3F_4 - y^1G_4^\circ$
3560.077	−1	gb	28081.29	−08	$z^7F_3^\circ - e^7F_4$
3559.465	1	g	28086.12	−12	$z^7F_1^\circ - e^7F_2$
3558.211	−2	g	28096.01	−05	$b^3F_2 - v^3D_3^\circ$
3554.649	1	g	28124.17	+01	$z^7D_2^\circ - e^5F_2$
3554.453	2	fb	28125.72	−09	$z^7P_4^\circ - e^5P_3$
3551.113	1	g	28152.17	−01	$z^7F_4^\circ - e^7F_3$
3544.860	−3	g	28201.83	+15	$z^7D_1^\circ - e^5F_1$
3543.102	−3	fb	28215.82	−08	$a^3H_6 - v^5F_5^\circ$
3542.572	−1	g	28220.04	−10	$z^7F_3^\circ - e^7F_2$
3541.243	−1N	fb	28230.63	−23	$a^3F_4 - y^7P_3^\circ$
3528.242	0	f	28334.66	−03	$a^3H_4 - v^5F_3^\circ$
3527.901	0	f	28337.40	−03	$a^3G_3 - z^1F_3^\circ$
3526.975	0	f	28344.83	−11	$z^5P_2^\circ - i^5D_3$
3523.185	0	f	28375.33	+01	$a^3D_3 - t^3G_3^\circ$
3522.738	−2	f	28378.92	−07	$a^1G_4 - s^3D_3^\circ$
3515.410	0	f	28438.08	00	$b^3F_2 - z^1D_2^\circ$
3514.469	0	f	28445.69	+06	$a^3F_4 - y^7P_4^\circ$
3513.605	−2	gb	28452.69	−12	$z^7F_6^\circ - f^5F_5$
3512.812	−1	f	28459.11	−10	$z^7F_3^\circ - e^7S_3$
3512.731	−1	f	28459.77	+04	$b^3H_5 - w^1G_4^\circ$
3510.193	1	fb	28480.35	−07	$z^5P_1^\circ - 4_2$
3509.732	1	g	28484.09	−05	$z^7F_0^\circ - f^5F_1$
3507.146	2	f	28505.09	−04	$z^5P_2^\circ - i^5D_2$
3506.595	1	f	28509.57	−12	$z^7F_1^\circ - f^5F_1$
3502.470	−1	f	28543.14	−06	$z^5D_3^\circ - f^3F_4$
3498.184	1N	gb	28578.11	−06	$z^7F_6^\circ - e^7G_6$
3494.263	−3	f	28610.19	−10	$a^3H_5 - w^3G_5^\circ$
3493.583	−1	g	28615.75	−08	$z^7F_5^\circ - f^5F_4$
3490.490	0	f	28641.10	−15	$z^5P_3^\circ - i^5D_4$
3477.986	−2	f	28744.07	−07	$z^5P_2^\circ - 4_2$
3473.228	−3	fb	28783.45	+06	$z^5D_4^\circ - f^3F_4$
3473.011	−2	f	28785.25	−02	$z^5D_2^\circ - f^3F_3$
3459.281	−2	fb	28899.49	+04	$z^5D_1^\circ - f^3F_2$
3450.141	−1	g	28976.05	−04	$b^3F_3 - v^3G_3^\circ$
3449.052	−3	f	28985.20	+07	$b^3G_5 - w^3H_4^\circ$
3448.206	0Nd?	fb	28992.31	−12	$a^3H_6 - z^1H_5^\circ$
3440.740	−1N	f	29055.22	+03	$a^3G_3 - v^3F_4^\circ$
3438.101	−2	f	29077.52	+01	$a^3G_4 - t^5D_4^\circ$
3434.967	−1	f	29104.04	−11	$a^1D_2 - t^3F_2^\circ$
3430.886	−2	gb	29138.67	−05	$b^3H_5 - v^3H_6^\circ$
*3429.818	−2	f	29147.74	−11	$b^3F_2 - w^3P_2^\circ$
		f		−01	$a^1G_4 - y^1H_5^\circ$
3428.021	0	f	29163.02	−06	$b^3H_4 - w^1F_3^\circ$
3426.674	2	f	29174.48	−01	$b^3H_5 - x^1H_5^\circ$
3426.093	−2	f	29179.43	−06	$c^3P_0 - t^5P_1^\circ$
3411.878	−2	f	29301.00	+05	$a^3G_5 - u^5F_5^\circ$
3410.565	−1	f	29312.28	−07	$b^3F_3 - w^3P_2^\circ$
3409.399	−2N	fb	29322.30	−03	$b^3G_3 - y^1F_3^\circ$
3407.562	3	gb	29338.11	−26	$a^5P_3 - u^5D_2^\circ$
3406.172	−1	g	29350.08	−02	$b^3P_1 - u^3D_2^\circ$
3398.111	−3N	f	29419.70	+07	$b^3H_6 - x^1H_5^\circ$
3391.842	0	f	29474.08	−04	$a^3D_2 - s^3G_3^\circ$
3369.152	0	gb	29672.56	−08	$a^3H_4 - v^3G_5^\circ$
3368.248	−2	f	29680.53	00	$a^3D_3 - s^3G_4^\circ$
3344.084	−2N	f	29894.99	+06	$b^3G_4 - 12_6^\circ$
3342.761	−1	g	29906.82	+01	$z^7P_2^\circ - g^7D_3$
3336.549	−1	f	29962.50	−06	$b^3G_3 - 13_4^\circ$
3315.176	0	f	30155.66	−03	$b^3H_4 - u^3F_3^\circ$
3308.761	2	gb	30214.13	−12	$a^3H_6 - y^3H_5^\circ$
3305.751	0	fb	30241.63	−04	$b^3H_5 - u^3F_4^\circ$
3304.366	0	g	30254.32	−05	$z^5F_2^\circ - i^5D_3$
3291.431	−1	f	30373.20	+10	$b^1G_4 - r^3G_4^\circ$
3274.227	1	g	30532.79	+08	$a^5P_1 - x^3P_0^\circ$
3272.607	1	g	30547.91	−08	$a^3F_3 - z^5H_4^\circ$
3270.672	0	fb	30565.98	+18	$b^1G_4 - r^3G_3^\circ$
3269.433	0	g	30577.56	−13	$a^5P_2 - x^3P_2^\circ$
3265.557	2	gb	30613.85	−06	$a^3G_4 - w^3H_5^\circ$
3263.467	−3	f	30633.46	−16	$a^3D_3 - u^3F_2^\circ$
3258.632	1N	g	30678.91	−11	$z^7D_1^\circ - f^5D_2$
*3256.497	1N	f	30699.02	+23	$z^7D_2^\circ - e^7P_3$
3254.471	2	g	30718.13	−15	$z^7D_3^\circ - e^7P_4$
3240.122	−1	g	30854.16	−10	$z^7D_3^\circ - e^7P_3$
3238.319	−1	g	30871.34	−03	$a^1G_4 - v^3H_4^\circ$
3235.328	0	f	30899.88	−01	$a^3G_4 - y^3I_5^\circ$
3230.098	0	g	30949.91	−09	$a^5F_3 - y^3D_2^\circ$
3227.177	−3	f	30977.92	−08	$b^3F_4 - u^5F_3^\circ$
3223.100	−2	f	31017.10	−23	$a^3D_2 - t^3F_3^\circ$
3219.370	−2N	g	31053.04	+03	$a^3G_5 - w^3H_4^\circ$
3216.050	−2	f	31085.10	+07	$a^3D_2 - t^3F_2^\circ$
3202.668	0	f	31214.98	−09	$a^3F_2 - w^5D_2^\circ$
3193.735	−1N	fb	31302.29	+02	$a^3D_1 - t^3F_2^\circ$
3191.415	0	fb	31325.03	−03	$a^3D_3 - t^3F_4^\circ$
3187.169	0	f	31366.76	−11	$z^7F_1^\circ - g^7D_2$
3183.582	0	gb	31402.11	−04	$a^3H_5 - x^3H_6^\circ$
3174.222	0	g	31494.70	−07	$z^5D_1^\circ - i^5D_2$
3172.298	0	f	31513.80	−02	$a^3G_3 - x^1F_3^\circ$
3169.076	−2	f	31545.84	+12	$a^1H_5 - t^3H_6^\circ$
3166.596	0	f	31570.54	−10	$a^5P_2 - v^3D_1^\circ$
3166.256	1	gb	31573.94	−11	$z^7D_3^\circ - e^7F_2$
3165.158	1	fb	31584.89	+02	$a^5P_3 - v^3D_3^\circ$
3165.085	−2	g	31585.62	−06	$a^3H_4 - u^3G_5^\circ$
3161.554	−1	f	31620.89	−01	$a^3H_4 - 4_4^\circ$
3160.924	0	g	31627.19	−07	$z^7D_2^\circ - e^7G_2$
3159.262	0	gb	31643.83	−08	$b^3F_2 - t^3D_2^\circ$
3157.144	1	gb	31665.06	+03	$a^3P_2 - w^3P_1^\circ$
3155.796	0	gb	31678.58	+01	$a^3H_6 - x^3H_5^\circ$
3154.121	−1	fb	31695.41	−15	$a^3F_2 - v^5D_2^\circ$
3149.498	−1	fb	31741.93	00	$b^3G_5 - x^1H_5^\circ$
3144.926	0	f	31788.08	−06	$a^3H_5 - 4_4^\circ$
3139.107	−2	f	31847.00	−11	$z^7D_3^\circ - f^5F_3$
3138.407	−2	f	31854.10	−08	$a^3F_3 - v^5D_4^\circ$
3136.086	−2	g	31877.67	−04	$z^7D_5^\circ - e^7G_6$
3133.967	−1d?	fb	31899.23	−02	$z^7D_4^\circ - f^5F_4$
3125.054	2	fb	31990.21	−28	$a^3F_3 - v^5D_3^\circ$
3119.036	−1	f	32051.92	+02	$a^3G_3 - 13_4^\circ$
3113.667	0	f	32107.19	−01	$z^7D_2^\circ - e^5P_1$
3102.644	0	g	32221.26	−06	$a^5F_3 - y^7P_3^\circ$
3097.491	−1	f	32274.86	+04	$z^7D_4^\circ - e^5P_3$
3094.069	−3	f	32310.55	+14	$z^7D_3^\circ - e^5P_2$
3081.841	0	f	32438.75	−12	$a^3F_4 - v^5D_4^\circ$
3079.826	−2	f	32459.96	+11	$a^5P_2 - w^3P_2^\circ$
3057.802	1N	g	32693.75	−07	$a^5F_3 - y^7P_4^\circ$
3035.238	0	f	32936.78	+08	$c^3P_1 - t^3F_2^\circ$
3018.258	−1	f	33122.08	−06	$b^3F_4 - x^1F_3^\circ$
*2990.336	0	f	33431.35	−02	$z^7F_4^\circ - 1$
		f		+11	$b^3G_4 - t^3F_4^\circ$
2980.586	−1	f	33540.69	+10	$a^3H_4 - 9_4^\circ$

* Fe I blend. Designations listed only if grade is "good" or "fair."
b (Column three) Solar line not entirely due to Fe I.

PART II

AN ANALYSIS OF THE ZEEMAN PATTERNS OF THE SPECTRUM OF Fe I

DOROTHY W. WEEKS

(27) OUTLINE OF THE WORK

This analysis of the Zeeman patterns of the spectrum of Fe I is based upon photographs taken at the Massachusetts Institute of Technology with the assistance of Works Progress Administration workers under the supervision of Dean George R. Harrison, who generously placed them at the disposal of the author. To obtain the Zeeman effect, the Bitter[33] electromagnet was used, with field strengths ranging from 83,000 to 87,000 oersteds for 5 sets of spectrograms containing approximately 12 plates each. The technique used in taking these spectrograms is described in papers dealing with other elements[34] that have been analyzed at the Massachusetts Institute of Technology. The plates were measured upon Harrison's automatic measuring machine,[35] which recorded on a film the wave-lengths and relative photographic intensities of the components. From the film-records thus produced the Zeeman patterns were studied according to the method developed by Russell.[36] These permanent film-records are noted for the high degree of accuracy of the measurements[35] and provide a convenient means for reexamination of the patterns.

The strong magnetic field resulted in a large number of resolved patterns with their components well separated. The magnification of these separations afforded by the automatic comparator made the identification of unblended patterns a comparatively easy task, but the spectrum is so rich that loss by overlapping of neighboring patterns was serious. Many lines originally measured were rejected for underexposure, and a few for overexposure, so that, of some 1250 originally recorded from the films, between λ6494 and λ2272, 1038 were adopted as a basis for the final study. Of this number, 519 were resolved completely enough to permit the determination of both the g-values involved, from the observed pattern alone; for 163 others, the pattern, though resolved, was so much blended that this was impracticable, though the difference $g_1 - g_2$ could be found from the separation of the components. There were also 345 lines with unresolved patterns, sufficiently clear of blends to permit the determination of one g-value when the other was known—as can also be done for the preceding group. The remaining 11 lines are of the "unaffected" type for which both g-values are very nearly zero. From these data, g-values were derived for 130 of the 184 known even levels, and 242 of the 280 odd levels. There are also 8 even and 12 odd levels for which $J = 0$, and g is effectively zero, though theoretically indeterminate—raising the whole number of available determinations to 392 levels out of 464. These values were computed by the usual process—starting with the values determined from fully resolved patterns, and using the mean g's so obtained for the even terms to obtain additional values for the odd terms from the unresolved or partially obscured patterns. A second approximation (as usual) was found to give stable values.

There were about 90 blended patterns for which only an unresolved maximum on one side of the mean position could be observed. These could be utilized by assuming that the correction to the wave-length recorded by the automatic machine was the same as for neighboring lines. These corrections are almost always small, but they vary from one well-observed line to another (perhaps because of Paschen-Back effect); enough to introduce quite sensible errors into work of this refinement. Such patterns were therefore rejected, except when they gave the only determination of g for the level involved—in which case the result is given to two instead of three decimals and marked with a colon. Almost all of these rejected observations were fully consistent with the pattern resulting from the g-values found from better data. Five lines, however, show patterns inconsistent with them and are evidently accidental blends.

(28) OBSERVED ZEEMAN PATTERNS OF FE I

Table D (pp. 184–202) contains the observed Zeeman patterns for the 1038 lines and the g-values deduced from each. Column 2 gives the number of different π-components observed (those equidistant from the middle being counted but once, no matter whether one or both were measured). Columns 3, 4, and 5 give the positions, in units of 0.001a, of the strongest of these (indicated by heavy type), and two adjacent ones. Others are omitted to save space. The letters B and D represent unresolved patterns, respectively, strongest at the outer edges and at the center.[37] The readings of the films correspond very

[33] G. R. Harrison and F. Bitter, *Phys. Rev.* **57**: 15, 1940.
[34] G. R. Harrison and J. Rand McNally, Jr., *Phys. Rev.* **58**: 703, 1940.
[35] G. R. Harrison, *Jour. Opt. Soc. Amer.* **25**: 169, 1935.
[36] H. N. Russell, W. Albertson, and D. N. Davis, *Phys. Rev.* **60**: 641, 1941.
[37] W. F. Meggers and H. N. Russell, *Bur. Standards Jour. Research* **17**: 131, 1936.

TABLE 25

AVERAGE ERRORS OF g-VALUES

N	Type of Level	Number of Levels	$\bar{N}$	$\bar{D}$	$\bar{D}'$
51 to 10	Low Even	30	21.5	5.4	5.5
10 to 5	Low Even	12	7.1	5.6	6.0
	Odd	86	6.3	6.4	7.0
	High Even	10	5.8	8.6	9.4
4 to 3	Odd	28	3.4	6.2	10.8
	High Even	14	3.3	9.0	7.4
2	Odd	84	2	5.6	7.9
	High Even	22	2	8.5	12.0

closely to the maximum of the actual pattern. Components masked by other lines or not capable of interpretation are indicated by "m". A dagger (†) denotes that the measures are insufficient to establish the position of the center, which has been estimated by comparison with the instrumental corrections for other lines (see above).

Similar data for the σ components are given in the next four columns. Unresolved patterns, strongest at the inner edge, the outer edge, or the center, are denoted [37] by A, B, or C. Sharp components are denoted by S, and "unaffected" lines (for which both g's are unresolvably small), by U. The values of J and g for the two terms involved in each line are given in the last four columns—g_1 corresponding to the level with larger J. These values are derived from the data for the individual line for completely resolved patterns; otherwise the mean value of g found from other lines for one term is used to find g for the other. The assumed values of g are given in parentheses, both for unresolved patterns and for those so badly blended that only Δg can be determined. A few values of lower accuracy are marked with colons. A double colon is used throughout the table to denote unusually doubtful entries.

(29) OBSERVED AND CORRECTED g-VALUES

A comparison of the observed and theoretical g-values for the principal groups of terms indicates that the observed values are systematically too large. Similar discrepancies have been found in other cases

TABLE 26

RESIDUALS OF g-VALUES FROM THEORY (UNIT 0.001a)

Term	J=6	5	4	3	2	1	All
d^6s^2							
a^5D			−4	−3	−6	−2	−15
a^3P					+6	0	+6
b^3D				−7	...	...	−7
b^3F			−15	−10	−4		−29
b^3G		0	−2	+11			+9
a^3H	−4	+5	+11				+12
b^1G			−21				−21
a^1I	+14						+14
Sum	+10	+5	−31	−9	−4	−2	−31
d^7s							
a^5P				−1	−13	−1	−15
a^5F		+4	−1	−2	−5	−14	−18
b^3P					−2	−11	−13
c^3P					−16	−34	−50
a^3D				+2	+11	+231	+244
a^3F			+4	+3	+3		+10
a^3G		−3	+1	+6			+4
b^3H	−2	−1	+11				+8
a^1P						−183	−183
a^1D					+28		+28
a^1G			+1				+1
a^1H		0					0
Sum	−2	0	+16	+8	+6	−12	+16
d^8							
c^3F			+14	−17	+10		+7
All low even terms	+8	+5	−1	−18	+12	−14	−8

Term	J=6	5	4	3	2	1	All
$z^7P°$			−3	−9	0		−12
$z^7D°$		−3	−8	−4	+8	−1	−8
$z^7F°$	−2	−2	−7	+13	+4	+49	+55
$z^5P°$				−10	+2	−13	−21
$z^5D°$			+2	0	+3	−5	0
$z^5F°$		−1	+5	0	+4	−12	−4
Sum	−2	−6	−11	−10	+21	+18	+10
$y^5D°$			−4	−8	−5	−8	−25
$y^5F°$		+17	−6	−6	−2	−16	−13
$z^5G°$	−1	−49	−47	−30	+2		−125
$y^3D°$				−9	−16	−7	−32
$y^3F°$			−4	+3	+21		+20
$z^3G°$		+48	+50	+41			+139
$y^5P°$				−6	+3	+2	−1
$x^5D°$			−11	+4	+1	−2	−8
$x^5F°$		−10	−22	+4	−2	−6	−36
$z^3P°$					−7	−4	−11
$z^3D°$				−12	+1	+13	+2
$z^3F°$			0	+3	+15		+18
$y^7P°$			0	−9	+7		−2
$z^5S°$					−15		−15
Sum	−1	+6	−44	−25	+3	−28	−89
Odd terms	−3	0	−55	−35	+24	−10	−79
e^7D		−15	+5	+5	+9	+2	+6
e^5D			+2	+8	+3	+18	+31
e^5F		+21	−19	−14	−9	+7	−14
e^3F			+38	+24	−45		+17
Sum		+6	+26	+23	−42	+27	+40

and probably arise from an error in the calibration constant, which was determined from a few persistent lines of other elements which appeared on the plate. It is not certain that the mean field was the same for the regions where these lines and those of iron were emitted. A close agreement with theory can be obtained by multiplying the observed g-values by the factor 0.987.

Table E (pp. 203–206) gives the observed and corrected g-values for all levels of Fe I for which they have been determined. The first column gives the designation, the second, the mean g-value resulting from the second approximation described above (giving half weight to a few weak determinations). The third and fourth give the numbers R of resolved patterns, and U of unresolved patterns or patterns disturbed by blending, upon which this mean is based, and the fifth (A. D.), the average residual, without regard to sign, for the individual determinations, in units of 0.001a. The sixth column gives the g-value corrected as just described, and the last, the residual excess of this above the predictions of Lande's theory. The uncorrected g-values should be used for comparison with table D; the corrected values, for all other purposes.

The means, without regard to sign, of the individual discordances D, grouped according to the number of determinations N for the various levels, are given in table 25. The last column gives $\bar{D}' = \bar{D}\sqrt{\bar{N}}/(\bar{N} - 1)$, which should be a fair, though not an exact, approximation to the average error of one determination of g. The weighted means of this (still in units of 0.001a) are 5.6 for the low even levels, 7.3 for the odd, and 11.1 for the high even levels. If two wildly discordant values are rejected, the last becomes 9.7. Multiplying by 0.85, the probable error of an average determination of g may be estimated as ±0.005a for the low even levels, ±0.006 for the odd, and ±0.009 for the high even levels.

The better-determined g-values in table E should have probable errors of 0.001 to 0.003a. The deviations from the theoretical values for most of the higher levels are very much greater, indicating the existence of large perturbations.

Perturbations may be anticipated between neighboring levels of the same parity, and with the same J —irrespective of multiplicity, L-value, and electron configuration. Within a group of isolated terms the g-sums for each J should be unaffected. The even levels below 34000 form such a group. There are two groups of odd terms, between 19000 and 30000, and 33000 and 42000, and one of high even terms, between 42000 and 49000. The residuals for these terms are collected in table 26. The theoretical g-sum for all the low even terms is 61.333; for the odd terms, 101.000; for the high even terms, 24.000. The observed sums are 61.225, 100.901, and 24.040. Taken separately, these indicate additional calibration factors of 0.9999, 0.9991 and 1.0017. Together they give 0.9995—a negligible correction.

Some interesting perturbations appear in this table. The terms involved are:

Desig	Level	Residual	Desig	Level	Residual
c^3P_1	24772	− 37	$z^5G_5^\circ$	34782	−48
a^3D_1	26406	+228	$z^3G_5^\circ$	35379	+51
a^1P_1	27543	−181			
Sum		+ 10			+ 3
$z^5G_4^\circ$	35257	− 43	$z^5G_3^\circ$	35611	−31
$z^3G_4^\circ$	35767	+ 53	$z^3G_3^\circ$	36079	+42
Sum		+ 10			+11

Perturbations of the g-values are accompanied by mutual repulsion of the levels involved and by strong intersystem combinations. The repulsion of a^1P_1 lowers a^3D_1 below a^3D_2, and that of $z^3G_5^\circ$ puts $z^5G_5^\circ$ below $z^5G_6^\circ$, interrupting the usual sequence of the components for both terms. The combination of z^5G° with a^3F produces the very strong line at λ 4383, and other intersystem lines are exceptionally strong; and a^1P and a^3D show a marked tendency to combine strongly with the same terms.

The higher levels, both odd and even, are closely packed, and the perturbations are great. For example, the configurations $(a^6D)4d$ and $(a^4D)5s$ give rise to 50 levels, lying between 50342 and 52257, all but one of which are known. For the higher J-values all the g's are known and the sum-rule holds—the algebraic sums and arithmetic means of the residuals being +18, ±13 for J = 6; +5, ±40 for J = 5; +8, ±64 for J = 4; for smaller J's some g's are missing. The identity of e^5G_2 appears to be conclusively settled by multiplet intensities, despite the enormous perturbation of +620, but in a case like this the levels of the same J share all their properties and an exact specification is illusory.

In conclusion, it is a pleasure to express gratitude to Dean George R. Harrison and to the Massachusetts Institute of Technology Works Progress Administration assistants for the photographs of the Zeeman patterns upon which the work of this paper has been based; to Professor Henry Norris Russell for his guidance in the preparation of the data and for preparing the final form of the manuscript for the press while the author has been engaged in war work; to Miss Helen P. Beard and Miss Margaret Aylesworth for assisting with the computations; to the American Philosophical Society for a grant from the Penrose Fund for an assistant to help with the computations; to the American Academy of Arts and Sciences for a grant from the Permanent Science Fund, also for computational assistance; and to Wilson College for co-operating in the support of the assistants.

TABLE D

OBSERVED ZEEMAN PATTERNS OF Fe I

λ	π Components				σ Components				J_1	Obs. g_1	J_2	Obs. g_2
	No. Meas.	π			No. Meas.	σ						
6494.985	1	0	D		1	975	A		6	(1178)	5	1219
6430.851	4	0	168	341	5	1011	1180	1348	4	1521	3	1692
6421.355	1	0			1	1522	C		2	(1526)	2	1518
6419.982	2		106	177	5	1282	1332	1475	3	1334	3	1275
6411.658	3	0	334	668	4	857	1193	1544	3	1528	2	1863
6408.031	2	0	1005		2	512†	1464†		2	1515	1	(2520)
6400.010	4	0	165	340	5	1021	1187	1321	4	1502	3	1665
6393.605	3	0	060	132	1	818	m	m	5	1058	4	1118
6380.748	1	0			1	687			2	(682)	2	692
6355.038	2	0	332		2	852	1178		2	1180	1	1509
6336.835	1	995			2	1532	2507		1	2517	1	1532
6335.335	3	023	332	661	3	791	1125	1517	3	1501	2	1848
6322.693	4	0	172	354	3	1493:	1609	1784	4	1257	3	1081
6318.022	3	0	068	157	2	622	691		4	826	3	894
6315.316	2			302	5	1292	1393	1425	3	1079	3	1180
6302.507	1	0	S		1	2507	S		1	2507	0	0
6301.515	2		340	664	4	1535	1897:	2179	2	1861	2	1528
6297.800	2	0	1009		2	509	1536		2	1518	1	2527
6270.238	1	0	S		1	509	S		1	509	0	0
6265.140	2		346	518	6	1517	1683	1850	3	1685	3	1515
6256.370	3	588	873	1166	5	825	1111	1421	4	1113	4	824
6254.262	1	0	D		1	1530	B		2	(1526)	1	1522
6252.561	6	0	083	167	1	758			6	(1178)	5	1261
6246.334	3	186	340	498	6	1518	1678	1841	3	1684	3	1528
6232.661	2	0	335		2		1876	2208	2	1875	1	1542
6230.728	1	047	B		1	1265	C		4	1271	4	1259
6219.290	2	332	665		4	1519	1842	2177	2	1847	2	1517
6215.152	3	0	103	179	1	940:	B		3	(768)	2	679
6213.438	1	1014			2	1517	2528		1	2528	1	1517
6200.323	3	0	430	848	3	1101	1521	1952	3	1100	2	675
6191.562	1	0	D		1	831	A		5	(1052)	4	1107
6180.212	2	0	283						4	(1065)	3	1348
6173.343	1	0	S		1	2531	S		1	2531	0	0
6170.492	1		621						2	1476	2	(1166)
6165.366	3	0	172	331	3	445	607	749	4	920	3	1080
6157.734	1			327	4	1022	m	1288	4	1202	4	1287
6141.734	2	0	155						3	(1677)	2	1522
6137.696	1	0			1	1102			3	(1087)	3	1117
6136.999	3	0	649	1355	1	2179:	B		2	(1844)	1	1509
6136.620	1	0	D		1	881	B		4	(822)	3	802
6102.178	2	0	166		1	879	B		2	682	1	(505)
6078.496	1	0			4	946	1022	m	3	1092	2	(1166)
6065.487	1	0			1	690			2	(672)	2	708
6055.987	4	0	185	396	2	615	m	m	4	1156	3	(1341)
6027.057	5	0	96	196	2	781†	942†		5	(1176)	4	1275
6024.066	1	0	D		1	1131	A		5	1237	4	(1263)
6008.577	1	0	D		2	1171	1226		4	1304	3	(1348)
6003.033	1	0			1	1288			4	(1267)	4	1309
5987.057	2	0	312						2	(1166)	1	1478
5984.805	3	0	126	268	3	1089	1250	1348	3	1349	2	1479

TABLE D—(*Continued*)

λ	π Components				σ Components				J_1	Obs. g_1	J_2	Obs. g_2
	No. Meas.	π			No. Meas.	σ						
5976.799	1	0			1	1087	1135	1211	3	1089	3	1130
5952.749	1	0			1	655	C		2	(690)	2	620
5934.658	1	0	D		1	1020	A		3	1129	2	(1183)
5930.173	2	0	208:		1	1166	B		3	853	2	(696)
5914.16	1	0	D		1	1176	B		4	1120	3	(1100)
5709.378	2		468	602					4	1520	4	(1370)
5701.553	1	0	D		1	998	A		4	(1251)	3	1335
5662.525	3	0	088	158	1	1203	A		5	(1442)	4	1512
5658.826	3	232	506	754					3	1519	3	(1267)
5624.549	2	523	1005		4	1028	1530	1960	2	1528	2	1023
5615.652	1	0	D		1	1035	A		5	(1418)	4	1514
5602.955	1	1509			1	60:	1545		1	1527	1	(−12)
5586.763	3	0	148	292	2	944	1058		4	1382	3	1528
5576.097	1	U			1	U			1	0	0	0
5572.849	3	0	250	498	4	780	1026	1273	3	1274	2	1522
5569.625	2	0	503		2	520	1042		2	1033	1	1546
5565.708	1	0			1	1089			3	(1100)	3	1078
5563.604	1	0			1	1519			3	1515	2	(1513)
5506.782	3	0	507	1008	2		2087	2521	3	1513	2	(1008)
5501.469	2	0	255						4	1519	3	(1264)
5497.519	2	0	1521		2		1559	3034	2	1514	1	−8
5476.571	1	0			1	1511			4	(1515)	4	1507
5455.613	1	1513			2	0	1522		1	1522	1	9
5446.920	2	508	1007		4	1021	1522	2028	2	1524	2	1020
5434.527	1	U			1	U			1	0:	0	0:
5429.699	3	252	506	753	6	1278	1525	1770	3	1523	3	1274
5424.072	1	0	D		2	1127	A		7	1315	6	(1346)
5415.201	1	0	D		1	1127	A		6	1247	5	(1264)
5410.913	1	0	D		1	1127	B		4	882	3	(801)
5405.778	2	0	504		3	514	1021	1522	2	1019	1	1523
5404.144	1	0	D		1	1151:	B		5	1124	4	(1117)
5397.131	4	309	456	600	7	1361	1519	1665	4	1520	4	1372
5393.174	1	0			1	1520			4	1518	3	(1517)
5383.374	1	0	D		1	1075	A		6	1207	5	(1233)
5371.493	4	0	249	499	5	773	1026	1271	3	1274	2	1524
5369.965	1	0	D		1	1115			5	1117	4	(1117)
5367.470	1	0	D		1	958†	B		4	913:	3	(898)
5365.403	1	0	D		1	955	A		5	(1013)	4	1027
5364.874	1	0	D		1	793	B		3	490	2	(339)
5341.026	2	494	983		4	671	1186	1692	2	1183	2	684
5328.534	3	205:	476:	720:	6	622†	861†	1147:†	3	1347	3	(1100)
5328.042	4	0	149	295	6	930	1072	1207	4	1365	3	1510
5324.185	1	0			1	1535			4	(1519)	4	1551
5302.307	1	0	D		1	1539	B		2	(1519)	1	1529
5283.628	1	0			1	1530			3	(1517)	3	1543
5281.796	4	0	574	1214	3	597	1188	1778	3	1779	2	2370
5273.379	1	0	S		1	511	S		1	511	0	0
5273.176	1	0	S		1	1542	S		1	1542	0	0
5270.360	2	0	165		3	533	680	848	2	687	1	527
5269.541	2	0	187		1	1064	A		5	(1422)	4	1512
5266.562	4	0	269	540	3	893	1159		4	1672	3	1936
5263.314	1	0			1	1525			2	(1521)	2	1529
5250.650	3	0	158	320	2	1366	1522		3	1681	2	1839
5242.495	1	0	D		1	991			6	1024	5	(1029)
5232.946	4	0	148	296	3	1045	1167	1275	5	1605	4	1751

TABLE D—(*Continued*)

λ	π Components				σ Components				J_1	Obs. g_1	J_2	Obs. g_2
	No. Meas.	π			No. Meas.	σ						
5227.192	1	0	D		1	991:	A		3	(1100)	2	1155::
5226.868	2	292	664		4	2043	2368	2717	2	2376	2	2033
5217.395	1	0	D		1	1523	A		4	(1519)	3	1518
5216.278	1	0			1	688	C		2	(679)	2	697
5208.601	1	0	D		1	1528	A		3	(1517)	2	1512
5202.339	1	0			2	1693:	m		3	(1688)	3	1698
5198.714	2	0	668		2	1191	1854		2	1855	1	2520
5194.943	1	0			1	1102			3	(1100)	3	1104
5192.350	2		364	503	2	m	2096	2232	3	1937	3	1785
5191.460	2	0	665		1	1705	A		2	2370	1	3035
5171.599	1	0			1	1273			4	(1271)	4	1275
5167.491	1	0	D		1	1101	A		4	(1271)	3	1328
5150.843	3	0	255	512	1	1791†	B		3	1268	2	(1008)
5142.932	3	0	111	219	2	1682†	B		4	1374	3	(1264)
5141.747	1	1017			2	501	1513		1	1513	1	501
5139.468	2		295	380	1	1695:	C		4	(1774)	4	1689
5137.388	1	0	D		2	1395	A		5	(1442)	4	1454
5133.692	1	0	D		5	m	1007	1110	6	1340	5	1452
5131.475	1	0			1	2527			1	(2526)	1	2528
5127.363	2	0	050		1	m	1566†	m	5	1416	4	1366
5123.723	1	U			1	U			1	0	1	0
5110.414	3	307	459	607	1	m	m	m	4	1669	4	(1516)
5107.645	3	0	407	820	3	1084:	1514	1918	3	1105	2	698
5107.452	1	0			1	1015			2	(1008)	2	1022
5098.703	3	0	164	332	4	1357	1517	1674	3	1678	2	1840
5096.998	4	0	142	306					3	1157	2	(1010)
5083.342	1	0			1	1269	C		3	(1264)	3	1274
5079.742	2	0	1010		2		1136	2022	2	998	1	−12
5079.226	2	0	686		2	1171	1849		2	1851	1	2533
5074.757	1	0	D		1	878	A		5	1264	4	(1361)
5068.774	1	0	D		1	1782			4	(1774)	3	1771
5065.020	3	0	169	321					4	1103:	3	(1263)
5051.636	1	0			1	1370	C		4	(1367)	4	1373
5049.825	3	0	181	361	4	987	1162	1341	3	1341	2	1520
5041.759	4	0	159	316	4	1431	1586	1736	4	1270	3	1113
5041.074	3	0	262	533	3	1238	1518	1775	3	1261	2	1004
5028.129	2	0	m	122	1	m	810		5	(1013)	4	1074
5022.244	1	0	D		1	582	B		2	(690)	1	798
5014.950	2	0		129:	1	991	A		3	(1101)	2	1156
5012.071	1	0			1	1422			5	(1422)	5	1422
5006.126	1	452	B		1	m	m	1779:	5	1608	5	(1518)
5001.871	2	0	117		2	986	1071		4	(1267)	3	1363
4994.133	3	0	102	195	3	1455	1572	1678	4	1364	3	1262
4985.261	1		B	123	1	1255	C		2	(1183)	2	1121
4973.108	1	303			2	507	840		1	825	1	522
4966.096	1	0			1	1434			5	(1418)	5	1450
4957.603					1	1135	A		6	(1522)	5	1600
4957.302	3	303	406	674					4	(1515)	4	1675
4891.496	4	0	262	501	4	769	1035	1297	4	1522	3	1773
4890.762	2	560	964		2	1560	2084	2561	2	2050	2	1538
4871.323	3	0	492	1009	4	546	1015	1533	3	1533	2	2031
4859.748	2	0	1511		3	0	1538	3000	2	1513	1	3028
4789.654	1	0			1	987	C		2	(1041)	2	933:
4786.810	3	0	127	248	2	1135	1226		3	1375	2	1500
4741.533	1	0	D		1	1493	A		3	1510	2	(1518)

TABLE D—(*Continued*)

λ	π Components				σ Components				J_1	Obs. g_1	J_2	Obs. g_2
	No. Meas.	π			No. Meas.	σ						
4736.780	3	0	092	198	1	1096	A		5	1429	4	(1519)
4710.286	1	0			1	774	C		3	(774)	3	774
4707.487	1	0	D		1	1566	B		2	1551	1	(1509)
4707.281	4	0	170	314	3	841	991	1147	4	1348	3	1517
4691.414	1	0			1	1068			4	(1068)	4	1068
4678.852	3	0	127	260	3	1167	1264	1448	4	1549	3	1677
4668.142	3	0	266	547	3	742	978	1249	3	1253	2	1510
4654.501	3	0	265	529	3	1597:	1890	2159	4	1359	3	1093
4647.437	1	0			1	1222			5	1216	5	1228
4638.016					1	1697	C		3	1717	3	(1677)
4637.512	2	0	521		1	504	A		2	1004	1	(1519)
4630.125	1	0			1	1536			3	1529	2	(1526)
4625.052	2		521	788	1	1267†	m		3	(1517)	3	1250
4619.294	1	0	D		1	1754	B		3	(1677)	2	1638
4613.210	1	U			1	U			1	0	0	0
4611.285	1			230					2	(1860)	2	1975
4602.944	5	0	148	298	6	1712	1858	2006	5	1420	4	1274
4598.122	1	1515			2	0	1522		1	1522	1	7
4592.655	1	158	335	496	6	1105	1265	1429	3	1264	3	1102
4587.132	1	0	D		1	1090	B		5	(1013)	4	994
4556.129	2	0	162:		3	1312†	1498†	1702†	3	(1677)	2	1860
4547.851	1	0	D		1	1013	A		3	(1032)	2	(1041)
4531.152	2	m	298	389	4	1266†	1355†	1454†	4	1362	4	(1271)
4528.619	4	0	168	336	6	1024	1187	1357	4	1519	3	1684
4525.142	1	473	B		1	1201	m	m	3	(1677)	3	1519
4517.530	1	142			1	1573	C		1	1627	1	(1485)
4494.568	3	0	322	645	5	891	1215	1536	3	1537	2	1859
4490.084	2	0	413		3	1098	1508	1859:	2	1509	1	1920
4489.741	1	0	S		1	1573	S		1	1573	0	0
4484.227	4	0	178	369	5	967†	1146†	1314†	4	1503	3	1677
4482.257	2	0	1003		3	528	1526	2522	2	1525	1	2525
4482.171	1	0			1	1525			2	1522	1	(1518)
4480.142	3	731	1054	1416	5	1044†	1399†	1625:†	4	1360	4	(1013)
4479.612	2	0	892						2	1628	1	(2520)
4476.021	2	0	293		3	931	1225	1513	2	1223	1	1515
4469.381	3	0	173	351	3	1336	1465	1647	3	1686	2	1861
4466.554	3	0	145	288	4	1096	1226	1361	3	1369	2	1510
4461.654	1	0			1	1571	B		3	1533	2	(1514)
4459.121	3	174	332	492	6	1521	1664	1822	3	1679	3	1521
4454.383	2	306	607		4	1225	1520	1822	2	1523	2	1224
4447.722	1	1003			2	1520	2523		1	2523	1	1520
4443.197	1	0	S		1	572	S		1	572	0	0
4442.343	2	328	650		4	1526	1842	2172	2	1847	2	1526
4433.223	2	0	591		3	1260†	1865†	2549:†	2	(1860)	1	2460
4430.618	1	0	S		1	2528	S		1	2528	0	0
4430.197	2	0	113		1	1481:	A		2	(1504)	1	1617
4427.312§	{1	0			1	1532			4	1522	3	(1518)
	{2		305	597	2	1895†	2200†		2	(1860)	2	1560
4422.570	1	942			2	576	1508		1	1508	1	576
4415.125	3	0	213	446	5	912	1133	1374:	3	909	2	687
4408.419	2	0	332		3	1523	1855	2178	2	1852	1	1522
4407.714	3	0	160	326	4	1695	1873	2015	3	1693	2	1531
4404.752	1	0	D		1	1166	B		4	1116	3	(1100)
4401.293	2		371	521					3	(1677)	3	1510
4390.954	4	0	132	264	5	1038	1172	1301	4	905	3	773

§ Two superposed Zeeman Patterns.

TABLE D—(*Continued*)

λ	π Components				σ Components				J_1	Obs. g_1	J_2	Obs. g_2
	No. Meas.	π			No. Meas.	σ						
4388.412	1	0			1	1702	C		3	1727	3	(1677)
4387.897	1	430			2	1489	1933		1	1933	1	1489
4383.547	1	0	D		1	1161	A		5	1249	4	(1271)
4377.793	1	0	D		1	1193	B		2	967	1	(741)
4375.932	1	0			1	1524			5	1518	4	(1516)
4373.563	1	0			1	790			3	765	2	(752)
4369.774	1	99	B		1	1044	C		4	1031	4	1056
4367.581	1	0			1	1126	B		5	1075	4	(1062)
4358.505	5	0	141	282	3	652	794	935	5	1220	4	1362
4352.737	2	0	516		3	1496	2007	2518	2	2007	1	2520
4337.049	3	209	403	598	6	910	1108	1301	3	1103	3	908
4325.765	3	0	122	248	3	808	928	1049	3	806	2	684
4315.087	2	164	319		4	1842	2004	2148	2	2000	2	1847
4309.380	1	0			1	1213	A		6	1215	5	(1216)
4309.036	1	56	B		1	1034	C		6	1039	6	(1030)
4307.906	1	0	D		1	1133	B		4	(1117)	3	1112
4305.455	2	0	404		2	1110	1515		2	1514	1	1918
4302.191	1	0:	B		2	m	1132	1189	4	1070:	4	(1013)
4298.040	5	0	210	402	3	1631	1830	2035	5	1217	4	1013
4294.128	4	293	435	581	7	1132	1274	1406	4	1271	4	1129
4288.148	1	0			1	778	C		3	(766)	3	790
4285.445	1	0			1	1191			6	(1180)	6	1202
4282.406	3	0	320	639	5	1060	1379	1700	3	1695	2	2013
4271.764	1	0			1	1262			5	1269	4	(1271)
4271.159	3	0	097	203	3	1377	1478	1575	4	1675	3	1774
4267.830	1	0	S		1	1570	S		1	1570	0	0
4266.968	1	0			1	1063			4	(1065)	4	1061
4260.479	1	0			1	1617			5	(1621)	5	1613
4250.790	3	292†	595†	904†	6	812	1109	1405	3	1108	3	806
4250.125	3	0	285	545	4	1259	1533	1784	3	1783	2	2045
4247.432	1	0			1	1388	B		5	1374	4	(1370)
4246.090	3	0	160	324	1	1235:†	A		3	1343	2	(1504)
4245.258	1	0	S		1	1911	S		1	1911	0	0
4239.847	1	0			1	1222			5	(1213)	5	1231
4238.816	1	0	D		1	1215	A		4	1254	3	(1267)
4235.942	1	0			1	1680	C		4	(1661)	4	1699
4233.608	2	0	1005		3	1026	2032	3037	2	2032	1	3037
4227.434	1	0	D		1	1125	A		6	1369	5	(1418)
4225.460	3	0	307	559	2	1305†		1918†	3	1306	2	(1017)
4224.176	3	0	130	287					5	1517	4	(1370)
4222.219	1	0			1	1786	C		3	(1771)	3	1801
4220.347	1	0	S		1	1478	S		1	1478	0	0
4219.364					1	1074	B		6	1023	5	(1013)
4216.186	3	512	764	1016					4	(1774)	4	1522
4210.352	1	0			1	3050			1	(3033)	1	3067
4207.130	2	0	395		1	1138	m	m	2	1533	1	1928
4203.987	1	0	D		1	1445	A		2	1477	1	(1509)
4202.031	4	305	451	599	8	1123	1274	1419	4	1266	4	1121
4199.098	1	0	D		1	1075	B		5	1026	4	(1013)
4198.310	1	0	D		1	1498	A		5	(1621)	4	1652
4195.337	1	194	B		2	1382	m	1466	5	1418	5	1378
4191.436	2	0	1008		2	1032	2030		2	2032	1	3035
4187.802	2	0	96		2	1379	1464	m	4	(1661)	3	1756
4187.044	3	0	252	497	4	1283	1535	1774	3	1779	2	2027
4184.895	1	124	B		1	1552:†	C		2	1456	2	(1518)

TABLE D—(*Continued*)

λ	π Components				σ Comopnents				J_1	Obs. g_1	J_2	Obs. g_2
	No. Meas.	π			No. Meas.	σ						
4181.758	3	**0**	097	207	3	**1210**	1317	1436	3	1422	2	1526
4175.640	2	**0**	225		3	**1058**	1295	1471	2	1288	1	1511
4172.126	3	**0**	132	262	3	**1107**	1210	1316	3	1355	2	1486
4156.803	2	202	**476**		4	**1300**	**1559**	1778	2	1527	2	1289
4154.812	1	**0**	D		1	**1482**	B		5	1392	4	(1370)
4147.673	1	**0**	D		3	1732	2190	**2658**	4	1267	3	804
4143.871	4	**0**	164	332	6	1443	1594	**1751**	4	1268	3	1105
4137.002	2	**0**	209		2	1039	**1246**		2	1038	1	830
4134.681	3	**0**	156	300	2	**1066**	1202		3	1364	2	1517
4132.903	2	**0**	304		2	**956**	1226		2	1231	1	1511
4132.060	3	**0**	425	851	4	1109	1532	**1958**	3	1107	2	682
4127.612	1	**0**	S		1	**781**	S		1	781	0	0
4118.549	1	**0**	D		1	**1074**			6	1023	5	(1013)
4114.449	1	**590**	B		3	938	**1223**	**1516**	2	1518	2	1229
4109.808	1	**748**			2	**779**	**1514**		1	1514	1	779
4107.492	1	m	92		1	**1610**	B		2	1518	1	1426
4095.975					2	**828**	953		3	1078	2	1203
4085.011	2	**0**	354		2	**803**	1146		2	1151	1	1505
4084.498					1	**1106**	A		5	(1418)	4	1496
4079.848	1	**0**	S		1	1902	S		1	1902	0	0
4078.365	2	**0**	110		2	667	**784**		2	669	1	556
4076.636	1	**93**	B		1	**1514**	C		4	(1519)	4	1509
4074.794					5	1016	**1185**	1388	4	1197	4	1011
4071.740	1	**0**			1	692	C		2	(696)	2	688
4067.984	1	**313**	B		1	**1575**	C		4	(1519)	4	1597
4067.275	1	**0**	D		3	**869**	978	1097	4	(1251)	3	1378
4066.979	1	**734**	B		4	**1154**	**1507**	1868	2	1511	2	1148
4063.597	1	**0**			1	**1094**			3	(1100)	3	1088
4062.446	1	**406**			2	**1506**	**1922**		1	1915	1	1509
4055.039					1	**1123**	A		5	1226	4	(1251)
4045.815	1	**0**			1	**1263**			4	(1272)	4	1254
4021.869	4	**0**	135	270	6	1027	1160	**1286**	4	895	3	762
4017.156					5	1475	1636	**1791**	5	1167	4	1012
4014.534	1	**0**			1	**1014**	C		5	(1013)	5	1015
4009.714	2	**0**	**841†**		3	**1163†**	1854†	2520†	2	1848	1	(2526)
4007.277	1	**0**	D		1	**767**	B		3	(766)	2	765
4006.631	1	**0**	S		1	**1405**	S		1	1405	0	0
4005.246	3	**0**	421	842	5	1097	1506	**1921**	3	1095	2	678
4003.764	1	**122**			2	**716†**	**828†**		1	(830)	1	708
4001.666	1	**0**			1	**1684**	C		3	(1688)	3	1680
4000.466	4	298	399	**527**	8	**1061**	**1193**	1313	4	1190	4	1062
3998.054	4	**0**	145	282	7	**648**	789	927	5	1214	4	1356
3997.394	1	**0**	D		1	**1107**	B		5	1073	4	(1065)
3996.968					1	**1004**			4	(1006)	4	1002
3995.996	3	**0**	303	612	5	**166**	**464**	761	4	1063	3	1364
3994.117	4	**0**	77	148	1	**1379**	B		5	1079	4	1004
3990.379	1	**315**	B		3	861:	**925:**	**993:**	4	998	4	926
3986.176	4	**0**	218	440	7	**489**	708	921	4	1136	3	1353
3983.960	4	**0**	118	234	4	690	818	952	4	1072	3	1198
3981.775	3	325	492	**679**	6	**898**	**1031**	1227	4	1063	4	894
3977.743	1	**0**			1	**1846**			2	(1844)	2	1848
3976.615	2	**0**	351		2	1157	**1503**		2	1157	1	811
3973.655	1	**0**	D		1	**1197**	B		3	1093	2	(1041)
3971.325	4	**0**	154	298	7	**617**	761	912	5	1217	4	1370
3970.391	1	**0**	S		1	**1477**	S		1	1477	0	0

TABLE D—(*Continued*)

λ	π Components				σ Components				J_1	Obs. g_1	J_2	Obs. g_2
	No. Meas.	π			No. Meas.	σ						
3969.261	4	0	165	330	7	1425	1586	1747	4	1264	3	1101
3967.423	1	0	D		1	842	B		4	(822)	3	815
3966.066	2	0	958		3	1345†	2014†	2683†	3	(1341)	2	672
3965.511					1	1397	A		4	1487	3	(1517)
3964.522	2	0	266		3	1535†	1823†	2060†	2	1771	1	(1509)
3963.108					3	483	959	1382:	2	980	1	(1519)
3961.147	1	0	S		1	2240	S		1	2240	0	0
3956.681	1	0	D		1	1220	B		6	1214	5	(1213)
3956.459	1	0	D		1	1325	B		6	(1180)	5	1151
3955.956	1	061			1	1436			1	1467	1	1406
3953.861	2	0	280		3	698†	962†	1235†	3	1244	2	(1518)
3953.156	4	0	173	343	3	1096	1262	1429	4	932	3	764
3952.702	2	0	219		3	1024	1273	1503	2	1273	1	1503
3952.606	3	432	569	706	8	1068	1209	1310	5	1208	5	1069
3951.164	2	0	302		3	731	1036	1335	2	1034	1	732
3949.954	3	0	170	338	5	1353	1519	1683	3	1684	2	1852
3948.779	1	0	D		1	949	A		5	(1046)	4	1070
3948.105					2	857	1090		4	1373	3	(1520)
3947.533	1	502	B		4	1146	1522	1898	2	1769	2	(1518)
3945.119	4	0	407	806	7	1894	2273	2638	4	1145	3	755
3944.890	4	0	104	208	6	1397	1509	1625	5	1187	4	1076
3943.339	2	0	643		3	1216	1852	2465	2	1844	1	2481
3942.443	2	0	222		3	1060	1286	1558:	2	1285	1	1509
3940.882	2	0	257		3	1772	2014	2266	4	1517	3	1267
3937.329	4	0	324	650	7	1841	2158	2479	5	1203	4	883
3935.815	1	461	B		4	1059	1285	1516	2	1516	2	1287
3932.629	5	0	255	512	7	1819	2086	2317	5	1313	4	1059
3930.299	1	0			1	1515			2	(1514)	3	1514
3927.922	1	0			1	1514			2	1516	1	(1518)
3925.946	1	0	S		1	1578	S		1	1578	0	0
3925.646	1	500	B		4	1283	1518	1755	2	1519	2	1276
3922.914	1	0			1	1517			4	1516	3	(1516)
3920.260	1	0	S		1	1518	S		1	1518	0	0
3919.069	2		412	543	3	921	1054	1182	4	1052	4	921
3918.644	1	0			1	771			3	(771)	3	771
3917.185	3	0	513	1045	4	1517	2018	2496:	3	1514	2	1006
3916.733	2	0	111		3	1503	1618	1719	6	1175	5	1065
3914.273	1	952†			1	1518	C		1	2470::	1	(1518)
3913.635	1	0	D		1	1477	A		3	1510	2	(1526)
3909.830	1	64			1	1541	C		1	1573	1	1509
3907.937	2	0	302		3	794	1069	1352	3	780	2	489
3906.748	3	0	147	275	4	1342†	1493†	1644†	3	1341	2	(1194)
3903.902	3	256	368	488	7	937	1055	1176	4	1054	4	935
3902.948	3	254	501	750	6	1096	1342	1583	3	1341	3	1095
3899.709	1	0			1	1518	C		2	(1514)	2	1522
3898.012	2	0	1525		3	0	1512	3022	2	1518	1	−10
3897.896	5	0	110	228	4	1651		1858	6	1323	5	1213
3895.658	1	0	S		1	1518	S		1	1518	0	0
3893.924	3	502†	682†	855†	6	m	1378	1554	5	1218	5	1047
3893.391	1	207	B		1	1205†	C		5	1223	5	1182
3891.928	1	458			2	1275	832		1	1277	1	830
3890.844	4	0	124	252	6	1183	1306	1424	4	1065	3	943
3888.517	2	508	1019		4	678	1177	1666	2	1172	2	674
3887.051	4	311	461	611	8	1367	1514	1667	4	1517	4	1367
3886.284	1	0			1	1519			3	(1517)	3	1521

TABLE D—(*Continued*)

λ	π Components				σ Components				J_1	Obs. g_1	J_2	Obs. g_2
	No. Meas.	π			No. Meas.	σ						
3885.512	2	0	307		3	899†	1213†	1512†	2	1210	1	(1520)
3884.359	2	0	174		5	1563†	1726†	1895†	5	(1213)	4	1040
3883.282	1	67	B		1	1335	C		3	1346	3	1324
3878.575	2	0	74		1	1522	B		2	(1514)	1	1506
3878.021	3	258	509	765	6	1262	1512	1759	3	1511	3	1260
3876.043	2	0	1535		2	0	1516	m	2	1516	1	−19
3873.763	1	0	B		1	1004	A		5	(1052)	4	1064
3872.504	2	513	1019		4	1009	1507	2004	2	1508	2	1006
3871.750	2	m	520	630	7	1080	1208	1329	5	1208	5	1083
3869.562					7	1505	1664	1792	5	1217	4	1069
3865.526	1	1536			2	0	1510		1	1510	1	−26
3861.341	2	0	463		3	743	1168	1600	2	1170	1	740
3859.913	1	0			1	1515			4	(1516)	4	1514
3859.214	1	0	D		1	1040	A		6	(1178)	5	1206
3856.373	1	0			1	1520			3	(1517)	2	1515
3852.574	4	0	178	359	5	979	1153	1335	4	1511	3	1689
3850.820	2	523	1024		4	1016	1505	2021	2	1511	2	1014
3849.969	1	U			1	U			1	0	0	0
3846.803	1	0			1	1340			3	(1353)	3	1327
3846.412					1	1017			5	(1013)	4	1012
3845.170	1	973			2	564	1514		1	1520	1	558
3843.259	1	0	D		1	992	A		4	(1013)	3	1020
3841.051	2	0	172		3	497†	679†	850†	2	(679)	1	506
3840.439	2	0	514		3	486	1009	1514	2	1006	3	1516
3839.259	1	81	B		1	1044	C		4	993	4	(1013)
3837.132	3	0	506	1013	3	1164	1658	2155	3	1158	2	657
3836.332	1	0			1	1178			2	(1194)	2	1162
3834.225	3	0	254	506	5	749	1014	1261	3	1258	2	1511
3833.311	2	m	333	432	6	1230	1339	1461	4	1348	4	1246
3829.458	1	0			1	732	C		1	(741)	1	723
3827.825	1	0	D		1	989	A		3	(1100)	2	1155
3825.884	4	0	152	301	5	938		1363	4	1368	3	1512
3824.306	1	0			1	822			4	(822)	4	822
3821.834	1	168	B		2	665†	743†		2	753	2	(672)
3821.181	1	0	D		1	951	A		6	1030	5	(1046)
3816.340	1	0	D		4	792†	1133†	1495†	3	1493	2	(1844)
3812.964	3	0	255	507	5	764	1015	1265	3	1262	2	1513
3811.892	2		344	542	4	228†	414†	603†	3	(766)	3	586
3810.759	1	0	D		1	1140	A		2	(1194)	1	1248
3808.731	2	345	467		6	1247	1358	1462	4	1359	4	1244
3807.534	2	0	962		3	585	1552	2519	2	1552	1	2517
3806.697	1	0	S		1	1046	S		5	(1046)	5	1046
3805.345	1	0	D		1	932	B		5	844	4	(822)
3802.283	2	308	638		4	1201	1514	1801	2	1516	2	1197
3801.681	3	0	296	589	4	633	923	1223	3	1224	2	1520
3799.549	2	m	95	194	2	1643	m	1465	4	1358	3	(1264)
3797.948	1	0	D		2	1091†	1431†	m	3	(1087)	2	747
3797.517	1	0			1	1184			6	(1180)	6	1188
3795.004	3	0	254	507	5	1263	1515	1762	3	1264	2	1012
3794.340	1	0	D		1	930	B		5	844	4	(822)
3792.156	3	241	366	471	6	939	1059	1185	4	1063	4	948
3790.095	2	0	512		3	507	1012	1515	2	1011	1	1519
3789.178	1	0	D		1	909	A		5	1034	4	(1065)
3787.883	2	0	1030		3	0	1012	2022	2	1016	1	−14
3786.678	1	U			1	U			1	0	0	0

TABLE D—(*Continued*)

λ	π Components				σ Components				J_1	Obs. g_1	J_2	Obs. g_2
	No. Meas.	π			No. Meas.	σ						
3785.950	1	0			1	1037			6	1054	5	(1052)
3781.188	2	0	489		3	415	890	1364	3	1365	2	1840
3779.444	1	534			2	743	1264		1	1264	1	743
3778.509	3	0	202	411	2		1522	1720	3	(1353)	2	1161
3777.061	1	0	D		1	1126	B		4	1054	3	(1030)
3776.454	3	0	231	470	4	777	992	1232	4	1462	3	1693
3774.823	1	1206			2	1334	2521		1	2521	1	1334
3773.699	1		1059†		4	1856	2348	2843	2	2350	2	1862
3770.305	1			364	3	1115†	1203†	1314†	5	(1213)	5	1143
3769.995	3	0	385†	698†	4	1931	2237	2521	3	1934	2	1632
3768.030	1	0	S		1	2521	S		1	2521	0	0
3767.194	1	U			1	U			1	0	1	0
3765.542	1	0	D		1	1054	A		7	1162	6	(1180)
3763.790	1	0			1	1009			2	(1008)	2	1010
3760.534	2	0	624		3	1304	1910	2512	2	1907	1	2517
3760.052	1	0	D		1	1077	m	m	7	1164	6	(1178)
3758.235	1	0			1	1263	C		3	(1264)	3	1262
3756.069	3	311	651	958	5	1355	1682	1980	3	1680	3	1364
3753.610	3	0	130	253	5	1693	1818	1940	3	1691	2	1564
3749.487	1	55	B						4	(1367)	4	1354
3748.264	2	0	518		3	512	1080	1512	2	1012	1	1521
3745.901	1	U			1	U			1	0	0	0
3745.561	3	0	258	520	5	768	1015	1264	3	1265	2	1518
3743.468	1	85	B		1	1022	C		5	1014	5	1031
3743.364	2	0	1021		3	0	1010	2024	2	1008	1	−10
3737.133	2	m	307	460	5	932	m	1360	4	1362	3	1513
3734.867	1	68	B		1	1435	C		5	1428	5	1442
3733.319	1	1542			1	0	1504†		1	1518	1	−24
3732.399	1				1	1871	C		2	1898	2	(1844)
3731.374	2	365†	562†		4	485	668	890	2	675	2	472
3730.945	1	0	D		1	708	B		3	684	2	(672)
3730.386	5	0	139	278	8	1435	1556	1692	5	1144	4	1007
3728.668	3	450†	671†	885†	3	817†	1026†	1258†	4	(1251)	4	1033
3737.621	3	0	259	512	5	1264	1518	1765	3	1265	2	1011
3724.380	3	0	154	306	5	1071	1235	1393	3	1377	2	1530
3722.564	2	515	1032		4	1013	1517	2019	2	1516	2	1014
3719.935	5	0	104	206	4	1004	1105	1201	5	1407	4	1508
3718.407	1	1169	B		5	311†	768†	1223†	3	1226	3	(766)
3716.442	1	0	D		2	1937	B		4	(1774)	3	1720
3715.911	1	638	B		4	906	1207	1522	2	1524	2	1213
3711.411	2	0	446		2	605	1031		2	1041	1	1487
3711.225	1	0	D		1	1046	A		4	1077	3	(1087)
3709.246	4	m	97	208	2	m	m	1682	4	1367	3	1262
3707.918	3	0	234	462	5	650	776	904	3	1686	2	1919
3707.824	2	0	1523		3	0	1475:	3047	2	1514	1	− 9
3705.567	3	258	512	763	6	1263	1511	1762	3	1514	3	1264
3704.463	5	0	140	280	7	1480	1620	1738	5	1210	4	1076
3702.033	1	0	S		1	1504	S		1	1504	0	0
3701.086	4	0	310	620	7	731	1022	1315	4	1626	3	1932
3698.611	2	0	403		4	313	713†	1105	3	1110	2	1510
3697.426	3	671	1298	1924	3	1308	1929	2578	3	1935	3	1304
3695.054	1	0	D		1	1290	B		5	1068	4	(1013)
3694.005	3	0	444	878	5	558	788	1018	3	1920	2	2355
3690.730	1	734	B		4	817†	963†	m	5	(1013)	5	1159
3690.450	1	0	S		1	820	S		1	820	0	0

TABLE D—(*Continued*)

λ	π Components				σ Components				J_1	Obs. g_1	J_2	Obs. g_2
	No. Meas.	π			No. Meas.	σ						
3689.457	4	229	427	643	4	1592	1776	1961	4	1776	4	1581
3687.656	3	297†	416†	530†	3	802†	931†	m	4	1196	4	(1065)
3687.458	1	0	D		1	1594	B		5	(1422)	4	1379
3686.260	1	0	S		1	1512	S		1	1512	0	0
3685.998	5	0	261	518	6	492	751	m	5	1534	4	1794
3684.108	4	0	163	319	7	563	747	906	4	1060	3	1220
3683.054	3	0	513	988	4	1507	2019	2554	3	1521	2	1013
3682.226	1	66	B		1	1015	C		2	1032	2	999
3679.915	4	310	460	609	8	1366	1515	1662	4	1517	4	1368
3677.630	1	0	D		1	912	B		3	(766)	2	693
3677.309	3	0	129	243	2	672†	782†	m	3	921	2	(1041)
3676.314	1	0	D		1	1108	A		5	1222	4	(1251)
3672.722	4	0	254	526	6	1598	1834	2085	5	1078	4	823
3670.810	1	0	S		1	789	S		1	789	0	0
3670.071	1	0	D		1	1060	A		6	1190	5	(1216)
3670.028	1	0	D		1	1480	A		2	1495	1	(1509)
3663.458	1	96	B		1	1267	C		4	1279	4	1255
3659.516	1	285	B		1	820	890		4	892	4	(822)
3657.139	2	0	242		3	1032	1278	1543	2	1277	1	1521
3655.465	1	62	B		1	1501	C		2	1517	2	1486
3653.763	4	0	174	352	7	1759	1941	m	6	1223	5	1046
3651.469	4	0	170	338	7	1095	1263	1420	4	931	3	764
3649.508	1	0	D		1	1267	B		5	(1213)	4	1200
3649.304	3	0	264	540	2		2040†	2292†	4	(1576)	3	1252
3647.844	4	m	129	266	5	711	m	1226	5	1229	4	1360
3645.822	1	0	S		1	714	S		1	714	0	0
3640.388	4	0	103	204	4	1390	1484	1565	5	1172	4	1069
3638.296	4	0	183	364	7	1120	1299	1475	4	942	3	763
3636.995	4	0	146	291	3	509†	650†	810†	4	941	3	(1087)
3636.650	2	0	673		3	−577†	m	820†	3	821	2	1504
3636.234	1	0	D		1	523	A		3	868	2	(1041)
3636.186	3	0	474	940	3	412	900	1354	3	1365	2	1841
3632.042	2	0	312		3	885†	1175†	1479†	2	1182	1	(1485)
3631.464	2	m	m	294	5	692	m...	1113	4	1113	3	1256
3625.140	1	0	D		1	1440	A		5	(1518)	4	1538
3623.187	1	186	B		1	1198	C		6	1214	6	1183
3622.001	1	0	B		1	768	C		3	766	3	(770)
3621.463	1	0	D		1	1151	B		5	1082	4	(1065)
3618.769	3	0	116	230	2	677	788		3	902	2	1015
3618.392	2	m	394†	526†	3	801†	931†	1075†	4	(1065)	4	930
3617.788	3	0	179	358	4	981†	1153†	1327†	3	1328	2	(1504)
3610.159	1	0	B		1	1511	C		6	(1522)	6	1500
3608.861	2	0	351		3	0	341	680	2	338	1	− 7
3606.679	1	0	D		1	927::	A		6	1161::	5	(1213)
3605.450	3	257	388	490	7	949	1067	1185	4	1062	4	948
3603.828	1	787			2	717	1496		1	717	1	1496
3603.205	1	210	B		1	1201	C		5	1180	5	1222
3599.624	5	0	151	301	6	431	557	710	5	1006	4	1154
3594.632	1	295	B		1	1543	C		4	1580	4	1506
3592.486	1	0	D		1	1062	A		4	1081	3	(1087)
3590.086	3	439†	569†	701†					5	(1216)	5	1085
3589.456	3	0	283	593	6	1320	1620	1911	4	1059	3	773
3589.107	1	937	B						5	(1422)	5	1235
3586.985	2	681	1360		3	338	1006	1690	2	1011	2	333
3585.708	4	490	754	998	4	1123	1357	1847	4	1361	4	1115

TABLE D—(*Continued*)

λ	π Components				σ Components				J_1	Obs. g_1	J_2	Obs. g_2
	No. Meas.	π			No. Meas.	σ						
3585.320	3	360	774	**1122**	6	**887**	**1265**	1632	3	1262	3	890
3584.663	4	332†	464†	**585†**	8	**1107†**	**1233†**	1357†	5	(1213)	5	1088
3582.201	6	0	192	381	7	**168†**	359†	575†	6	(1180)	5	1368
3581.195	1	0	D		1	**987:**	A		6	1350	5	(1422)
3575.374	2	337	**667**		4	**1172**	**1504**	1828	2	1504	2	1172
3573.896	2	0	126		1	**508**	A		4	(822)	3	930
3571.995	1	0	B		1	**1523**	C		5	(1518)	5	1528
3570.100					1	**849**	A		5	1263	4	(1367)
3568.977	4	0	273	560	6	1806†	2070†	**2341†**	5	(1213)	4	940
3567.038					1	**919**			3	1322	2	(1524)
3565.381	4	**0**	151	302	7	**331**	406	474	4	1120	3	1272
3559.506	1	**217**			2	**1204†**	**1428†**		1	(1485)	1	1265
3558.518	3	**0**	217	431	5	**373**	584	793	3	793	2	1006
3556.877	2	**0**	132		2	**946**	1056		5	1402	4	(1515)
3554.922	1	m	88		1	**1049**	A		6	1440	5	(1518)
3554.122	1	**0**	D		3	1262†	2183†	**3094†**	3	(1264)	2	348
3552.112	1	**0**	D		1	**1518**	B		2	1502	1	(1485)
3549.868	2	**0**	682		3	0	677	**1356**	2	676	1	− 6
3547.203	3	385	588	**780**	4	**833**	**1017**	1197	4	826	4	1016
3545.639	1	**517**	B						4	1644	4	(1515)
3543.669	2	**0**	172		3	842	1011	**1167**	2	1007	1	840
3542.076	3	**0**	148	305	1	**1378†**	A		4	1383	3	(1534)
3541.083	1		102		1	**1005**	A		5	1414	4	1516
3540.709	1	m	474†		3	1834†	2309†	m	4	(1367)	3	889
3540.121	1	**0**	D		1	**1496**	A		4	1524	3	(1534)
3537.729	1	**0**	D		1	775	B		2	(672)	1	569
3536.556	5	**0**	238	502					3	1262	2	(1524)
3529.818	1	**1942**	B		1	**1188**			1	1565	1	−378
3527.792	1	**637**	B						4	(1515)	4	1355
3526.465	1	B			1	**1500**			2	(1526)	2	1474
3526.167	3	480	945	**1420**	6	**798**	**1264**	1760	3	1267	3	797
3526.039	3	**0**	171	336	4	1685	1860	**2008**	3	1685	2	1519
3524.236	2	**0**	114		2	**1201**	1295		3	1413	2	1523
3521.833					1	**1166**	A		2	1846	1	(2526)
3521.264	4	518	768	**1018**	8	1112	**1363**	1614	4	1364	4	1111
3518.86	3	0	171	332	3	**1358**	1514	1672	3	1675	2	1837
3516.403	1	**0**	D		1	**957**	B		4	817	3	(771)
3513.820	5	468	621	**766**	10	**1267**	**1416**	1568	5	1416	5	1263
3511.748	2			**207**					4	(1251)	4	1199
3510.446	1	**0**	S		1	**1571**	S		1	1571	0	0
3509.870	1	**47**			1	**2493**			1	2516	1	2470
3508.494	1	**0**			1	**1018**			5	1053	4	(1062)
3506.498	1		**501**		4	**1286**	**1526**	1764	2	1525	2	1278
3505.065	2	**0**	246		3	1316†	1531†	**1768†**	2	(1504)	1	1268
3504.859	1	**0**	D		1	1440	A		2	(1526)	1	1612
3500.564					3	1099	1490	**1900**	3	1086	2	676
3497.843	2	**0**	342		3	1523	1856	**2188**	2	1855	1	1518
3497.110	1	**0**			1	**1687**			3	(1688)	3	1686
3490.575	3	191	347	**513**	6	**1513**	**1681**	1846	3	1682	3	1512
3489.670	1	**0**	D		1	**1059**	A		6	1190	5	(1216)
3485.342	2	**0**	624		3	**1233**	1849	2460	2	1847	1	2466
3476.853					5	**428**	571	726	4	933	3	(1087)
3476.704	1	**0**	S		1	**2525**	S		1	2525	0	0
3475.450	2	345	**687**		4	**1516**	**1849**	2184	2	1852	2	1513
3471.350	1	**0**	D		1	**1617**	B		2	(1526)	1	1435

TABLE D—(*Continued*)

λ	π Components				σ Components				J_1	Obs. g_1	J_2	Obs. g_2
	No. Meas.	π			No. Meas.	σ						
3471.27					3	**380**	1463	2530	2	1458	1	2530
3469.012	1	**0**			1	**824**			4	(822)	4	826
3468.849	1	**0**	D		1	**875**	A		5	1176	4	(1251)
3465.863	1	**1036**			2	**1516**	**2524**		1	2534	1	1512
3463.305	1	**0**			1	**1263**			4	(1271)	3	1274
3462.353	1	**0**	D		1	**1795**	A		2	(1844)	1	1893
3458.304	1	**0**	S		1	**1517**	S		1	1517	0	0
3453.022	1	**0**	D		1	**670**	A		3	(766)	2	814
3452.273	1	**0**	D		1	**1255**			4	1262	3	(1264)
3451.915	2	**0**	1265		3	0	1272	2527	2	1272	1	2527
3451.628	1	**44**			1	**1541**	C		1	1563	1	1519
3450.328	1	**926**			2	**1615**	**2522**		1	2527	1	1610
3447.278	2	407	**799**		4	**1448**	**1843**	2238	2	1845	2	1447
3445.151	3	**0**	442	882	5	**544**	978	1390	3	1408	2	1845
3443.878	2	**0**	1029		3	**508**	1511	2525	2	1515	1	2525
3442.364	1		**752**		2	1010	1395		2	(1526)	2	(1148)
3440.989	3	**0**	343†	680†	5	**844**	1180	1512	3	1513	2	1850
3440.610	4	**0**	165	350	7	**1020**	1172:	1344	4	1512	3	1679
3439.039	1	**0**	D		1	**1103**	B		5	1073	4	(1065)
3437.046	3	m		217	5	**m**	m	281	4	(1013)	3	1223
3431.815					3	**931**	1115	1323	3	1315	2	1507
3428.192	2	578	**1163**		4	**1266**	**1840**	2399:	2	1842	2	1265
3427.121	4	**0**	338	673	7	**355**	**686**	1019	4	1354	3	1689
3424.284	3	285	560	**834**	6	**1416**	**1689**	1963	3	1690	3	1416
3422.656	2	0	1298		3	**−59**	1237	2507	2	1238	1	2535
3419.706	1	**712**			2	**809**	**1509**		1	1512	1	806
3418.507	1	**0**	S		1	**2525**	S		1	2525	0	0
3417.842	1	**1115**			1	**1424**			1	2539	1	1424
3413.135	3	**0**	493	986	5	**389**	875	1361	3	1363	2	1853
3411.353	1	**274**	B		1	**1104**	C		4	1138	4	1070
3410.171	2	**0**	135		3	**554**	693	826	2	691	1	828
3407.461	4	**0**	328	661	7	**387**	707	1029	4	1357	3	1684
3404.357	3	**0**	689	1374	5	−186	486	1157	3	1160	2	1840
3402.256	1	**0**			1	**1182**			6	(1180)	6	1184
3401.521	3	**0**	327	649	5	**396**	723	1045	4	1364	3	1686
3399.336	2	627	**1249**		4	**1229**	**1843**	2458	2	1845	2	1227
3396.978	3	**0**	618	1212	5	**80**	677	1261	3	1270	2	1870
3392.652	3	323	665	**991**	6	**1358**	**1680**	2018	3	1684	3	1356
3392.304	1		**2176**		4	**759**	**1841**	2880	2	1845	2	762
3392.014	1	0			1	1551			2	(1504)	1	1457
3389.748	2	**0**	1408		3	**−210**	1161	2552	2	1161	1	2532
3387.410	3	**0**	144	282	5	**479**	611	751	3	754	2	895
3383.981	2		1060	**1555**	6	**1195:**	**1693**	2210	3	1690	3	1170
3383.692	2	**0**	1102		3	769	1855	**2913**	2	1846	1	763
3380.111	1	**113**	B		1	**781**	C		3	800	3	762
3379.017	2	**0**	469		2	2150	**2597**		3	1686	2	1228
3378.676	1	**0**	D		1	**1444**	B		5	(1213)	4	1155
3373.874	1	**0**	D		1	**1100**	B		5	1072	4	(1065)
3372.070	3	**0**	953	1877	4	1681	2619	**3554**	3	1681	2	744
3370.786	1	**315**	B		1	**1185**	C		5	1217	5	1154
3369.549	1	**45**	B		1	**1073**	C		4	1079	4	1068
3366.867	2	693	**1407**		4	**1167**	**1850**	2533	2	1852	2	1159
3366.789	5	**0**	247	515	8	**1712**	1934	**2184**	5	1212	4	966
3356.407	2	256	**503**		4	**1533**	**1762**	1993	2	1768	2	1526
3355.228	1	**183**	B		2	**809**	**854**		4	854	4	809

TABLE D—(*Continued*)

λ	π Components				σ Components				J_1	Obs. g_1	J_2	Obs. g_2
	No. Meas.	π			No. Meas.	σ						
3354.064	1	0	S		1	1261	S		1	1261	0	0
3351.750	4	0	270	531	6	1330	1597	1828:	4	1070	3	807
3351.529	1	0	D		1	1808	A		2	(1844)	1	1880
3347.927	1	480	B		4	1290	1518	1752	2	1520	2	1283
3346.936	2	0	547		3	1692	2219	2743	3	1682	2	1146
3342.298	1	235			2	1280	1503		1	1506	1	1277
3342.216	2	0	732		2	799	1492:	m	2	1531	1	2263
3341.906	2	m	617†	728†	5	933†	1099:†	1221:†	5	(1213)	5	1083
3340.566	1	491	B		3	1289	1524	1757	2	1523	2	1284
3337.666	4	0	128	264	6	1478	1608	1728	5	1219	4	1090
3336.254	3	708	1060	1418	6	798	1162	1527	4	1163	4	814
3335.776	1	0	D		1	1503			2	1506	1	(1509)
3334.223	1	171	B		1	1088	C		5	1105	5	1071
3331.778	1	0	S		1	1409	S		1	1409	0	0
3331.612	2	0	124		6	1274	1383	1511	5	1033	4	913
3328.867	1	0			1	1045			5	(1046)	5	1044
3327.498	1	0	D		1	1180	A		4	1445	3	(1534)
3325.468	1	0	D		1	934	B		4	(822)	3	785
3324.541	1	0	D		1	1220	B		6	(1180)	5	1172
3323.737	1	0			1	1507			2	(1518)	2	1496
3319.258	1	504	B		6	1083	1196	1303	4	1199	4	1076
3317.121	1	0	D		1	1499	A		2	(1526)	1	1553
3314.742	1	0			1	911			3	1100	2	(1194)
3310.496	3	m	476	981	5	−587	−90	354	4	866	3	1351
3310.347	1	210	B		1	1220	C		5	1241	5	1199
3306.356	2	0	762		3	1004	1763	2565:	2	1764	1	2525
3305.971	3	0	160	323	2	1344	1506		3	1667	2	1828
3301.227	1	234			2	1294	1509		1	1514	1	1289
3298.133	2	0	1259		2	33	1282		2	1284	1	2538
3292.590	1	290			2	2232	2516		1	2516	1	2232
3292.022	4	0	192	382	4	600	789	978	4	1171	3	1362
3290.988	2	0	1253		3	0	1246:	2530	2	1267	1	2520
3288.967	2	0	611		3	0	640	1243	3	1245	2	1852
3288.651	1	0	S		1	1520	S		1	1520	0	0
3286.755	1	0			1	1678			3	(1688)	3	1668
3284.588	1	152	B		1	1791	C		2	1829	2	1753
3282.891	1	0	D		1	658	A		2	700	1	(741)
3280.261	1	0	D		1	926	B		5	843	4	(822)
3278.741	1	0			1	1099	C		3	(1087)	3	1111
3276.468	2	574	1148		4	1288	1846	2402	2	1846	2	1288
3274.453					1	1389	C		4	(1370)	4	1408
3271.002	2	0	397		3	1448	1840	2230	2	1839	1	2233
3268.234	1	970			2	1573	2521		1	2526	1	1568
3265.616	1	0	D		1	1559	A		3	(1688)	2	1752
3265.046	2	0	162		3	990	1156	1342	3	1339	2	1515
3264.512	2	0	1602		2	1850	3450		2	1850	1	249
3263.378	1	0	D		1	1465	A		2	1492	1	(1520)
3260.261	1	446	B		1	1119†	m	m	4	(1251)	4	1139
3257.594	2	0	402		3	1698	2084	2479	3	1691	2	1295
3254.363	1	0	D		1	1009	A		6	1040	5	(1046)
3253.610					4	510	800	1080	4	1067	3	(1353)
3248.206	1	425	B		1	1776	C		3	1776	3	1634
3246.962	2	0	278		2		1834	2129	2	1839	1	1553
3244.190	2	0	142†		1	1011†	A		5	1519	4	(1661)
3243.109	2	m	514:†	678:†	1	776:†	m	m	4	(822)	4	987:

TABLE D—(*Continued*)

λ	π Components				σ Components				J_1	Obs. g_1	J_2	Obs. g_2
	No. Meas.	π			No. Meas.	σ						
3239.436	1	483	B		1	1563	C		4	(1661)	4	1540
3236.223	4	0	256	511	5	507	762	1015	4	1268	3	1522
3234.614	1	541			4	1224†	1390†	1572†	3	(1517)	3	1337
3233.967	1	255	B		1	1427:			4	1597	4	(1661)
3233.053	1	0	D		1	1066	A		7	1164	6	(1180)
3229.123	1	0	S		1	510	S		1	510	0	0
3228.254	2	0	353		3	1679†	2060†	2401†	2	(2041)	1	1684
3227.798	1	0	D		1	1712	B		4	(1661)	3	1644
3225.789	1	0	114		1	m	1045:		6	1519	5	(1621)
3222.069	1	315	B		1	1574	C		5	1606	5	1542
3219.806	1	0	D		1	1631	A		4	(1661)	3	1671
3219.581	2	0	174		2	1075:†	1254:†		4	1595	3	(1771)
3215.940	1	306	B		1	m	2058:†		2	(2041)	2	1888
3214.396	3	0	436	872					3	(1514)	2	1078
3211.989	1	0	D		1	1677	B		5	(1621)	4	1607
3211.683					1	1417†			6	1418::	5	(1418)
3208.470	2	0	343		2	m	360†	690†	2	348	1	(−12)
3205.400	1	526			2	2525	3029		1	3033	1	2521
3202.562	1	0	D		1	1268	B		4	(1013)	3	928
3200.475	1		127	B					2	(2041)	2	1978
3199.530	1	219	B		1	1574:	m	m	4	(1661)	4	1606
3196.930					1	1003	A		5	1529	4	(1661)
3193.228	3	512	780	1024	6	1272	1506	1771	4	1511	4	1256
3191.659	4	0	183	361	6	1703†	1875†	2048†	4	(1516)	3	1339
3188.819	2	0	2154			−1182	972	2837:	2	970	1	3022
3188.567	1	1190	B						5	(1621)	5	1383
3184.896	2		827	1245	6	1098	1519	1915	3	1515	3	1104
3182.970					4	56†	671†	1258†	3	1236	2	(1844)
3181.522	1	0	D		1	947	A		3	(1087)	2	1157
3180.756	1	835†	m		3	678	1522	2378	2	1526	2	678
3180.223		D			1	1292	A		4	1651	3	(1771)
3178.967	1	0			1	1052			5	(1052)	5	1052
3178.545	1	454	B						3	(771)	3	922
3178.015	1	0	D		1	1659	B		5	(1621)	4	1612
3176.366					1	665			2	(672)	1	679
3175.447	1	495	B		3	1489:†	1637†	1739†	5	(1621)	5	1521
3171.353	4	0	146	291	4	1160	1305	1433	4	1016	3	873
3167.907					1	926:	A		4	1369:	3	(1517)
3166.435					2	1025	m	m	4	(1251)	3	1326
3165.860	4	0	420	844	5	578	899	1349	4	1331	3	1754
3161.949	6	0	196	393					6	1428	5	(1621)
3160.658	1	90	B		1	1642	C		4	1653	4	1631
3160.344	1	0			1	1170			6	(1178)	6	1162
3157.88	1	0	D		1	1801	A		3	1967	2	(2041)
3157.040	5	0	319	550	6	325	568	839	5	1381	4	1648
3156.275	1	244	B		1	m	m	m	3	(1517)	3	1436
3155.293	1	0	D		1	794	A		5	(1052)	4	1116
3151.353	1	0	D		1	1116	B		5	1075	4	(1065)
3142.888	1	68	B		1	1500	C		2	1517	2	1483
3134.111	3	0	254	512	4	1772	2026	2272	4	1514	3	1260
3129.334	4	0	250	484	6	557	793	1015:	4	1266	3	1502
3125.653	2	m	516	1043	3	1534	2032	2543	3	1523	2	1010
3120.435	1	0	D		1	846	B		4	(822)	3	814
3119.495	1	0	D		1	965	A		5	(1052)	4	1074
3116.633	2	0	1538		3	21	1518	3036	2	1513	1	−15

TABLE D—(*Continued*)

λ	π Components				σ Components				J_1	Obs. g_1	J_2	Obs. g_2
	No. Meas.	π			No. Meas.	σ						
3112.079	1	271	B						5	(1216)	5	1162
3100.666	2		308:	783	6	1264	1521	1772	3	1519	3	1261
3100.304	2	463	1036		4	1008	1520	2027	2	1008	2	1522
3099.971	1	605	B		1	m	1503†		4	1518	4	(1367)
3099.897	1	1542			2	51	1524		1	1524	1	−18
3098.192	1	0			1	1236			5	(1213)	5	1259
3095.270	4	451†	610†	742†	6	1073	1233	1377	5	1226	5	1379
3093.883	1	0	D		1	1263	B		4	(1251)	3	1247
3093.806	2	537	1087		4	675	1211	1763	2	1218	2	678
3092.778	3	0	1122	2225	3	196†	1286†	2400†	3	(1264)	2	2371
3091.578	1	0	U		1	0	U		1	0	0	0
3090.209	1	528	B		5	944	1116	1291	3	939	3	763
3083.742	2	0	513		3	551	1048	1550	2	1010	1	1517
3078.436	1	0	S		1	713	S		1	713	0	0
3075.721	3	0	258	514	5	756	1007	1260	3	1262	2	1518
3073.982	1	0	D		1	1279	B		5	(1213)	4	1197
3067.244	5	0	157	304	1	909	A		4	1371	3	1525
3063.933	1	727	B		2	814	1523		1	1528	1	810
3060.984	3	286	571	853	6	1099	1372	1654	3	1378	3	1097
3060.545	1	0	D		1	1003	A		4	(1062)	3	1082
3059.086	1	0			1	1514			4	1516	3	(1517)
3057.446					1	1052			5	(1422)	4	1515
3055.263	1	0	D		1	892	A		3	(1100)	2	1204
3053.065	2	0	349		2	826	1175		2	1175	1	1524
3047.605	1	0			1	1514			3	1514	2	(1514)
3045.077	4	0	543	1165	5	307:	793	1358	4	1362	3	1927
3042.666	3	0	260	516	5	1265	1523	1780	3	1267	2	1011
3042.020	2	0	1018		2	951:	2050		2	1032	1	−14
3041.745					1	1582	B		4	1344	3	(1264)
3040.428	1	0	D		1	1560	B		5	1406	4	(1367)
3039.322	1	0	D		1	950	A		6	1035	5	(1052)
3037.388	1	0			1	1516			2	1517	1	(1518)
3033.101	1	832			2	715	1519		1	1526	1	708
3031.638	1	0	U		1	0	U		1	0	1	0
3031.213	1	0			1	822			4	(822)	4	822
3030.149	1	0			1	1046			5	(1052)	5	1040
3029.237	3	335	698	1024	6	766	1107	1436	3	1107	3	768
3025.843	1	0	S		1	1510	S		1	1510	0	0
3025.638	1	0			1	1188			6	(1178)	6	1198
3024.033	1	0			1	1514			2	1516	1	(1518)
3021.074	1	0			1	1521			3	(1517)	3	1525
3020.640	1	0			1	1513			4	(1516)	4	1510
3020.487	1	0			1	1516			2	(1514)	2	1518
3018.983	1	0			1	1265			3	(1264)	3	1266
3016.186	2	0	1018		3	0	1010	2018	2	1008	1	− 5
3015.913	3	m	238	m	6	1526	1729	1954	5	1049	4	822
3014.175	3	0	760	1513	3	−224	521	1274	3	1273	2	2022
3011.482	1	0	D		1	974	B		4	818	3	(766)
3009.570	1	0			1	1364			4	(1367)	4	1361
3009.098	3	m	243	377	3	m	1703:	1802	6	(1178)	5	1045
3008.139	1	0	S		1	1518	S		1	1518	0	0
3007.281	1	0			1	1508			2	(1514)	2	1502
3005.302	1	0	D		1	1066	A		7	1162	6	(1178)
3004.119	4	615	809	1024	10	856	1053	1235	5	1049	5	848
3003.031	3	0	253	534	4	1266	1518	1757	3	1267	2	1012

TABLE D—(*Continued*)

λ	π Components				σ Components				J_1	Obs. g_1	J_2	Obs. g_2
	No. Meas.	π			No. Meas.	σ						
3000.950	1	0			1	1519			2	(1514)	1	1509
2999.512	1	0			1	1416			5	1422	5	1410
2996.386	1	0			1	1496			2	(1526)	1	1511
2994.507	1	0	S		1	1507	S		1	1507	0	0
2994.427	1	0			1	1530			3	(1517)	2	1510
2990.392	1	0	D		1	1095	B		5	1071	4	(1065)
2988.468	3	402	617	817	8	1072	1268	1464:	4	1274	4	1071
2987.292	1	0	D		1	1613	B		4	(1367)	3	1285
2983.574	1	0	D		1	1556	B		4	(1516)	3	1503
2981.852	1	0	D		1	958	A		4	1506	3	(1688)
2981.446	1	0			1	1520			3	(1517)	2	1516
2980.532	2	m	322†	459†	6	765†	910†	1055†	3	910	3	(766)
2976.126	3	0	203	416	5	928	1122	1309	3	1325	2	1525
2973.237	2	0	266		4	734	992	1273	4	1255	3	1517
2970.106	2	0	514		3	503	1010	1516	2	1009	1	1519
2968.481	1	245			2	1309†	1531†		1	(1520)	1	1283
2966.901	1	0	D		1	1151	A		5	1443	4	(1516)
2965.255	1	U			1	U			1	0	0	0
2960.299					1	1425			1	1425	0	0
2959.992	1	0	D		1	1069	A		6	1189	5	(1213)
2957.365	1	1542			1	0	1517		1	1517	1	−25
2954.651	3	0	187	372	4	979†	1162†	1342†	3	1342	2	(1526)
2953.940	2	504	1037		4	1004	1516	2012:	2	1519	2	1009
2947.877	3	265	530	782	6	1260	1515	1773	3	1517	3	1258
2947.363	2	379	756		3	1185	1547	1902	2	1543	2	1172
2941.343	2	0	1531		3	24	1517	3026	2	1506	1	−14
2936.904	2		507	625	6	1364	1514	1662	4	1516	4	1361
2929.618	3	0	525†	993†	3	1167†	1658†	2129†	3	1179	2	(679)
2929.008	3	0	513	1028	4	1518	2025	2528	3	1518	2	1009
2925.899	2	562	1144		4	668	1225	1784	2	674	2	1228
2925.359	1	0	D		1	1140	B		4	860	3	(766)
2923.288	1	0:			1	1046			5	(1046)	5	1046
2920.691	1	163	B		1	703	C		2	744	2	662
2918.354	1	89	B		1	1474			1	1519	1	1430
2918.023	1	0:			1	1178			6	(1180)	6	1178
2914.305	1	0			1	586	A		2	(679)	1	772
2912.158	4	0	258	519	6	1775	2027	2276	4	1519	3	1267
2908.864					1	1387†	A		4	1675::	3	(1771)
2907.518	1	0	D		1	1020	A		2	(1844)	1	2668
2901.910	1	0			1	1614			5	(1621)	5	1607
2901.381	3	283	549	810	6	1098	1361	1624	3	1359	3	1095
2899.416	2	0	273		3	1252	1523	1785	2	1520	1	1250
2895.035	1	206	B		1	1127	C		3	1161	3	1093
2894.505	1	0:			1	1514			2	(1526)	2	1502
2893.882	2	0	217		3	298†	501†	713†	4	893	3	(1100)
2887.806	1	0	D		1	1063	A		6	1188	5	(1218)
2886.316	3	0	371	697	4	1091	1434	1777	3	1089	2	744
2877.300	1	340	B		1	1334	C		4	1376	4	1292
2874.172	5	0	288	578	6	93	379	667	5	1238	4	1525
2872.333	3	440	844	1232	5	1264	1663	2084	3	1674	3	1264
2869.308	3	0	402	805	6	69	316	729	4	1116	3	1508
2866.624	2	844	1676		3	168	987	1813:	2	1829	2	995
2858.896	2	0	1184		2	−958†	m	1635:	2	334	1	(1518)
2851.798	2	0	345		3	28	343	679	2	338	1	− 5
2851.52					3	880†	1073†	1263†	3	864	2	(672)

TABLE D—(*Continued*)

λ	π Components				σ Components				J_1	Obs. g_1	J_2	Obs. g_2
	No. Meas.	π			No. Meas.	σ						
2848.713	2	0	1510		3	**−381**	1018	2493	2	1009	1	2502
2846.830	1	**0**	D		1	**1514**	B		4	(1271)	3	**1190**
2845.595	3	**0**	599	1184	4	**84**	677	1264	3	1267	2	1859
2843.977	1	**0**	D		1	**754**	A		3	923	2	(1008)
2843.631	4	**0**	312	624	6	**446**	748	1060	4	1370	3	1680
2838.120	2	679	**1363**		3	**334**	**1010**	1681:	2	1011	2	333
2835.457	3	810†	1224†	**1623†**	6	**1127†**	**1520†**	1924†	4	(1516)	4	1115
2832.436	4	**0**	228	456	6	**366**	589	810	4	1039	3	1265
2828.808	3	**0**	486	981	5	**−441:**	0	441	3	511	2	1014
2827.892	3	**0**	409	819	5	m	m	721†	4	1115	3	(1517)
2825.557	4	0	387	777	6	**−243**	145	495	4	884	3	1261
2823.276	3	358	718	**1077**	6	911	**1258**	1608	3	1262	3	908
2820.801	3	**0**	1203	2329	5	1519	2687	**3888**	3	1520	2	342
2819.286	1	**0**	D		1	**884**	B		3	(766)	2	707
2817.505	3	**0**	953	1871	5	1266	2185	**3104**	3	1260	2	330
2815.506	1	0	D		1	**433**	A		3	597	2	(679)
2813.288	5	0	146	281	5	**595**	758	917	5	1219	4	1371
2808.328	3	764	1506	**2234**	6	**535**	**1266**	2011	3	1267	3	522
2806.984	5	**0**	294	588	8	**−75**	163:	497	5	1074	4	1361
2804.521	3	673	1002	**1332**	8	**1039**	**1366**	1694	4	1369	4	1039
2803.613	1	**0:**			1	**812**			4	(822)	4	802
2797.775	3	1385	1906	**2439**	5	**880**	**1353**	1834	4	1358	4	881
2796.871	2	**0**	178						3	(1100)	2	1278
2795.540	2	m	m	924	5	1818†	2313†	**2694†**	4	(1367)	3	920
2792.397	4	**0**	155	299	4	**503**	647	802	4	950	3	1100
2791.786	1	**111**	B		1	**1060**	C		5	1071	5	1049
2789.803					1	1078			5	(1213)	4	1247
2788.106					1	**1048**	A		6	1360	5	(1422)
2781.835	3	0	506	1003	5	1507	1992	**2477**	3	1501	2	1005
2778.221	5	617	820	**995**	10	1219	**1414**	1607	5	1416	5	1219
2774.730	2	**0**	1588		3	10	1562	**3117**	2	1554	1	−17
2772.113	2	m	166	348	5	1655	1839	**2003**	3	1666	2	1496
2772.083	4	1040	1399	**1732**	7	740	**1089:**	**1471:**	5	1417	5	1063
2769.670	4	m	399†	805†	6	2185†	2567†	**2960†**	5	1422	4	1024
2769.297	1	0:			1	**1178**			6	(1178)	6	1178
2766.909	2	0	1173		3	50	1146	2266	2	1133	1	−13
2764.323	3	0	342	684	4	**782**	1153	1466	2	1837	2	1495
2763.108	3	**0**	364	730	5	1376	1713	**2064**	3	1364	2	1006
2762.027	2	m	496	**730**	6	1262	**1505**	1734	3	1494	3	1253
2761.780	2	602	**1124**		4	**997**	**1553**	2116	2	1565	2	1002
2759.814	1	**301**			2	**24**	**287**		1	287	1	− 7
2754.427	4	**0**	202	400	3	1468	1670	1870	4	1469	3	1269
2754.030	1	**244**	B						2	(1008)	2	1130
2750.140	1	529	B		2	m	m	1977†	3	1686	3	(1517)
2744.526					1	**692**	A		2	(1008)	1	1324
2743.564	1	**313**	B		1	**1300**	C		3	1368	3	(1264)
2742.406	2	351	**707**		4	**1520**	m	2211	2	1869	2	1521
2742.256					3	691†	979†	1267†	3	(1264)	2	1552
2737.310	1	**1061**			2	**1520**	**2535**		1	2547	1	1509
2735.475	1	m	127		1	**979**	A		4	(1367)	3	1495
2734.002	2	**0**	400		4	1384†	1782†	**2190†**	3	1410	2	(1008)
2733.581	1	**0**	D		1	**1097**	A		5	(1422)	4	(1503)
2728.819	1	**180**	B		1	**863**	C		4	885	4	841
2728.020	1	**348**	B		1	m	**1456†**		4	1454	4	(1367)
2723.577	2	**0**	1017		3	**505**	1515	2528	2	1516	1	2530

TABLE D—(*Continued*)

λ	π Components				σ Comopnents				J_1	Obs. g_1	J_2	Obs. g_2
	No. Meas.	π			No. Meas.	σ						
2720.902	1	m	m	709	5	**798**	1220	1525	3	1520	2	1867
2719.418	1	**0**			1	**1046**			5	(1052)	5	1040
2718.435	2	0	399		2	690†	m	**1480†**	3	1407:	2	(1008)
2717.786	3	**0**	669	1389:	5	**0**	626	1286	3	(1264)	2	1929:
2714.868	1	**421**	B		1	1299:†			3	1404	3	(1264)
2711.655	1	**0**	D		1	**1514**	B		5	1396	4	(1367)
2710.543	1	**0**	D		1	**936**	B		3	765	2	(679)
2708.570	1	**0:**			1	**1253**			4	(1251)	4	1255
2706.581	2	**0**	158		1	**1000**	A		3	(1264)	2	1396
2706.012	1	**0:**			1	**1178**			6	(1178)	6	1178
2699.107	1	**148**	B		1	**1390**	C		4	1372	4	1408
2697.019	3	**0**	170	322	4	**435**	577	750	4	926	3	1093
2695.032	1	**0**	D		1	**1302**	A		5	(1422)	4	1452
2690.067	4	**0**	423	**868**	5	1943	2329	**2773**	4	1508	3	1086
2689.827	2	**0**	142	**m**	3	**445†**	580†	761†	4	942	3	(1100)
2689.212	1	0	D		1	**1294**	A		4	(1367)	3	1391
2680.452	2	201	**416**		4	**1016**	**1215**	1412	2	1214	2	1011
2679.062	1	**0:**			1	**1413**			5	(1422)	5	1404
2673.213	1	**580**	B		1	**0**	**564**		1	564	1	−16
2669.492	1	**0**	D		1	**1006**	A		6	1045	5	(1052)
2667.912	3	**0**	175	376	3	**1002**	1163	1322	3	1322	2	1482
2666.398	1	**331**	B		1	**1294**	C		3	1349	3	1239
2662.056	1	**0**	D		1	**1330**	B		3	(1264)	2	1231
2661.196	2	**0**	456		3	547	1007	**1465**	2	1006	1	548
2660.396	3	**0**	253	507	5	**267**	515	767	3	767	2	1019
2656.792	5	**0**	179	368	5	**388**	554	735	5	1099	4	1279
2656.145	1	**0**	D		1	**1069**	A		7	1162	6	(1178)
2651.706	2	m	409	597	3	**467**	668	m	4	1070	3	1271
2647.558	1			**534**	6	**1344**	**1518**	1684	3	1517	3	1348
2645.422	1	**0**	D		1	**820**	A		2	1169	1	(1518)
2643.997	2	**0**	353		3	0	350	690	2	343	1	−10
2641.645	1	**0:**			1	**1352**			4	(1368)	3	1372
2636.477	3	**0**	m	296	4	**641**	770	m	5	1227	4	1376
2635.808	1	0	D		1	**804**	A		3	940	2	(1008)
2632.593	2	331	**726**		3	**m†**	**1489†**	1861†	2	(1514)	2	1167
2632.238	2	698	**1373**		4	**330†**	**990†**	1694†	2	(1008)	2	325
2623.532	1	**0**	D		1	**901**	A		4	1173	3	(1264)
2618.708	3	**0**	180	338	1	**1716†**	m	m	4	(1516)	3	1346
2618.018	3		528	**856**	5	**924**	**1275**	1589	3	1269	3	937
2614.494	1	m	**768†**		3	1107†	2045†	3009†	3	(1264)	2	313
2612.771	3	**0**	351	722					3	(1517)	2	1156
2610.750	2	**0**	1027		3	**431†**	1522†	**2538†**	2	(1514)	1	492
2606.826					1	**985**	A		5	1291	4	(1367)
2594.150	4	**0**	444	876	3	**1809†**	2226†	**2632†**	4	(1367)	3	942
2584.536	1	**0**	D		1	**1013**	A		6	1354	5	(1422)
2580.062	1	**1653**			1	**0**	**1640**		1	1640	1	−13
2576.688	1	680	B		3	**996†**	**1144†**	m	5	(1422)	5	1280
2564.555	2	**0**	1270		3	0	1217†	**2454†**	2	1241	1	(−14)
2561.262	2	**0**	369						3	1377	2	(1008)
2560.556	2	**0**	773†		2		750	**1519**	2	752	1	−14
2556.298					2	1075			4	1094	3	(1100)
2552.827	1	0	D		1	**1266**			3	(1264)	2	1263
2549.612	1	**0**	D		1	**1378**	A		4	1482	3	(1517)
2545.977	1	**0**	D		1	**1527**	B		3	1518	2	(1514)
2544.706	1	**0**	D		1	**1028**	A		5	1206	4	(1251)

TABLE D—(*Continued*)

λ	π Components				σ Components				J_1	Obs. g_1	J_2	Obs. g_2
	No. Meas.	π			No. Meas.	σ						
2543.920	1	0	D		1	913	A		4	1043	3	(1087)
2542.101	1	0	D		1	1026	B		3	790	2	(672)
2540.971					1	1521			2	1520	1	(1518)
2535.604	1	0	S		1	1514	S		1	1514	0	0
2530.694	3	0	418	845	3	1926†	2340†	2775†	3	1936	2	(1514)
2527.433	1	0			1	1519			3	(1517)	3	1521
2524.290	1	0	S		1	1515	S		1	1515	0	0
2522.848	1	0			1	1512			4	(1516)	4	1508
2519.628	2	0	472		3	0	468	937	2	467	1	− 3
2518.100	1	0	D		1	1516			2	(1514)	1	1512
2516.569	4	0	236	471	5	268:	532	797	4	1025	3	(1264)
2510.833	1	0	D		1	1527†	B		3	(1517)	2	1512
2508.751	3	0	337	648	3	0	352	694	3	670	2	(1008)
2507.899	1	0	D		1	902	A		4	1173	3	(1264)
2501.130	1	0	D		1	1508			4	(1516)	3	1519
2496.532	1	0	D		1	1169	A		5	1327	4	(1367)
2493.998	2	0	1320		3	0	1297	2582	2	1287	1	−18
2490.642	1	m	m	512	4	608	896	1164	3	1258	2	(1514)
2487.368	2	508	1030						2	(1514)	2	2011
2486.690	1	0	D		1	1294	B		4	1272	3	(1264)
2486.372	1		421		1	1207			4	(1516)	3	1937
2485.989	3	383†	575†	816†	5	1141†	1320†	1533†	4	(1367)	4	1161
2479.775	2	534	975		4	1013	1507	2005	2	1508	2	998
2476.654	2	0	797		3	1007	1780	2533	2	1001	1	216
2473.156	2	m	694†	947†	1	1342:†			4	(1516)	4	1769
2468.878	1	473	B		1	1381†	C		5	(1422)	5	1327
2467.730	1	0	D		1	1247	A		3	(1264)	2	1273
2463.728	1	0	D		1	1234	A		3	(1264)	2	1279
2462.645	1	588	B		1	1405	C		4	1479	4	1331
2462.178	2	0	518		2	1499†	m	m	3	1506	2	1003
2457.596	1	375	B		1	1372	C		5	1410	5	1334
2453.475	1	0	D		1	1666	B		4	(1367)	3	1267
2443.871	5	659	849	1041	9	1208:	1409:	1647	5	1435	5	1221
2442.567	1	0			1	1057			5	(1052)	5	1062
2440.106	1	0			1	819			4	(822)	4	816
2439.743	1	0			1	1178			6	(1178)	6	1178
2438.181	1	0	D		1	1940	B		5	(1422)	4	1292
2389.971					1	1938			3	1655	2	(1514)
2371.428	2	373	689		4	1517	1851	2140	2	1843	2	1505
2369.454	1	988			1	1974::			1	(1518)	1	2506
2320.356	1	0	D		1	1499			4	1512	3	(1517)
2313.102	1	0			1	1491			3	1506	2	(1514)
2308.997	1	0	D		1	1579	B		2	1548	1	(1518)
2300.140	1	0	D		1	1117	A		3	1382	2	(1514)
2299.218	1	0			1	1531			2	(1514)	2	1548
2298.175	1	0			1	1518			4	(1516)	4	1520
2297.785	1	0			1	1504			3	(1517)	3	1491
2294.406	1	0	S		1	1523	S		1	1523	0	0
2292.523	1	0	D		1	1352	A		4	1476	3	(1517)
2287.248	1	0	D		1	1690	B		2	(1514)	1	1338
2284.087	1	0	D		1	1473	A		3	(1517)	2	1539
2276.025	1	0	D		1	1540	B		4	(1516)	3	1508
2272.067					1	1169			4	1430	3	(1517)

TABLE E

OBSERVED AND CORRECTED g-VALUES

Desig	Observed g	R	U	A.D.	Corrected g	Res
a^1P_1	0.828	5	1	±6	0.817	−.183
a^1D_2	1.041	1	2	7	1.028	+.028
a^1G_4	1.014	6	7	6	1.001	+.001
b^1G_4	0.992		1		0.979	−.021
a^1H_5	1.013	3	3	4	1.000	.000
a^1I_6	1.027	1	1	3	1.014	+.014
a^3P_2	1.526	12	9	6	1.506	+.006
a^3P_1	1.520	12	4	4	1.500	.000
b^3P_2	1.518	12	5	8	1.498	−.002
b^3P_1	1.509	16	4	4	1.489	−.011
c^3P_2	1.504	5	4	6	1.484	−.016
c^3P_1	1.485	6	3	8	1.466	−.034
e^3P_2	1.478	1	1	2	1.459	−.041
e^3P_1	1.478		1		1.459	−.041
a^3D_3	1.353	6	2	6	1.335	+.002
a^3D_2	1.194	2	3	4	1.178	+.011
a^3D_1	0.741	4	3	7	0.731	+.231
b^3D_3	1.343		1		1.326	−.007
e^3D_3	1.363		1		1.345	+.012
e^3D_2	1.140		2	16	1.125	−.042
e^3D_1	0.812	1	1	14	0.801	+.301
f^3D_3	1.275	1			1.258	−.075
a^3F_4	1.271	11	8	6	1.254	+.004
a^3F_3	1.100	16	11	6	1.086	+.003
a^3F_2	0.679	10	8	7	0.670	+.003
b^3F_4	1.251	4	11	4	1.235	−.015
b^3F_3	1.089	4	8	6	1.073	−.010
b^3F_2	0.672	4	6	6	0.663	−.004
c^3F_4	1.281	1	1	6	1.264	+.014
c^3F_3	1.080	2		1	1.066	−.017
c^3F_2	0.686		2	6	0.677	+.010
e^3F_4	1.305		2	0	1.288	+.038
e^3F_3	1.122		2	8	1.107	+.024
e^3F_2	0.630		2	10	0.622	−.045
f^3F_4	1.156		1		1.141	−.109
f^3F_3	1.085		2	7	1.071	−.012
f^3F_2	0.685		1		0.676	+.009
a^3G_5	1.213	11	15	6	1.197	−.003
a^3G_4	1.068	16	10	5	1.051	+.001
a^3G_3	0.766	7	11	6	0.756	+.006
b^3G_5	1.216	3	5	9	1.200	.000
b^3G_4	1.062	5	5	8	1.048	−.002
b^3G_3	0.771	2	4	4	0.761	+.011

Desig	Observed g	R	U	A.D.	Corrected g	Res
e^3G_5	1.264		1		1.248	+.048
e^3G_4	1.110		2	8	1.096	+.046
e^3G_3	0.853		1		0.842	+.092
a^3H_6	1.178	2	12	5	1.163	−.004
a^3H_5	1.052	8	12	8	1.038	+.005
a^3H_4	0.822	5	9	4	0.811	+.011
b^3H_6	1.180		7	3	1.165	−.002
b^3H_5	1.046		6	4	1.032	−.001
b^3H_4	0.822	1	7	4	0.811	+.011
e^3H_6	1.241		1		1.225	+.058
e^3H_5	1.124		1		1.109	+.076
e^3H_4	0.882		1		0.871	+.071
e^5S_2	1.978		1		1.952	−.048
a^5P_3	1.688	23	6	5	1.666	−.001
a^5P_2	1.844	32	7	5	1.820	−.013
a^5P_1	2.526	26	2	5	2.499	−.001
e^5P_3	1.686	1			1.664	−.003
e^5P_1	2.464		2	4	2.432	−.068
a^5D_4	1.516	3	18	5	1.496	−.004
a^5D_3	1.517	13	19	6	1.497	−.003
a^5D_2	1.514	17	19	7	1.494	−.006
a^5D_1	1.518	10	10	5	1.498	−.002
e^5D_4	1.522	2	3	13	1.502	+.002
e^5D_3	1.528	3	3	6	1.508	+.008
e^5D_2	1.523	4	4	5	1.503	+.003
e^5D_1	1.538	5		7	1.518	+.018
f^5D_4	1.534	1	3	12	1.514	+.014
f^5D_3	1.636		2	6	1.615	+.115
f^5D_2	1.635	1	1	3	1.614	+.114
f^5D_1	1.684		1		1.662	+.162
g^5D_4	1.507	1	5	11	1.487	−.013
g^5D_3	1.512		2	3	1.492	−.008
g^5D_2	1.59:		2	34	1.57:	+.07:
h^5D_4	1.454		1		1.435	−.065
i^5D_4	1.402		3	17	1.384	−.116
i^5D_3	1.433		2	3	1.415	−.085
a^5F_5	1.422	6	16	5	1.404	+.004
a^5F_4	1.367	18	20	5	1.349	−.001
a^5F_3	1.264	30	21	6	1.248	−.002
a^5F_2	1.008	33	8	5	0.995	−.005
a^5F_1	−0.014	21		6	−0.014	−.014
e^5F_5	1.440		2	10	1.421	+.021
e^5F_4	1.348	1			1.331	−.019
e^5F_3	1.252	1	1	2	1.236	−.014
e^5F_2	1.004		1		0.991	−.009
e^5F_1	0.007	1			0.007	+.007

TABLE E—(*Continued*)

Desig	Observed g	R	U	A.D.	Corrected g	Res
f^5F_5	1.402		1		1.384	−.016
f^5F_4	1.373		1		1.355	+.005
f^5F_2	0.980		1		0.967	−.033
e^5G_6	1.369		1		1.351	+.018
e^5G_5	1.378		3	3	1.360	+.093
e^5G_4	1.254		1		1.238	+.088
e^5G_3	1.311	1	2	8	1.294	+.377
e^5G_2	0.970	1			0.953	+.620
f^5G_6	1.340	1			1.323	−.010
f^5G_5	1.237		1		1.221	−.046
f^5G_3	1.157		1		1.142	+.225
g^5G_6	1.42:		1		1.40:	+.07:
g^5G_2	0.348		1		0.343	+.010
e^5H_7	1.32:		1		1.30:	+.01:
e^5H_6	1.207		1		1.191	−.023
e^5H_5	1.117		1		1.102	+.002
e^5H_4	0.91:		1		0.90:	.00:
e^5H_3	0.490		1		0.484	−.016
e^7S_3	1.94:	1	1	24	1.92:	−.08:
e^7P_4	1.606	1	2	6	1.585	−.165
e^7P_3	1.709		3	16	1.687	−.230
e^7D_5	1.606	1	3	4	1.585	−.015
e^7D_4	1.677	2	4	11	1.655	+.005
e^7D_3	1.778	3	4	10	1.755	+.005
e^7D_2	2.035	5		6	2.009	+.009
e^7D_1	3.041	3	1	13	3.002	+.002
f^7D_5	1.530	1	1	12	1.510	−.090
f^7D_4	1.595	2	3	12	1.574	−.076
f^7D_2	1.868	1	2	14	1.844	−.156
g^7D_5	1.607		1		1.586	−.014
g^7D_4	1.68:		1		1.65:	.00:
e^7F_6	1.510		3	7	1.490	−.010
e^7F_5	1.525	1	4	6	1.505	+.005
e^7F_4	1.638	2	2	10	1.617	+.117
e^7F_3	1.519		1		1.499	−.001
e^7F_1	2.521	1			2.490	+.990
e^7G_6	1.434		3	4	1.415	+.010
e^7G_5	1.397	1	2	14	1.379	+.012
e^7G_4	1.356	2	1	18	1.338	+.038
e^7G_3	1.262		1		1.244	+.077
e^7G_1	−0.378	1			−0.374	+.126
$z^1P_1^\circ$	1.283	2	1	4	1.266	+.266
$z^1D_2^\circ$	0.93:		1		0.92:	−.08:
$y^1D_2^\circ$	1.038	3		2	1.025	+.025
$x^1D_2^\circ$	0.895	1			0.883	−.117
$w^1D_2^\circ$	1.003	2		4	0.990	−.010
$z^1F_3^\circ$	1.031		3	7	1.018	+.018
$y^1F_3^\circ$	1.22:		1		1.21:	+.21:

Desig	Observed g	R	U	A.D.	Corrected g	Res
$x^1F_3^\circ$	1.093		1		1.079	+.079
$w^1F_3^\circ$	0.920		4	5	0.908	−.092
$z^1G_4^\circ$	1.038	1	4	8	1.025	+.025
$y^1G_4^\circ$	1.077	1	2	3	1.063	+.063
$x^1G_4^\circ$	0.991		3	3	0.978	−.022
$w^1G_4^\circ$	1.014	1	1	2	1.001	+.001
$v^1G_4^\circ$	1.067		1		1.053	+.053
$z^1H_5^\circ$	1.031		3	4	1.018	+.018
$y^1H_5^\circ$	1.04:		2	30	1.03:	+.03:
$x^1H_5^\circ$	1.031	1			1.018	+.018
$z^1I_6^\circ$	1.023		1		1.010	+.010
$z^3S_1^\circ$	1.913	3	1	11	1.888	−.112
$y^3S_1^\circ$	1.909	5	1	7	1.884	−.116
$z^3P_2^\circ$	1.513	3	4	4	1.493	−.007
$z^3P_1^\circ$	1.516	2	1	6	1.496	−.004
$y^3P_2^\circ$	1.463	2	3	11	1.444	−.056
$y^3P_1^\circ$	1.621	2	3	10	1.600	+.100
$x^3P_2^\circ$	1.280	5	2	6	1.263	−.237
$x^3P_1^\circ$	1.567	7	1	8	1.547	+.047
$w^3P_2^\circ$	1.488	5		4	1.469	−.031
$w^3P_1^\circ$	1.407	3		2	1.389	−.111
$v^3P_2^\circ$	1.505	1	6	5	1.495	−.005
$v^3P_1^\circ$	1.437	2	1	16	1.418	−.082
$z^3D_3^\circ$	1.338	1	3	5	1.321	−.012
$z^3D_2^\circ$	1.183	1			1.168	+.001
$z^3D_1^\circ$	0.520	3		6	0.513	+.013
$y^3D_3^\circ$	1.341	5	3	6	1.324	−.009
$y^3D_2^\circ$	1.166	2	4	8	1.151	−.016
$y^3D_1^\circ$	0.504	4	1	6	0.493	−.007
$x^3D_3^\circ$	1.370	6	2	6	1.352	+.019
$x^3D_2^\circ$	1.216	6	3	7	1.200	+.033
$x^3D_1^\circ$	0.563	6		8	0.556	+.056
$w^3D_3^\circ$	1.364	4	2	4	1.346	+.013
$w^3D_2^\circ$	1.232	6	1	5	1.216	+.049
$w^3D_1^\circ$	0.777	4	1	7	0.767	+.267
$v^3D_3^\circ$	1.227	2	2	5	1.211	−.122
$v^3D_2^\circ$	0.967		1		0.954	−.213
$v^3D_1^\circ$	0.569		1		0.562	+.062
$u^3D_3^\circ$	1.323	3	1	4	1.306	−.027
$u^3D_2^\circ$	1.171	4	2	5	1.156	−.011
$u^3D_1^\circ$	0.709	4	3	9	0.700	+.200

TABLE E—(*Continued*)

Desig	Observed g	R	U	A.D.	Corrected g	Res
$t^3D_3^\circ$	1.334		4	7	1.317	−.016
$t^3D_2^\circ$	1.160	1	2	2	1.145	−.022
$t^3D_1^\circ$	0.812	3		5	0.801	+.301
$s^3D_3^\circ$	1.247		1		1.231	−.102
$z^3F_4^\circ$	1.266	2	1	7	1.250	.000
$z^3F_3^\circ$	1.100	2	2	11	1.086	+.003
$z^3F_2^\circ$	0.691	2	1	9	0.682	+.015
$y^3F_4^\circ$	1.262	3	2	6	1.246	−.004
$y^3F_3^\circ$	1.100	4	2	8	1.086	+.003
$y^3F_2^\circ$	0.697	1	2	13	0.688	+.021
$x^3F_4^\circ$	1.362	3	1	4	1.344	+.094
$x^3F_3^\circ$	1.174	5	2	13	1.159	+.076
$x^3F_2^\circ$	0.753	5	3	7	0.743	+.076
$w^3F_4^\circ$	1.197	3	3	3	1.181	−.069
$w^3F_3^\circ$	1.180	1			1.165	+.082
$w^3F_2^\circ$	0.686	1	2	6	0.677	+.010
$v^3F_4^\circ$	1.137	2	3	9	1.122	−.128
$v^3F_3^\circ$	1.110		1		1.096	+.013
$v^3F_2^\circ$	0.814		1		0.803	+.136
$u^3F_4^\circ$	1.163	2	1	6	1.148	−.102
$u^3F_3^\circ$	1.091		2	9	1.077	−.006
$u^3F_2^\circ$	0.696	1	1	4	0.687	+.020
$t^3F_4^\circ$	1.251		2	4	1.235	−.015
$t^3F_2^\circ$	0.707		1		0.698	+.031
$z^3G_5^\circ$	1.264	1	3	3	1.248	+.048
$z^3G_4^\circ$	1.114	4	3	4	1.100	+.050
$z^3G_3^\circ$	0.801	5	1	4	0.791	+.041
$y^3G_5^\circ$	1.223	2	4	7	1.207	+.007
$y^3G_4^\circ$	1.067	2	3	3	1.053	+.003
$y^3G_3^\circ$	0.775	2	2	8	0.765	+.015
$x^3G_5^\circ$	1.219	2	2	2	1.203	+.003
$x^3G_4^\circ$	1.075		2	2	1.061	+.011
$x^3G_3^\circ$	0.677	1	1	7	0.668	−.082
$w^3G_5^\circ$	1.29:		1		1.27:	+.07:
$w^3G_4^\circ$	0.946	3		4	0.934	−.116
$w^3G_3^\circ$	0.592	1	1	6	0.584	−.166
$v^3G_5^\circ$	1.178	4	3	7	1.163	−.037
$v^3G_4^\circ$	0.926	7	2	5	0.914	−.136
$v^3G_3^\circ$	0.773	1	4	5	0.763	+.013
$u^3G_5^\circ$	1.155	2	2	8	1.140	−.060
$u^3G_4^\circ$	1.081	2	3	8	1.067	+.017
$u^3G_3^\circ$	0.811	2	3	6	0.801	+.051
$t^3G_5^\circ$	1.250	1	1	9	1.234	+.034
$t^3G_4^\circ$	1.198	1	1	1	1.183	+.133
$t^3G_3^\circ$	0.934	1	1	10	0.922	+.172
$s^3G_5^\circ$	1.160		2	2	1.145	−.055
$s^3G_3^\circ$	0.868	1	2	3	0.857	+.107
$r^3G_5^\circ$	1.206		1		1.190	−.010
$r^3G_4^\circ$	1.043		1		1.030	−.020
$r^3G_3^\circ$	0.790		1		0.780	+.030

Desig	Observed g	R	U	A.D.	Corrected g	Res
$z^3H_6^\circ$	1.216	2	2	3	1.200	+.033
$z^3H_5^\circ$	1.074	2	3	3	1.060	+.027
$z^3H_4^\circ$	0.892	4	2	5	0.880	+.080
$y^3H_6^\circ$	1.18:		2	20	1.17:	.00:
$y^3H_5^\circ$	1.089	4	2	8	1.075	+.042
$y^3H_4^\circ$	0.941	3	2	3	0.929	+.129
$x^3H_6^\circ$	1.176		2	14	1.161	−.006
$x^3H_5^\circ$	1.062		2	10	1.038	+.005
$w^3H_6^\circ$	1.192		3	4	1.177	+.010
$w^3H_5^\circ$	1.047		4	4	1.033	.000
$w^3H_4^\circ$	0.821	1	3	2	0.810	+.010
$v^3H_6^\circ$	1.184		3	4	1.169	+.002
$v^3H_5^\circ$	1.071		2	0	1.057	+.024
$v^3H_4^\circ$	0.815		3	9	0.804	+.004
$u^3H_6^\circ$	1.181		3	4	1.166	−.001
$u^3H_5^\circ$	1.043		2	3	1.029	−.004
$u^3H_4^\circ$	0.870		2	10	0.859	+.059
$t^3H_6^\circ$	1.178		2	0	1.163	−.004
$t^3H_5^\circ$	1.054		2	8	1.040	+.007
$t^3H_4^\circ$	0.816		1		0.805	+.005
$z^3I_7^\circ$	1.164		1		1.149	+.006
$z^3I_6^\circ$	1.054		1		1.040	+.016
$z^3I_5^\circ$	0.844		1		0.833	.000
$y^3I_7^\circ$	1.162		2	0	1.147	+.004
$y^3I_6^\circ$	1.032	1	3	5	1.019	−.005
$y^3I_5^\circ$	0.841	1	1	2	0.830	−.003
$x^3I_7^\circ$	1.160		2	2	1.145	+.002
$x^3I_6^\circ$	1.042		2	2	1.028	+.004
$x^3I_5^\circ$	0.843		1		0.832	−.001
$z^5S_2^\circ$	2.011	4	1	6	1.985	−.015
$y^5S_2^\circ$	1.913	2	2	11	1.888	−.112
$z^5P_3^\circ$	1.679	6	1	5	1.657	−.010
$z^5P_2^\circ$	1.859	6		7	1.835	+.002
$z^5P_1^\circ$	2.520	6		5	2.487	−.013
$y^5P_3^\circ$	1.683	3	2	8	1.661	−.006
$y^5P_2^\circ$	1.860	5		10	1.836	+.003
$y^5P_1^\circ$	2.535	3	1	6	2.502	+.002
$x^5P_3^\circ$	1.672	2	2	9	1.650	−.017
$x^5P_2^\circ$	1.846	4	2	7	1.822	−.011
$x^5P_1^\circ$	2.496	1	2	10	2.464	−.036
$w^5P_3^\circ$	1.680	1	1	6	1.658	−.009
$w^5P_2^\circ$	1.846		1		1.822	−.011
$w^5P_1^\circ$	2.468	2		2	2.436	−.064
$v^5P_3^\circ$	1.668	1	1	1	1.646	−.021
$v^5P_2^\circ$	1.763	3	3	7	1.740	−.093
$v^5P_1^\circ$	2.242	4		10	2.213	−.287
$u^5P_1^\circ$	2.668		1		2.633	+.133
$t^5P_2^\circ$	1.72:		1		1.70:	−.13:

TABLE E—(*Continued*)

Desig	Observed g	R	U	A.D.	Corrected g	Res
$z^5D_4^\circ$	1.522	1	6	9	1.502	+.002
$z^5D_3^\circ$	1.520	2	6	6	1.500	.000
$z^5D_2^\circ$	1.523	3	5	5	1.503	+.003
$z^5D_1^\circ$	1.515	3	2	7	1.495	−.005
$y^5D_4^\circ$	1.516	3	2	3	1.496	−.004
$y^5D_3^\circ$	1.512	5	3	5	1.492	−.008
$y^5D_2^\circ$	1.515	5	3	4	1.495	−.005
$y^5D_1^\circ$	1.512	4	2	3	1.492	−.008
$x^5D_4^\circ$	1.509	3	3	9	1.489	−.011
$x^5D_3^\circ$	1.524	5	4	5	1.504	+.004
$x^5D_2^\circ$	1.521	6	2	5	1.501	+.001
$x^5D_1^\circ$	1.518	4	2	4	1.498	−.002
$w^5D_4^\circ$	1.512	1	3	4	1.492	−.008
$w^5D_3^\circ$	1.501	3	7	6	1.481	−.019
$w^5D_2^\circ$	1.553	5	3	6	1.533	+.033
$w^5D_1^\circ$	1.332	2	1	5	1.315	−.185
$v^5D_4^\circ$	1.419	1	1	11	1.401	−.099
$v^5D_3^\circ$	1.404		3	4	1.386	−.114
$v^5D_2^\circ$	1.396	1			1.378	−.122
$v^5D_1^\circ$	1.407		1		1.389	−.111
$u^5D_4^\circ$	1.359	5		8	1.341	−.159
$u^5D_3^\circ$	1.415	4		4	1.397	−.103
$u^5D_2^\circ$	1.277	6	1	7	1.260	−.240
$u^5D_1^\circ$	1.428	2	1	4	1.410	−.090
$t^5D_4^\circ$	1.506		1		1.486	−.014
$z^5F_5^\circ$	1.417	2	2	6	1.399	−.001
$z^5F_4^\circ$	1.373	3	3	5	1.355	+.005
$z^5F_3^\circ$	1.266	4	4	6	1.250	.000
$z^5F_2^\circ$	1.017	6	1	7	1.004	+.004
$z^5F_1^\circ$	−0.012	2	2	6	−0.012	−.012
$y^5F_5^\circ$	1.435	2	1	10	1.417	+.017
$y^5F_4^\circ$	1.362	2	4	3	1.344	−.006
$y^5F_3^\circ$	1.260	5	2	4	1.244	−.006
$y^5F_2^\circ$	1.011	5	1	2	0.998	−.002
$y^5F_1^\circ$	−0.016	3		6	−0.016	−.016
$x^5F_5^\circ$	1.408		2	2	1.390	−.010
$x^5F_4^\circ$	1.345	1	2	10	1.328	−.022
$x^5F_3^\circ$	1.270	1	4	8	1.254	+.004
$x^5F_2^\circ$	1.011	4		11	0.998	−.002
$x^5F_1^\circ$	−0.006	2		1	−0.006	−.006
$w^5F_5^\circ$	1.400		2	4	1.382	−.018
$w^5F_4^\circ$	1.463	2	3	8	1.444	+.094
$w^5F_3^\circ$	1.369	3	2	6	1.351	+.101
$w^5F_2^\circ$	1.132	1	1	2	1.117	+.117
$w^5F_1^\circ$	0.287	1			0.283	+.283
$v^5F_5^\circ$	1.334	1			1.317	−.083
$v^5F_4^\circ$	1.281	1	2	7	1.264	−.086
$v^5F_3^\circ$	1.252	1	2	10	1.236	−.014
$v^5F_2^\circ$	1.284	7	1	6	1.267	+.267
$v^5F_1^\circ$	0.233	2		16	0.230	+.230

Desig	Observed g	R	U	A.D.	Corrected g	Res
$z^5G_6^\circ$	1.350		1		1.332	−.001
$z^5G_5^\circ$	1.234	2	3	8	1.218	−.049
$z^5G_4^\circ$	1.117	5	2	3	1.103	−.047
$z^5G_3^\circ$	0.899	5	1	8	0.887	−.030
$z^5G_2^\circ$	0.339	4	1	5	0.335	+.002
$y^5G_6^\circ$	1.360		1		1.342	+.009
$y^5G_5^\circ$	1.219	2		0	1.203	−.064
$y^5G_4^\circ$	1.037	2	1	7	1.024	−.126
$y^5G_3^\circ$	0.917	1	2	6	0.905	−.012
$y^5G_2^\circ$	0.335	3		4	0.331	−.002
$x^5G_6^\circ$	1.354		1		1.336	+.003
$x^5G_5^\circ$	1.286		2	6	1.269	+.002
$x^5G_4^\circ$	1.173		1		1.158	+.008
$x^5G_3^\circ$	0.940	1	2	2	0.928	+.011
$x^5G_2^\circ$	0.327	1	2	11	0.323	−.010
$w^5G_6^\circ$	1.323		1		1.306	−.027
$w^5G_5^\circ$	1.322	1	2	6	1.305	+.038
$w^5G_4^\circ$	1.160	1	2	10	1.145	−.005
$w^5G_3^\circ$	0.943	1			0.931	+.014
$w^5G_2^\circ$	0.478	2		11	0.472	+.139
$z^5H_5^\circ$	1.068	2		6	1.054	−.046
$z^5H_4^\circ$	0.882	2		2	0.871	−.029
$z^5H_3^\circ$	0.516	2		5	0.509	+.009
$z^7P_4^\circ$	1.770	3	2	8	1.747	−.003
$z^7P_3^\circ$	1.933	4	1	2	1.908	−.009
$z^7P_2^\circ$	2.364	5		9	2.333	.000
$y^7P_4^\circ$	1.77:		1		1.75:	.00:
$y^7P_3^\circ$	1.933	1	2	4	1.908	−.009
$y^7P_2^\circ$	2.371		1		2.340	+.007
$z^7D_5^\circ$	1.618	1	5	9	1.597	−.003
$z^7D_4^\circ$	1.664		9	10	1.642	−.008
$z^7D_3^\circ$	1.769	2	5	15	1.746	−.004
$z^7D_2^\circ$	2.034	1	3	7	2.008	+.008
$z^7D_1^\circ$	3.038	3	1	11	2.999	−.001
$z^7F_6^\circ$	1.518		2	8	1.498	−.002
$z^7F_5^\circ$	1.518		4	2	1.498	−.002
$z^7F_4^\circ$	1.513	2	4	7	1.493	−.007
$z^7F_3^\circ$	1.533	1	1	0	1.513	+.013
$z^7F_2^\circ$	1.524	2	1	9	1.504	+.004
$z^7F_1^\circ$	1.569	2		4	1.549	+.049
1_2°	1.152	5	1	6	1.137	
4_4°	0.966	1			0.953	
6_5°	1.075	1	4	8	1.061	
8_1°	1.262	4	3	8	1.246	
10_3°	1.495	1			1.476	
12_6°	1.374	1	1	6	1.356	

INDEX

www.ingramcontent.com/pod-product-compliance
Lightning Source LLC
Chambersburg PA
CBHW081336190326
41458CB00018B/6023
* 9 7 8 1 4 2 2 3 7 7 2 5 3 *